AF740761

OXFORD MONOGRAPHS IN INTERNATIONAL LAW

General Editor: Professor Ian Brownlie QC, DCL, FBA
Chichele Professor of Public International Law in the University of Oxford and Fellow of All Souls College, Oxford.

THE INTERNATIONAL LAW OF MARITIME BOUNDARIES AND THE PRACTICE OF STATES IN THE MEDITERRANEAN SEA

OXFORD MONOGRAPHS IN INTERNATIONAL LAW

The aim of this series of monographs is to publish important and original pieces of research on all aspects of public international law. Topics which are given particular prominence are those which, while of interest to the academic lawyer, also have important bearing on issues which touch the actual conduct of international relations. None the less the series is wide in scope and includes monographs on the history and philosophical foundations of international law.

RECENT TITLES IN THIS SERIES

Land-Locked and Geographically Disadvantaged States in the International Law of the Sea
S. C. VASCIANNIE

Surrender, Occupation, and Private Property in International Law
NISUKE ANDO

The Human Rights Committee: Its Role in the Development of the International Covenant on Civil and Political Rights
DOMINIC McGOLDRICK

Human Rights in States of Emergency in International Law
JAIME ORAÁ

China's Practice in the Law of the Sea
JEANETTE GREENFIELD

The International Law of Maritime Boundaries and the Practice of States in the Mediterranean Sea

FARAJ ABDULLAH AHNISH

CLARENDON PRESS · OXFORD
1993

UNIV DUNELM

- 5 AUG 1993

Oxford University Press, Walton Street, Oxford OX2 6DP
Oxford New York Toronto
Delhi Bombay Calcutta Madras Karachi
Kuala Lumpur Singapore Hong Kong Tokyo
Nairobi Dar es Salaam Cape Town
Melbourne Auckland Madrid
and associated companies in
Berlin Ibadan

Oxford is a trade mark of Oxford University Press

Published in the United States
by Oxford University Press Inc., New York

© *Faraj Abdullah Ahnish 1993*

All rights reserved. No part of this publication may be reproduced, stored in a retrieval system, or transmitted, in any form or by any means, without the prior permission in writing of Oxford University Press. Within the UK, exceptions are allowed in respect of any fair dealing for the purpose of research or private study, or criticism or review, as permitted under the Copyright, Designs and Patents Act, 1988, or in the case of reprographic reproduction in accordance with the terms of the licences issued by the Copyright Licensing Agency. Enquiries concerning reproduction outside these terms and in other countries should be sent to the Rights Department, Oxford University Press, at the address above

British Library Cataloguing in Publication Data
Data available

Library of Congress Cataloging in Publication Data
Ahnish, Faraj Abdullah.
The international law of maritime boundaries and the practice of states in the Mediterranean sea/Faraj Abdullah Ahnish.
(Oxford monographs in international law)
"A revised and updated Ph.D. thesis submitted in 1989 to the University of Cambridge"—Acknowledgements.
Includes bibliographical references and index.
1. Territorial waters. 2. Baselines (Maritime law) 3. Continental shelf. 4. Territorial waters—Mediterranean Region. 5. Baselines (Maritime law)—Mediterranean Region. 6. Continental shelf—Mediterranean Region. I. Title. II. Series.
JX4131.A47 1993 341.4'48—dc20 92-25901
ISBN 0-19-825739-2

Typeset by Best-set Typesetter Ltd., Hong Kong
Printed in Great Britain by
Biddles Ltd
Guildford and King's Lynn

341.87
AHN
93/346
341.448091638
AHN

To
all members of my family,
for their love and support

Editor's Preface

THE work of Dr Ahnish originated as a doctoral thesis prepared at Cambridge under the supervision of Professor D. W. Bowett, CBE, QC, then the Whewell Professor of International Law. The author presents a general picture of the development of the law concerning delimitation of the territorial sea, continental shelf, and exclusive economic zone, together with the practice of delimitation by means of a single maritime boundary. Having set the scene, he embarks upon a monographic treatment of the practice of States in the Mediterranean Sea.

Such a regional study is relatively rare, and the results are impressive. Not only is it evidently useful to have a systematic analysis of Mediterranean practice, but the study illustrates the mode in which the available legal principles and optional constructions are applied to the circumstances of a region. Particular benefits include studies of the disputes relating to the Gulf of Sert and the Aegean Sea continental shelf.

All Souls College,
Oxford
3rd November 1992

IAN BROWNLIE

Acknowledgements

This study is a revised and updated Ph.D. thesis submitted in 1989 to the University of Cambridge. I would like to express my profound appreciation to several people for their help with this work. During my research I was fortunate in having the kind and expert guidance of Professor Derek Bowett, to whom I am deeply grateful.

Part II of this research raised topical problems relating to many Mediterranean countries. I am most grateful to the following academics who have assisted in various ways while I was collecting materials in Italy, Malta, Spain, and Libya: Professor T. Scovazzi (University of Parma), Dr G. Francalanci (State University, Milan), Dr M. Iovane (Second University, Rome), Dr D. Attard (attorney in Malta), Dr E. Mizzi (Chairman of the Law Commission, Malta), Professor R. Pastor (Madrid State University), Professor A. El-Murtadi-Suleiman (Gar Younis University, Libya), and Mr Ibrahim A. Omar (People's Bureau of Liaison, Tripoli).

I am also indebted to Professor B. Vukas (University of Zagreb, Yugoslavia) and Dr R. Versan (University of Istanbul, Turkey) for their invaluable notes and comment on certain parts of this work. A debt of gratitude is also owed to Professor I. Brownlie whose editorial suggestions so enriched certain chapters of my work.

It gives me great pleasure to recall with gratitude my friends who have provided encouragement and assistance in various ways, in particular D. Lambourn (Lady Margaret Hall, Oxford), S. Lees (University Library, Cambridge), A. El-Sheikh (Churchill College, Cambridge), Mrs S. Barkham (Computing Service, Cambridge), E. Ellafi (Libyan Interests Section, London), and Vittorio and Mauro Zingerelli (L'Aquila, Italy).

My thanks are due to the staff at all levels of the following institutions for their assistance over the years of research: Cambridge University Library, Squire Law Library, Churchill College Library, Cambridge, the Institute of Advanced Legal Studies, the International Maritime Organization, the Food and Agriculture Organization, and the International Ocean Institute (Malta).

Finally, I owe much to all members of my family, especially to Mohammed and Salih, whose support really made my study at Cambridge not only possible, but enjoyable, and to my wife for her unfailing support and encouragement, with devotion, throughout my work.

F. A. A.

Contents

PART II. PRACTICE OF STATES IN THE MEDITERRANEAN SEA

Maps

Tables

Abbreviations

AFDI	*Annuaire français de droit international*
AJIL	*American Journal of International Law*
AYBIL	*Australian Year Book of International Law*
BFSP	*British and Foreign State Papers*
BYIL	*British Yearbook of International Law*
CILJ	*Cornell International Law Journal*
CLP	*Current Legal Problems*
Cmnd.	Command Paper (United Kingdom)
1958 Conventions	1958 Conventions on the Law of the Sea
1982 Convention	United Nations Convention on the Law of the Sea (1982)
CYIL	*Canadian Yearbook of International Law*
CYILA	*Chinese Yearbook of International Law and Affairs*
Decisions	*ICJ and arbitral decisions on maritime boundary disputes*
ECCP	European Community Command Papers
EEC	European Economic Community
EESDLS	*The Eastern European States and the Development of the Law of the Sea*, 2 binders (1979)
EEZ	exclusive economic zone
EFZ	exclusive fishery zone
FAO	Food and Agriculture Organization
FZ	fishery zone
GFCM	General Fisheries Council for the Mediterranean
GJICL	*Georgia Journal of International and Comparative Law*
GYIL	*German Yearbook of International Law*
HILJ	*Harvard International Law Journal*
IBGT	*Institute of British Geographers: Transactions*
IBL	*International Business Lawyer*
IBS	*International Boundary Study* (Series, US Department of State)

ICJ	*International Court of Justice*
ICLQ	*International and Comparative Law Quarterly*
ICNT	*Informal Composite Negotiating Text*
IHR	*International Hydrographic Review*
IJECL	*International Journal of Estuarine and Coastal Law*
IJIL	*Indian Journal of International Law*
IL	*International Lawyer*
ILC	*International Law Commission*
ILM	*International Legal Materials*
IMO	International Maritime Organization
IYIL	*Italian Yearbook of International Law*
Keesing's	*Keesing's Contemporary Archives* (Longman, London)
LIS	*Limits in the Sea* (Series, US Department of State, Bureau of Oceans & International Environmental & Scientific Affairs)
LSB	*Law of the Sea Bulletin* (United Nations, Office of the Special Representative of the Secretary-General for the Law of the Sea)
MLR	*Maine Law Review*
MNRG	*Martens nouveau recueil général*
MP	*Marine Policy*
MPM	*Marine Policy Management*
MPR	*Marine Policy Reports*
New Directions	*New Directions in the Law of the Sea*, 11 vols. (1973–81)
n.m.	nautical miles
NRJ	*Natural Resources Journal*
NRL	*Natural Resources Lawyer*
ODIL	*Ocean Development and International Law (Journal of Marine Affairs)*
PCIJ	*Permanent Court of International Justice*
PYIL	*Polish Yearbook of International Law*
RDI	*Rivista di diritto internazionale*
REDI	*Revue égyptienne de droit international*
REs. DI	*Revista española de derecho Internacional*
RGDIP	*Revue général de droit international public*
RIAA	*Reports of the International Arbitral Awards (United Nations)*
RIRI	*Revue iranienne des relations internationales*
1958 Shelf Convention	*Geneva Convention on the Continental Shelf (1958)*

SI	*Lo spettatore internazionale*
SJILC	*Syracuse Journal of International Law and Commerce*
SLR	*San Diego Law Review*
SNT	*Single Negotiating Text (UNCLOS III)*
UNCLOS I	*United Nations First Conference on the Law of the Sea*
UNCLOS III	*United Nations Third Conference on the Law of the Sea*
UNLS	*United Nations Legislative Series*
UNTS	*United Nations Treaty Series*
WEDLS	*Western Europe and the Development of the Law of the Sea*, 4 binders (1983)
YJIL	*Yale Journal of International Law*
Yearbook of the ILC	*Yearbook of the International Law Commission*

Table of Cases

Introduction

This study is divided into two parts. Part I examines the various sources on the law of maritime boundaries. Part II focuses on the problem of boundary delimitation in the Mediterranean Sea, with the particular aim of exposing and analysing outstanding problems which highlight the increasing complexity of practical delimitation.

The general aim of the first part is twofold: (1) to provide evidence of the inception and gradual growth of a substantive rule of law, namely, the 'equidistance–special circumstances' rule, with the central thread being the discussions at the International Law Commission in 1951–6, the deliberations at the Geneva Conference on the Law of the Sea in 1958, and State practice; and (2) to describe the different view taken by adjudication and to speculate on the reasons behind this attitude.

This attempt stems from realizing the significant role played by adjudication in the formation of international law. It provides evidence of the views of international courts not only on the efficacy of existing rules, but, furthermore, on the need for modification or even creation of new rules. Another important factor in weighing the influence of the Courts' jurisprudence upon the formation of international law is the particular subject involved. In this regard the international law of maritime boundaries relates to an area of intense diplomatic activity, witnessed by the proliferation of settled/disputed boundary problems around the world. As of July 1988, it was estimated that some 412 maritime boundaries had to be delimited, of which only about 32 per cent had been settled.[1] Indeed, the vast majority of the work of the International Court of Justice during the last six years had been pre-occupied by this subject.[2]

The formulation of relevant rules that enhance both efficacy and predictability would seem to be a primary concern for an institution concerned with administering the law. The view taken by the ICJ, however, seems not only to give little regard to this concern but to put forward an elusive rule, briefly described as the 'equitable principles' rule, as the only 'primordial' rule of delimitation. The legal contortions resulting from this approach are reflected in the dissenting opinions of the members of the Court itself. Judge Gros recently observed that

[1] *LIS* 108: Table 1, 3. [2] See, Table of Cases, *supra*.

the course taken since [the 1982 *Tunisia–Libya* case] has been to indulge in an equity beyond the law, detached from any established rules, based solely on whatever each group of judges seised of a case declares itself able and free to appreciate in accordance with its political or economic views of the moment.[3]

It is also the primary concern of this study to show that the reasons given by the Court in refusing to recognize the existence in law of any other rule beside equity, though presentable, are refutable. The research utilizes State practice and the preparatory work of the ILC and UNCLOS I to demonstrate that a reasonable reading of such materials is sufficient to clear any confusion blurring the sound argument that a more substantive, efficient, and predictable rule of law exists.

Chapter 1 thus attempts to describe the various connotations of the concept of baselines and to discuss their relevance to inter-State delimitation. The aim is to demonstrate the confusion resulting from the specific formulation of this concept in the delimitation provisions of the 1958 Conventions on the Law of the Sea, which has partly affected the flexible quality of the substantive rule described above.

Chapter 2 examines the 'historical source' of this rule of delimitation, which was eventually embodied in the 1958 Conventions. The particular application of this rule to territorial sea boundaries in recent judicial and State practice, and finally its reaffirmation in the United Nations Convention on the Law of the Sea, will be examined. This is thought essential, for the validity of a rule applying to more extended maritime zones depends largely on obtaining a historical perspective demonstrating its flexible formulation.

Chapter 3 examines the adoption of the rule in the 1958 Geneva Convention on the Continental Shelf and its interpretation by two subsequent decisions, given before the adoption of the 1982 Convention.

Chapter 4 aims principally to examine the reasons behind the departure in the 1982 Convention and the subsequent arbitral and judicial decisions from the long-standing application of the rule of delimitation discussed in the previous chapters. It also attempts to examine whether contemporary State practice on shelf and EEZ boundaries has undergone any 'change of policy' as a result of this departure. Sections 1 and 2 analyse the new rule as embodied in the 1982 Convention. Section 3 elects to discuss the jurisprudence of the Decisions and detail the substantive elements of this jurisprudence. Section 4 analyses contemporary State practice and identifies the reasons underlying the Decisions' refusal to recognize the evidential value of that practice. Chapter 5 extends this discussion to the newly developed concept of a single maritime boundary.

[3] His dissent in the *Gulf of Maine* case, ICJ Reports (1984), para. 47. For similar comments by other judges see *infra*, Ch. 4 nn. 272, 277.

Obviously, some of these issues could form individual research topics on their own. This study, however, aims to devote only as much attention as is necessary for the basis of the central argument, that the legal sources establish the existence of a substantive rule of law.

The selection in Part II of the Mediterranean Sea[4] as a 'study field' is justified by the fact that the problems associated with the delimitation of extended maritime jurisdictions are particularly noticeable in enclosed and semi-enclosed seas. As one commentator indicates:

> The physical disposition of coastal States in and the geographical configuration of these enclosed and semi-enclosed seas make the delimitation of maritime zones especially difficult and the presence of features such as islands which perhaps constitute a characteristic of these maritime regions merely aggravates the problem.[5]

In this respect the political geography of the Mediterranean reveals characteristics which are rarely encountered in other semi-enclosed seas. Within the definition given above, the Mediterranean comprises a long and narrow corridor with a length of slightly over 2,000 nautical miles and, at its widest point between the Strait of Otranto and the Libyan coast, a width of just under 600 n.m. The coast of this limited area is divided between eighteen sovereign States as well as three dependent territories.[6] The coasts of these littorals vary dramatically in length, configuration, and direction.[7] Of particular significance is the fact that the coasts of three States, namely Spain, Italy, and Greece, form great peninsulas running into the sea and effectively dividing it.

The relationship between the continental coasts (resulting from this arrangement) is further complicated by the presence of many islands of contrasting size, location, economic significance, and political status;[8] and both of these factors give to the Mediterranean one of its most distinctive characteristics. They ensure that in no part of the Mediterranean can one travel more than 184 n.m. before meeting land. As a result, any unilateral claim of a 200-mile limit would overlap with the maritime jurisdiction of *at least two* or more States calculated in a similar fashion.

To these characteristics are added sensitive and complex political

[4] The term 'Mediterranean Sea' applies in this study to the area defined in Art. 1(1) of the Convention for the Protection of the Mediterranean Sea against Pollution of 1976. It provides that: 'For the purposes of this Convention, the Mediterranean Sea proper, including its Gulfs and Seas bounded to the west by the meridian passing through Cape Spartel lighthouse, at the entrance of the Strait of Gibraltar, and to the east by the southern limits of the entrance of the Straits of Dardanelles between Mehmetick and Kumkale lighthouse.'

[5] L. Nelson, 'The Emerging Law of the Sea', 42 *Modern Law Review* (1979), 42, at 58.

[6] See Table 1.

[7] See *infra*, Map 1, and *infra* Ch. 6 s.1.

[8] See Table 1; Map 1.

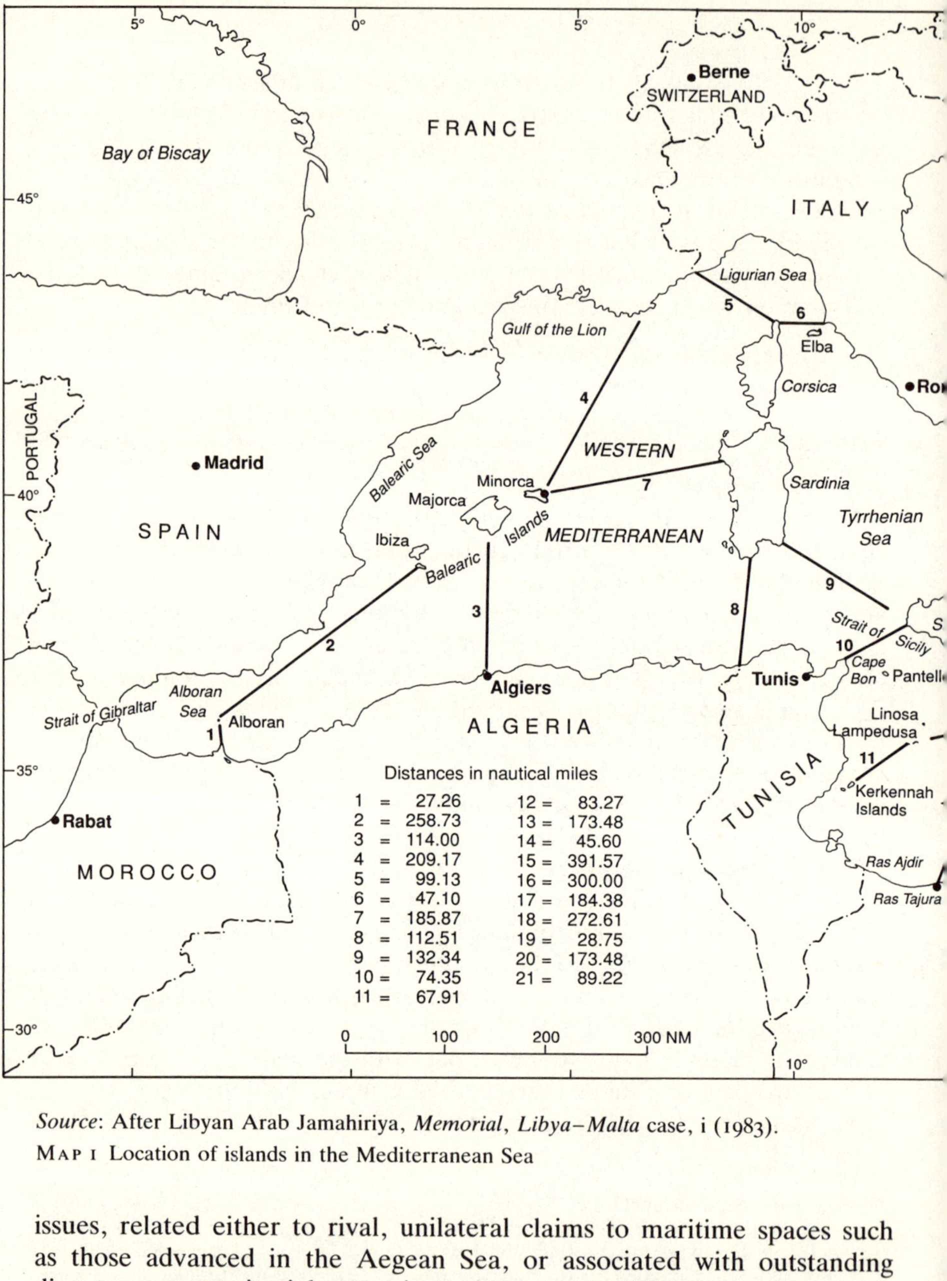

Source: After Libyan Arab Jamahiriya, *Memorial*, *Libya–Malta* case, i (1983).

MAP 1 Location of islands in the Mediterranean Sea

issues, related either to rival, unilateral claims to maritime spaces such as those advanced in the Aegean Sea, or associated with outstanding disputes over territorial sovereignty, such as the Gibraltar and Spanish enclaves questions.

These are the basic ingredients which give rise to so many complex and delicate legal/diplomatic problems in the Mediterranean boundary-making process. Since no comprehensive treatise on the problems of

maritime boundary delimitation in the Mediterranean yet exists, it is useful to speculate on these problems. Two chapters in Part II will therefore analyse the unilateral claims of Mediterranean States to baselines. This subject is relevant to boundary delimitation in the Mediterranean, partly because of the limitation of the area, making its littorals more aware of the 'encroachments' made by the 'neighbours' via baseline claims; partly because such claims are effected in areas where international straits are involved (consequently giving them world-wide importance); and, lastly, because such claims usually underpin allegedly

Table 1. *Mediterranean coastlines and seabed allocation (according to hypothetical equidistant boundaries)*

	Coastal length (n.m.)	Seabed area (sq. n.m.)
Albania	153	6,114
Algeria	596	32,461
Cyprus	290	26,000
Egypt	537	51,387
France[a]	491	25,787
Gaza Strip	21.6	—
Gibraltar (UK)	7	159
Greece	1,645	134,000
Israel	120	5,969
Italy[b]	2,450	155,000
Lebanon	105	4,586
Libya	910	93,311
Malta	45	17,815
Monaco	3	82
Morocco	190	6,332
Sovereign Base Area (UK)	86	4,360
Spain[c]	618	73,514
Syria	82	2,984
Tunisia	555	28,387
Turkey	973	17,323
Yugoslavia	426	20,453

[a] Including Corsica.
[b] Including Sardinia and Sicily.
[c] Including Balearic Islands.

Sources: 'Sovereignty of the Sea', *Geographic Bulletin*, 3 (US Department of State, 1969); G. Blake, 'Mediterranean Non-energy Resources; Scope for Cooperation and Dangers of Conflict', in G. Luciani (ed.), *The Mediterranean Region* (Croom Helm, London, 1984), 44–5 and other sources.

'exclusive' interests of coastal States that are opposed (and consequently challenged) by other States. The Libyan claim over the Gulf of Sert and the ensuing conflict between Libya and the United States provides a good example.

Chapter 6, therefore, studies the claims of Mediterranean States in relation to three kinds of baselines: low-water line, straight baselines, and lines closing juridical bays. Chapter 7 focuses on the Libyan claim over the Gulf of Sert as an historic bay, while outlining similar claims by other Mediterranean States.

Chapter 8 addresses the problems of territorial sea delimitation between the Mediterranean States. The Mediterranean's outstanding problems over, or related to, territorial sea boundaries are analysed in some detail.

Chapter 9 examines the patterns of existing, as well as future, shelf/ EEZ/single boundaries among the Mediterranean States. It aims primarily (1) to analyse existing shelf boundaries as part of the evidential value of State practice analysed in Part I; (2) to highlight the impact of the Mediterranean's geo-political characteristics on the jurisprudence of the International Court, two decisions of which were delivered on shelf boundary disputes in the Mediterranean; (3) to speculate on the future pattern of maritime boundary behaviour by the Mediterranean States. In this regard, the impact of the concept of a single boundary on the one hand and the relevance of the potential economic resources of the Mediterranean on the other is discussed.

Chpater 10 takes the Aegean Sea continental shelf dispute as a case-study, since it represents the most peculiar—and indeed complicated—picture in the geography of the Mediterranean. The aim is to attempt to resolve this peculiarity in order to make its analogy to State and judicial practice feasible.

The Conclusions at the end of this study summarize the substantive rule on boundary delimitation, borne out by international deliberations and long-standing application by States. The approach of adjudication and the legal consequences attributed to this approach are restated. This is followed by a brief restatement of the conclusions of the chapters dealing with the practice of the Mediterranean States, highlighting the potential of the Mediterranean's future resources and the pattern of 'boundaries' that may suit them.

Part I

The International Law of Maritime Boundaries

I

The Delineation of Baselines

1. INTRODUCTION

The very purpose of a baseline is to give juridical definition to the 'coast' of a coastal State from which that State's maritime zones are measured. Baselines normally coincide with the low-water line, as marked on large-scale charts, or, alternatively, are determined by geographical co-ordinates recognized by the coastal State.[1] Nevertheless, international law permits States—in specific circumstances—to measure their national maritime zones from straight or closing baselines drawn along parts of the coast. The original intention of this allowance is often framed in the need to avoid complicated applications of rules dealing with the 'normal' baseline when such applications would produce a complex pattern of State maritime jurisdiction.[2]

A proper examination of State and judicial practice, however, seems to disclose a more realistic rationale. It indicates that such an allowance has often been permitted as a necessary arrangement for recognizing a coastal State's exclusive interests. This is especially exhibited in the recent proliferation of enclosures to coastal configurations to protect security and economic interests which usually underlie direct reference to the concepts of 'historic bays' or straight baselines.[3] It will be recalled, however, that this rationale did not originate exclusively in State practice. In fact, States' claims have often been justified on 'judicial precedents' such as the *Gulf of Fonseca*, *Grisbadarna*, and *Anglo-Norwegian Fisheries* Decisions. They all placed special emphasis on socio-economic factors (normally manifested in a dynamic interaction between the forms of geography and human activity) in acknowledging not only the existence of pre-eminent exclusive State interests but also the validity of the companion, exclusive claims *vis-à-vis* other States.[4]

The recently adopted 1982 Convention seems to reflect this basic ration-

[1] See Art. 3 of the Territorial Sea Convention; Art. 16(1) of the 1982 Convention.

[2] See e.g. P. Beazley, 'Maritime Limits and Baselines: A Guide to Their Delineation', The Hydrographic Society, Special Paper Publication No. 2 (rev. Aug. 1978).

[3] For recent claims to 'historic bays', see *infra*, Ch. 7 s.3(*b*). For recent analysis of State practice on straight baselines, see J. Prescott, 'Straight Baselines: Theory and Practice', paper given at the 19th Annual Conference of the Law of the Sea Institute (Cardiff, 1985).

[4] See *infra*, Ch. 7 s.3(*a*).

ale. It relaxes the already flexible rules pertaining to the criteria for drawing straight baselines.[5] It further widens the concept of baselines by introducing new rules in respect of 'reefs' and archipelagic States.[6] This should not of course conceal the fact that this allowance has often been misused by States to increase the width of the combined internal and territorial waters, and effectively the rest of the zones of maritime jurisdiction. The attractiveness of this method is demonstrated by the fact that, at present, some fifty-seven States[7] have claimed straight baselines along their coasts. In many cases, straight lines are drawn to enclose large areas of high seas which cannot be said to reflect genuine State exclusive interests.[8]

The fact, however, remains that the primary function of a straight or closing baseline is to draw a 'politico-legalistic' boundary between internal (or archipelagic) waters over which a coastal State enjoys an almost absolute sovereignty and other maritime jurisdictional zones with varied emphases on State exclusive/international community inclusive usages.

This might explain, partly at least, the attitude, especially reflected in State practice, which tends to 'neutralize' States' official baselines for boundary purposes. As they normally relate to a subjective evaluation of a State's economico-political interests, baselines are often left aside and the negotiators proceed to determine the boundary from mutually agreed basepoints.[9] The rather interesting phenomenon, however, is that adjudication has explained the same attitude by reference to legal reasoning inherent in that body's general approach to the question of maritime boundaries.

The important remark in this respect is that this correspondence between adjudication and State practice seems to depart from the verbal meaning of the relevant delimitation provisions of conventional law, especially those embodied in the 1958 Geneva Conventions, whereby 'baselines' appear to be part of the delimitation provisions.

The third section of this chapter will state this departure and then attempt to consider its significance for the application of the relevant

[5] See *infra*, s.2(*b*).

[6] See *infra*, text at nn. 17, 32–3.

[7] *LIS*, *National Claims to Maritime Jurisdictions*, No. 36 (5th Revision), Country Profiles, 12–191; D. Francalanci, D. Romano, and T. Scovazzi (eds.), *Atlas of Baselines* (Giuffrè Editore, Milan, 1986).

[8] e.g. Bangladesh (Proclamation No. LT1.3.74) has drawn straight baselines based on the water depth criterion. The Maldives has drawn a rectangle which in effect creates baselines connecting geographical co-ordinates, not points on land (see analysis in *LIS* 78: 8–9). One segment of the straight baselines system of Cuba is depicted on only two points on land, which are about 400 n.m. apart. For further examples, see J. Prescott, *The Maritime Political Boundaries of the World* (Methuen, London, 1985), 163, 237, 313, 336.

[9] See *infra*, s.3(*b*)(iii).

rules on delimitation. This will be preceded by the next section, which gives a brief account of the various kinds of baselines and the various characteristics that coasts should exhibit in order to qualify for one kind or another of baselines. These are detailed in Articles 3–11 of the Territorial Sea Convention and Articles 4–13 and 47 of the 1982 Convention.

2. VARIOUS KINDS OF BASELINES

(*a*) *The general rule: the low-water line*

In normal circumstances, a coastal State jurisdiction extending from the low-water line[10] affords adequate protection of coastal interests. This was stated by the ICJ in the *Anglo-Norwegian Fisheries* case when it recognized that it had 'no difficulty in finding that for the purpose of measuring the breadth of the territorial sea, it is the low-water mark . . . which has generally been adopted in the practice of States. This criterion is the most favourable to the coastal State and clearly shows the character of territorial waters as appurtenant to the land territory.'[11] Thus, Article 3 of the Territorial Sea Convention (now Article 5 of the 1982 Convention) affirms that '[e]xcept where otherwise provided [in this Convention], the normal baseline for measuring the breadth of the territorial sea is the low-water line along the coast'.

One major question relating to the application of the low-water line, however, is whether the coastal State is entitled to employ any of the several low-water data;[12] and, if so, what the solution would be when two or more States involved in a maritime boundary delimitation use different low-water lines.[13] Neither the Conventions nor customary international

[10] There are several levels which would define 'low-water line'. These include: highest astronomical/lowest astronomical tides, mean high-water/low-water springs and mean high-water/low-water neaps. These are the results of changes in tidal levels in accordance with a set of periodic tides, superimposing one upon another. For various definitions, see the Hydrographer of the Navy, *Admiralty Tide Tables: Pacific Ocean and Adjacent Seas, Including Tidal Stream Tables*, iii (London, 1989), p. xxxviii.

[11] *Anglo-Norwegian Fisheries* case, *UK* v. *Norway*, ICJ Reports (1951), 116, at 128.

[12] See *supra*, n. 10.

[13] It is apparently a fact that some low-water lines might be located considerably further seaward than other possible low-water or high-tide lines. In the proceedings before the Special Master in *US* v. *California*, the latter stated that 'in many places' there was a 'substantial distance' between low-water and high-water lines (Brief of California in relation to Report of the Special Master, *US* v. *California*, 332 US 19 (1947), 92). The possibility of noticeable variations of this kind perhaps partly underlined the British objection to Egypt about use of lowest low-water line for defining an island (see the *Anglo-Norwegian Fisheries* case, Oral Pleadings, Documents, Correspondence, iv, ICJ Reports (1951), 579).

law seems to offer a positive rule,[14] and the matter is usually left to the States involved to agree either on a specific tidal line,[15] or, as is often the case, to ignore totally their proclaimed baselines and use 'chosen basepoints' for delimitation purposes.[16] However, the application of this 'normal' line in the specific cases of reefs and low-tide elevations is of some interest.

Article 6 of the 1982 Convention deals for the first time with the particular case of drawing normal baselines around 'islands situated on atolls or . . . islands having fringing reefs'.[17] The baseline for measuring the breadth of the territorial sea around such features 'is the seaward low-water line of the reef, as shown by the appropriate symbol on charts officially recognized by the coastal State'. Although the terms 'atoll' and 'reef' convey special scientific meanings,[18] there is little doubt that Article 6 is also concerned with features which, whilst formed in a different way, generally correspond to the 'popular view' of atolls.[19] Thus, structures such as 'almost-atolls',[20] 'horseshoe reefs',[21] and 'faros'[22] are similar features to which Article 6 also applies. It would also be reasonable to assume that the reference to fringing reefs in Article 6 can be applied without distinction to any reefs including barrier reefs whether their origin is biological, as in the case of coral reefs, or erosional, as in the case of wave-cut platforms common along many island and continental coasts. It is pertinent to realize in this respect that, unlike the provisions on archipelagos, Article 6 does not require that atolls or reefs should

[14] At the time of the 1958 Geneva Conference, the ILC stated that 'the traditional expression "low-water mark" may have different meanings; there is no uniform standard by which States in practice determine this line' (Report of the ILC to the General Assembly, *Yearbook of the ILC* (1956), ii. 267).

[15] e.g. the Iran–Saudi Arabia Continental Shelf Agreement of 1968, *infra*, n. 95.

[16] See *infra*, s.3(*b*)(iii).

[17] For a concise description of the natural formation and characteristics of atolls and reefs, see L. Alexander and D. Hodgson, *Toward an Objective Analysis of Special Circumstances: Bays, Rivers, Coastal and Oceanic Archipelagos and Atolls*, Occasional Paper No. 13 (Law of the Sea Institute, University of Rhode Island, 1972), 52–4.

[18] Geomorphologists reserve the term 'atoll' for reefs which surround a lagoon and are surmounted by one or more islands. The reef will usually be interrupted by channels generally on the leeward side of the atoll and the water in the lagoon will have a depth of at least 35 fathoms. See UN, 'Baselines: An Analysis of the Provisions of the United Nations Convention on the Law of the Sea Dealing with Baselines' (a study conducted by experts on the request of the UN Office for Ocean Affairs and the Law of the Sea) (unpublished document, second draft, 1 Sept. 1987).

[19] Ibid. 7–9.

[20] Such as Aitutaki in the Cook Islands. They consist of a small subsiding island which is surrounded by a circular reef with a lagoon in between.

[21] e.g. Pickersgill Reef off the Queensland coast. The limbs of the horseshoe are formed on the leeward side of the reef and may converge and join.

[22] Ebon in the Marshall Islands and Manikiki in the Cook Islands are called faros. They appear to be identical with atolls except that the water in the lagoon has a maximum depth of less than 35 fathoms.

comprise the whole political territory of a State. Mainland States with offshore reefs can therefore benefit from Article 6.

A low-tide elevation is an elevation of the seabed only above water at low tide. Article 11 of the Territorial Sea Convention (Article 13 of the 1982 Convention) states that the 'low-water line' of these features can be used to measure the outer limit of the territorial sea only if they are wholly or partly at a distance not exceeding the breadth of the territorial sea from the mainland or the island. Article 13 also applies, *a fortiori*, to islands located within the breadth of the territorial sea of the mainland or another island.[23]

(*b*) *Straight and closing baselines*

The system of straight baselines is widely employed around deeply indented coasts or those fringed with islands. But the system is also utilized to close mouths of bays, rivers, and archipelagic coasts.

Article 4 of the Territorial Sea Convention (7 of the 1982 Convention) provides that States are entitled to draw straight lines along localities which are 'deeply indented and cut into', or which have 'a fringe of islands' in the 'immediate vicinity'. Article 7(2) adds to the list coasts which are highly unstable because of the 'presence of delta and other natural conditions'. The system must be effected in accordance with requirements[24] related to the configuration of the coast,[25] its general direction,[26] the use of low-tide elevations,[27] and the avoidance of restricting the access of other States to their exclusive economic zone or the high seas.[28] However, the regime generally allows for a relaxation in the interpretation of these requirements in the case of the existence of 'econ-

[23] D. Bowett, *The Legal Regime of Islands in International Law* (Oceana Publications, New York, 1979), 10–17.

[24] For a detailed account of these requirements see Prescott, 'Straight Baselines', 3–31.

[25] That is the requirement of 'deeply indented and cut into' or 'a fringe of islands' (para. 1) which is derived from the language of the Court in the *Anglo-Norwegian Fisheries* case. See *Yearbook of the ILC* (1956), ii. 267.

[26] Para. 3 reads that the 'drawing of straight baselines must not depart to any appreciable extent from the general direction of the coast, and the sea areas lying within the lines must be sufficiently closely linked to the land domain'.

[27] Art. 4(3) of the Territorial Sea Convention states that straight baselines 'shall not be drawn to and from low-tide elevations, unless lighthouses or similar installations which are permanently above sea level have been built upon them'. Art. 7(4) of the 1982 Convention adds a new exception to this restriction. Permanent installations are not required in instances where the drawing of baselines to and from such elevations 'has received general international recognition'.

[28] Para. 6 states that the system of straight baselines 'may not be applied by a State in such a manner as to cut off the territorial sea of another State from the high seas or an exclusive economic zone'.

omic interests peculiar to the region concerned, the reality and importance of which are clearly evidenced by long usage'.[29]

According to Article 13 of the Territorial Sea Convention (Article 9 of the 1982 Convention) a coastal State is entitled to draw a closing line across the mouth of a river flowing 'directly into the sea', 'between points on the low-water line of its banks'. The provision seems to apply to all mouths of rivers whether they fall within the territory of one State or are shared by two or more States. Nor is there any restriction on the length of the line closing the river. In both these respects, the system is more liberal than that dealing with 'juridical bays'.

A juridical bay is defined in Article 7(1) of the Territorial Sea Convention, now Article 10(1) of the 1982 Convention, 'as a well-marked indentation in such a proportion to the width of its mouth as to contain land-locked waters and constitute more than a mere curvature of the coast'. This general definition is qualified by the precise geometrical criterion that an indentation is not regarded as a bay 'unless its area is as large as, or larger than, that of the semi-circle whose diameter is a line drawn across the mouth of that indentation'. However, for the purpose of the actual utilization of a closing line, the fourth and fifth paragraphs specify that the maximum length of any closing line is 24 miles. If the mouth of the bay exceeds that length the closing line may be drawn anywhere within the bay so as to close the maximum area of water possible.[30]

The first and last paragraphs of Articles 7 and 10 name three classes of bays which are not covered by the regime of juridical bays. Bays which are bounded by more than one State cannot be closed by a closing line; historic bays are exempted from the definitional/geometrical criteria for juridical bays; and finally bays which are converted to internal waters by the regime of straight baselines are not subject to the regime of juridical bays.

In the case of ports, Article 8 (11 of the 1982 Convention) provides that the outermost permanent harbour works which form an integral part of the harbour system are regarded as forming part of the coast, irrespective of its extension seawards.[31] Whilst the Articles do not mention baselines explicitly, it is obvious that harbour works could be tied to the mainland

[29] Art. 4(4) of the Territorial Sea Convention. See G. Fitzmaurice, 'Some Results of the Geneva Conference on the Law of the Sea', 8 *ICLQ* (1959), 73, at 77.

[30] The regime of bays has inspired several legal writings; and the meaning of the above-mentioned requirements was interestingly scrutinized in several judgments of the Supreme Court of the USA (the so-called Submerged Lands Cases). For a detailed examination of these requirements and relevant State practice, see G. Westerman, *The Juridical Bay: Its Designation and Delimitation in International Law* (Oxford University Press, New York, 1987).

[31] Art. 11 of the 1982 Convention adds a new sentence which states that 'Off-shore installations and artificial islands shall not be considered as permanent harbour works.'

or island by a straight baseline. This is explicitly mentioned in Article 50 of the 1982 Convention, which allows an archipelagic State to draw closing lines to create internal waters in accordance with Articles 9, 10, and 11.

The use of 'straight archipelagic baselines' to join 'the outermost islands and drying reefs' of an archipelagic State was codified for the first time in Articles 46 and 47 of the 1982 Convention. According to the definition of Article 46, archipelagos which are politically part of a continental State are excluded from the special archipelagic regime.[32] Article 47 sets out five tests which archipelagic baselines must satisfy. They must include the main islands; they must enclose an area of sea as large as the area of enclosed land but not more than nine times that land area; no segment of baseline may exceed 125 nautical miles in length; not more than 3 per cent of baseline segments may exceed 100 miles in length; and the baselines must not depart to any appreciable extent from the general configuration of the archipelago.[33] Under the ratio criteria of this Article only five States are said to have qualified for the special status of ocean-archipelagos.[34]

3. RELEVANCE OF BASELINES TO BOUNDARY DELIMITATION

(*a*) *Delimitation of the outer limit*

The term delimitation in a broader sense may also mean the determination of the boundary between the coastal State's internal (or archipelagic) waters and the rest of its national zones in the first place and between the latter and the high seas in the second place. The significance and relevance of baselines for this dual function is self-evident. The area landwards of a baseline is normally described as 'internal waters' upon which the State enjoys an unrestricted right of sovereignty. Article 5(1) of the Territorial Sea Convention (now Article 8(1) of the 1982 Convention) provides that 'waters on the landward side of the baseline of the territorial sea form part of the internal waters of the State'. Two caveats, however, should be

[32] e.g. the Galapagos belonging to Ecuador and the Andamans belonging to India. The 1974 SNT included an article which stated that provisions on archipelagic States were without prejudice to archipelagos belonging to continental States. This was deleted from the subsequent revised texts. For details, see C. Amerasinghe, 'The Problem of Archipelagos in the International Law of the Sea', 23 *ICLQ* (1974), 539; R. Anand, 'Mid-ocean Archipelagos in International Law', 19 *IJIL* (1979), 228.

[33] For a critical analysis of these conditions, see D. O'Connell, *The International Law of the Sea*, i (Clarendon Press, Oxford, 1982), Ch. 6.

[34] These are the Bahamas, Fiji, Indonesia, Mauritius, and the Philippines, ibid. 257–8.

mentioned. First, Article 5(2) (now Article 8(2) of the 1982 Convention) maintains the right of innocent passage in those *internal* waters enclosed by a straight baseline system if they have been previously considered part of the territorial sea or the high seas of the claimant State. Secondly, waters enclosed under the archipelagic baselines system are given a distinct title, namely 'archipelagic waters'. Under this regime, a right of passage is recognized through specified sea-lanes, and the archipelagic State is endowed with wide designatory and supervisory powers.[35]

In respect of the delimitation of the outer limit of national jurisdiction, a baseline is a determinative factor. Under Articles 6 and 24(2) of the Territorial Sea Convention (now Articles 4 and 33(2) of the 1982 Convention) the outer limit of the territorial sea and contiguous zone is respectively determined by distances from the baselines of the coastal State. The 1982 Convention provides for a further use of the baselines. Article 48 states that the maritime jurisdictional zones of an archipelagic State 'shall be measured from archipelagic baselines'. Articles 57 and 76(1) respectively state that the exclusive economic zone and (partly) the continental shelf extend to a distance of 200 miles from the baselines from which the breadth of the territorial sea is measured.[36] Article 76(6) provides that the outer limit of the continental shelf shall not exceed in any event 350 miles from these baselines.

Finally, paragraph 7 of the same Article introduces a new, *ad hoc* straight baseline system for the measurement of the outer limit of a continental shelf which extends beyond 200 miles. The coastal State 'shall delineate' the outer limit of that shelf 'by straight lines not exceeding 60 nautical miles in length'. The novelty of this system is exhibited in its imperative nature in the sense that the State is under an obligation to draw straight baselines, as well as in placing a maximum limit on the length of segments of the lines drawn.

(*b*) *Inter-State delimitation*

(i) Treaty law

Upon a superficial reading of the relevant provisions on delimitation, one would assume that the attitude of treaty law regarding the relevance of baselines depends on a method. The same reading would suggest an immediate conclusion: the Geneva Conventions, by adopting equidistance as the delimitatory rule, point out the relevance of a baseline very clearly. Article 12 of the Territorial Sea Convention employs a line 'equidistant from the nearest points on the baselines from which the breadth of the

[35] See Arts. 49–53 of the 1982 Convention.

[36] See *infra*, Ch. 5 s.2(*e*).

territorial seas of each of the two States is measured'.[37] Article 24(3) of the same Convention states in greater precision that neither State is entitled to extend its contiguous zone beyond the median line 'every point of which is equidistant from the nearest points on the baselines from which the breadth of the territorial seas of the two States is measured'. Article 6 of the Continental Shelf Convention repeats the latter and former expressions for the delimitation of the continental shelf between opposite and adjacent States, respectively.[38] Article 7(5) of the Convention on Fishing and Conservation of Living Resources of the High Seas adopts the same provision by referring back to Article 12 of the Territorial Sea Convention.

Article 15 of the 1982 Convention (delimitation of the territorial sea) adopts the same provision of Article 12 of the Territorial Sea Convention. Articles 74 (EEZ) and 83 (continental shelf), on the other hand, are different. Since they provide no 'substantive rule' of delimitation[39] they accord no role to the concept of baselines.

This is briefly the position of delimitatory conventional law on baselines. What seems to be worth pointing out, however, is the drastic impact that this absolute and strict formulation of the provisions on baselines may have on the proper understanding of the rule of equidistance. Taking them verbatim, the words 'every point of which' and the 'nearest points on the baselines from which' the 'breadth of the territorial seas of each of the two States is measured' imply that it is possible to apply equidistance only if: (1) the contracting States have already established baselines that (2) are identical and (3) that there exist no disputes between the States regarding the establishment of such baselines. Such an assumption cannot reflect the reality of the situation in the majority of cases, where at least one of these three conditions is absent.[40]

The preparatory work of the Geneva Conventions discloses very little of the reason behind this precise formulation on baselines. Yet it appears that this precise formulation was not intended to be the norm for the rule of equidistance. In this respect, one has to assume that the basic charac-

[37] *Infra*, Ch. 2 s.3(*a*).
[38] *Infra*, Ch. 3 s.1(*a*).
[39] *Infra*, Ch. 4 ss.1, 2.
[40] Not all States have predetermined baselines along all of their coasts before they negotiate or submit a dispute on a maritime boundary (see e.g. Greece and Libya, *infra*, Ch. 6 s.2). Even if one considers in such cases 'the low-water line' to be the normal applicable rule of law, the problem still arises of determining which of the several data available best serve the interests of that State (see *supra*, nn. 10, 13). A State normally uses different kinds of baselines along its coasts which cannot be assumed to be comparable with those adopted by the other neighbouring State. Finally, neighbouring States often oppose each other's claims on baselines and may even argue on these lines in order to attack each other's views on the delimitation solution (e.g. the arguments of the parties in *Tunisia–Libya* and *Libya–Malta* cases, *infra*, text at nn. 60–9).

teristics of a delimitatory rule were not absent from the thinking of the members of the International Law Commission. The rule has to be elastic enough to govern a diversity of geographical/socio-economic situations and to fit the basic contractual nature of the law on boundaries. A boundary can be established only by mutual agreement, either directly or by submission to a third party. Disagreements are likely to arise before reaching a final settlement. The determination of the baseline—or, rather precisely, the basepoints—is a very possible source of disagreement, for each State will be capable of modifying the extent of its maritime zone depending on its more or less restrictive way of drawing its baselines, and its opponents will likewise be entitled to deny the lawfulness of its action. The only way for all States concerned to eliminate conflicting claims will be to reach an agreement on this issue. Therefore, it must have been a basic understanding of the members of the Commission that, in respect of the 'basepoints' to be taken into account in the application of equidistance, the line will often not be measurable automatically and without any dispute; rather, the parties will have to make a settlement in order to avoid friction. Articles 12 (territorial sea) and 6 (continental shelf) must be intended to make such a basic requirement possible. The comments of the Commission in its final draft on these Articles respectively noted that 'the rule should be very flexibly applied'[41] and 'the rule adopted is fairly elastic'.[42]

This exposition may find some echo in the preparatory work of the Commission. For instance, in the discussion on the continental shelf a different proviso was proposed by Pal in 1953 (and defended by several members of the Commission).[43] He suggested a line equidistant from points along the seaward boundary line of the territorial seas of the States concerned, but this view was persuasively opposed by François:

> On the face of it, the idea was logical, but did not allow for the crucial difficulty that no unanimity obtained regarding the breadth of territorial waters. The consequence would therefore be that, by stating a claim for a wider territorial sea, a State would be able to secure a larger extent of continental shelf. That was why his proposal to draw the boundary equidistant from the coast was fairer.[44]

It was only at this point that François suggested the replacement of the term 'coast' by 'baselines of the territorial waters'.[45] Explaining the purpose of this amendment, François added that the term '"coast" was not an expression such as could be used in a legal text; it was necessary to define it, and that was what he had sought to do'.[46] François's amendment was adopted by the majority of the Commission[47] and was not questioned

[41] *Yearbook of the ILC* (1956), ii. 272. [42] Ibid. ii. 300.
[43] Ibid. (1953), i. 125–7. [44] Ibid. at 125. [45] Ibid. i. 128.
[46] Ibid. [47] Ibid. i. 129.

any more until UNCLOS I. The same amendment was adopted for the relevant provisions on the territorial sea in the 1954 Session.[48] At the Fourth Committee of UNCLOS I, the United Kingdom delegate convincingly pointed out the difficulties in applying in all cases a line equidistant from the low-water line.[49] He also argued that 'while the high-water line did not move as rapidly as the low-water line, it was nevertheless liable to move, and in certain places it had moved out to seaward by several miles in the course of fifty years'.[50]

This clearly shows that, firstly, the choice of the baseline from which the territorial sea is measured did not stem from a legal conviction that it was an essential part of equidistance; and, secondly, even that baseline suggested by François was meant to represent only one of several types of baseline to which equidistance could equally apply. The literal meaning which the provisions suggest cannot be reconciled with the Commission's awareness that even 'the traditional expression "low-water mark" may have different meanings'; and 'there is no uniform standard by which States, in practice, determine this line'.[51]

The codification of 'equidistance' as a general applicable rule can only make sense if the essential element of 'special circumstances' which is written into the text of the Geneva Conventions[52] allows the recognition of various elements and factors when determining the baselines from which the equidistance line is to be effected. The 'inherent logic of the 1958 Convention', as Judge Oda pertinently states,[53] should allow this understanding:

> While the sole use of the equidistance method can be expected to lead to an equitable result, this is on the understanding that the baseline to be employed for the purpose of the geometrical construction will vary from case to case, from the strict version used in measuring the limit of the territorial sea to the certain modified baselines employed because of special circumstances in the geography of the region.

The problem with adjudication is that while—as will be seen below—it *applies* this understanding, it refuses to attribute it to the existence of a more complicated rule of law established under the 'equidistance–special circumstance' formula. This, for instance, is manifested in the Court's statement in the *Tunisia–Libya* case that 'it is the virtue—though it may also be the weakness—of the equidistance method to take full account of

[48] Ibid. (1954), i. 103.
[49] UNCLOS I, *Official Records*, vi. 93, para. 21.
[50] Ibid. para. 2.
[51] *Supra*, n. 14.
[52] See *infra*, Ch. 2 s.3(*a*); Ch. 3 s.1.
[53] His dissent in the *Tunisia–Libya* case, ICJ Reports (1982), para. 168.

almost all variations in the relevant coastlines'.[54] This strict interpretation comes mainly as a result of the linguistic construction of the rule in the Geneva Conventions. Having been structured in such a strict, yet unintended, formula,[55] equidistance has been placed almost consistently under a 'judicial attack' that demeaned it to a mere strict, geometrical method. Further reasons behind this attitude in adjudication can also be identified.[56]

(ii) Adjudication

The Decisions are unanimous on the principle governing States' claimed baselines. This is perhaps best illustrated in the *Libya–Malta* case, when the Court stated that 'the baselines as determined by coastal States are not *per se* identical with the points chosen on a coast to make it possible to calculate the area of continental shelf appertaining to that State'.[57]

Perhaps there was a general principle in the mind of the Court when it delivered this definite statement. This was alluded to in the earlier Judgment of the Court in the 1969 *North Sea* cases. In that case, the Court declared that the appurtenance of a given area, considered as an entity, in no way governs the precise delimitation of its boundaries.[58] According to this principle, a baseline—from which the appurtenance of a maritime zone could be measured—is not regarded as a legal element in the consideration of the rules of law pertaining to maritime boundaries.

However, the more specific factor which seems to have shaped the judicial line of thought not only with regard to baselines but, more importantly, in the identification of applicable criteria and methods of delimitation is the introduction of 'equity' as the only 'primordial' rule of delimitation.[59] It is this notion which has enabled the Court to, *inter alia*, 'neutralize' the issue of baselines even when it is a matter of dispute between the parties to the case.

In the *Tunisia–Libya* case, the legality of the baselines system of Tunisia was challenged by Libya.[60] The relevance of the dispute on this point was particularly manifested in relation to whether the test of proportionality required the inclusion or otherwise of the entire seabed, including that enclosed by Tunisia as territorial and internal waters. Tunisia was of the view that such areas must be excluded from calculation.

[54] Para. 126. See also a similar statement by the Chamber in the *Gulf of Maine* case, ICJ Reports (1984), para. 210.

[55] For further problems associated with the construction of 'equidistance' in the Geneva Conventions, see *infra*, Ch. 2 s.3(*b*); Ch. 3 ss. 2, 3.

[56] see *infra*, Ch. 4 s.4(*c*).

[57] ICJ Reports (1985), para. 64.

[58] ICJ Reports (1969), para. 46.

[59] See *infra*, Ch. 4 s.2(*d*).

[60] See *infra*, Ch. 6 s.3(*j*).

Libya argued for the opposite view.[61] The Court stated that it was not convinced by either party's view. It did 'not consider that any general rule of law exists which requires the test of proportionality always to be applied by adopting one of the two methods'.[62] The Court then observed that in a situation when the adoption of either calculation would produce different results, it was the relevant circumstances of the area under consideration which would afford the basis for determining which of the two calculations would produce an equitable result;[63] for 'the element of proportionality is related to the lengths of the coasts of the States concerned, not to straight baselines drawn around those coasts'.[64] In the circumstances of the case, the Court assumed no differences[65] and felt accordingly that it was not necessary to make 'any ruling as to the validity or opposability to Libya of the straight baselines of Tunisia',[66] and, consequently, the Court did 'not find it necessary to pass on the question of historic rights [claimed by Tunisia] as justification for the baselines'.[67]

In *Libya–Malta*, the Court, complementing the statement quoted above, explained that '[i]n this case, the equitableness of an equidistance line depends on whether the precaution is taken of eliminating the disproportionate effect of certain islets, rocks, and minor coastal projections'.[68] In conformity with this requirement the Court found it reasonable to ignore the Filfla Rock (enclosed within the straight baselines system of Malta and challenged by Libya) without having to pronounce on the legality of this enclosure.[69]

The same reason—though explained differently—had been pointed out by the Court in the 1969 *North Sea* Judgment. In that case, the Court implicitly rejected a baseline following all the details of a coastal irregularity. The Court found that the application of such a line may often cause the equidistance line to be incompatible with the principle of non-encroachment. It recognized that the application of such a line in an equidistance boundary only emphasized the relative location of selected basepoints and took no account of the broad geographical relationship of neighbouring States.[70]

In the *Anglo-French* Arbitration the Court of Arbitration in delimiting the boundary in the Atlantic region found that Article 6 of the Continental Shelf Convention was applicable. In determining such a finding, however, the Court declared that it was modifying or varying the 'strict baseline system' of Article 6. It employed different baselines ignoring the detailed configuration of the coast, including islands, and proceeded to draw the line of equidistance.[71] In that case, however, the Court made a rather

[61] Para. 97. [62] Para. 103. [63] Ibid. [64] Para. 104. [65] Ibid.
[66] Ibid. [67] Para. 105. [68] Para. 64.
[69] See *infra*, Ch. 6 s.3(*d*); Ch. 9 s.2(*f*). [70] See *infra*, Ch. 3 s.3(*a*).
[71] See ibid. s.3(*b*).

casual reference to the low-water line in the Channel Islands region. Giving them a semi-enclave of 12 miles,[72] the Court stated that that radius should be 'delimited from the relevant base-points, at low-water, in the Bailiwick of Guernsey'.[73]

Another casual cognizance of a baseline was that given by the Court in the *Libya–Malta* case. In considering the hypothetical situation of a median line between Libya and the island of Sicily, which in effect determined the final adjustment of the boundary line between Libya and Malta,[74] the Court calculated this median line from 'the baselines on the coasts of Sicily established by the Italian Government'.[75]

These are the only exceptions to the general 'indifference' of adjudication toward States' established baselines. In fact, they seem to come as a consequential application of the principle of equity as envisaged by the Courts rather than a 'judicial' recognition of the legal relevance of States' established baselines to the rules on delimitation.

(iii) State practice

State practice pertaining to maritime boundaries reflects, as will be established,[76] a widespread application of the 'equidistance–special circumstances' rule. A common point in this practice is that States (including parties to the Geneva Conventions) do not follow the rigid language of the relevant provisions which, as shown earlier, identify 'equidistance' with the determined baselines. Instead, States opt for selecting specific basepoints for effecting the boundary line while ignoring their own official baselines. Reasons for this practice are related to a variety of relevant factors the assessment of which constitutes the essential operation for reaching an equitable boundary.[77] The wide range of such factors will be

[72] Ibid. The conformity of baselines with the rules of law was raised before the Court when the UK objected to the French Republic's claim to use a straight baseline across the Anse de Vauville situated on the coast of Normandy, east of the Channel Islands. The Court, however, declined to investigate the British contention, as it found that it did not have competence to delimit the boundary between the Channel Islands and the French coast. See 18 *RIAA*, paras. 15–22.

[73] Paras. 201–2. In the Second Decision of 14 Mar. 1978, the UK asked the Court for an interpretation of the Award of 1977 on the basis that the 12-mile boundary delimited by the Court in that Award had been delimited in the *dispositif* and the Boundary-Line Chart by ignoring certain low-tide elevations and dry-land low features as basepoints from which the baselines of the territorial sea of the Channel Islands had been determined. The Court, finding that 'there is no dispute between the parties' on this point, acknowledged that there was a 'discrepancy', but of a 'material error' that did not require the revision of the 1977 Award. Text in 18 *RIAA* 271–330.

[74] See *infra*, Ch. 9 s.2(*f*).

[75] Para. 72.

[76] See *infra*, Ch. 2 s.2; Ch. 4 s.4; Ch. 5 s.4.

[77] For a clarification of this point by adjudication, see *infra*, Ch. 4 s.3(*b*).

considered in detail elsewhere;[78] for it is only intended here to allude to some 'common factors' which would illustrate that 'equidistance' in practice normally assumes a departure from strict conformity with baselines.

Even within the narrow confines of the territorial sea, States feel free (despite their announcement that equidistance is the applicable rule) to ignore small islands, low-tide elevations, or rocks which, even if legitimately used as part of a baseline, would tend to produce an equidistance line which is 'distorted' thereby, that is, a line which does not reflect the true relationship of the mainland coasts or the narrowness of the area involved. Thus, in the Boundary Agreement between India and Sri Lanka which divides the historic waters of Palk Bay, selective basepoints are chosen which reflect a modified, median line, not based on the two States' established baselines.[79] Similarly, the Indonesia–Singapore Agreement of 1973 adopts a modified median line in which three points are equidistant from low-tide elevations of both coastlines, but another three turning-points are negotiated—non-strict equidistance—points. Indeed, one of these three points lies inside the Indonesian straight baselines.[80] The Denmark–Sweden Territorial Sea Agreement of 1932 used turning-points in the narrows of the Sound which ignore offshore islands and are measured from the mainland coasts.[81]

The extensive use of the 'equidistance–special circumstances' rule is perhaps most manifested in agreements delimiting extended zones such as the continental shelf or EEZ. In such boundaries, States depart from strict conformity with baselines and agree on modified equidistance in order to alleviate common problems arising from islands, coastal configurations, allocation of natural resources, or simply for administrative convenience. Additionally, they might agree on special alternative basepoints because the established baselines are disputed.

Examples of the first instance (islands) are abundant: if the island is smaller than the opposite foreign territory, equidistance is adjusted to reflect this circumstance. This is apparently the reason behind the departure from the strict median line between Indonesia and Malaysia in favour of a long Malaysian coast which extends opposite to small, scattered islands belonging to Indonesia.[82] Similarly, the 1973 shelf boundary between Thailand and Malaysia utilizes the respective baselines, except for a slight modification to give effect to a Malaysian island which is much larger than an opposite Thai group of islands.[83]

[78] See analysis of State practice in *infra*, Ch. 2 s.2; Ch. 4 s.4; Ch. 5 s.4.
[79] Agreement of 26–8 June 1974, *LIS* 66.
[80] Agreement of 25 May 1973, *LIS* 60.
[81] Declaration of 30 Jan. 1932, *LIS* 26.
[82] See *infra*, Ch. 4, text at n. 243.
[83] See *infra*, Ch. 4, text at n. 248.

The Italian–Spanish shelf boundary of 1974 reveals that the entire equidistance line may be tilted to reflect the dimension of the larger feature. The minutes of the meetings leading to that agreement state that 'the criterion applicable to the entire length of the line of delimitation [between Minorca and Sardinia] shall be that of equidistance from the respective baselines'.[84] The resulting boundary, however, does not conform to this statement: the line is actually drawn in a concave shape, apparently reflecting the concave shape of the western coast of Sardinia rather than following exactly a median from the two States' respective straight baselines along Minorca and the larger island of Sardinia.[85]

Islands—especially 'straddling' or 'on the wrong side' of the equidistance line—may be ignored as basepoints so as to reflect an 'equitable apportionment' of overlapping jurisdictions. Examples are found in Agreements in the Persian Gulf,[86] the North Sea,[87] the Caribbean Sea,[88] and in the Mediterranean Sea.[89] When sovereignty over an island is disputed or uncertain it is normal practice to ignore it for the delineation of the principal, mainland-to-mainland equidistance boundary. This is the solution adopted in the agreements between India and Sri Lanka,[90] Canada and the United States,[91] Canada and Denmark (Greenland),[92] Iran and Qatar,[93] Bahrain and Saudi Arabia,[94] and Iran and Saudi Arabia.[95]

Natural resources can also cause the strict equidistance line (drawn from the baselines claimed by States) to deviate. In the 1968 Shelf Agreement between Iran and Saudi Arabia, the equidistance line is modified in the north, apparently to allocate to Iran an area of seabed which had been known for its potential hydrocarbon resources.[96] In the Bahrain–Saudi Arabia Agreement, although the line is delimited 'on the basis of the median line', Points 12–14 are deviated towards Bahrain,

[84] Minutes of the Italian–Spanish Negotiations on Delimitation of the Continental Shelf between the Two Countries, Madrid, 4–5 Oct. 1973 (unpublished document).

[85] See *infra*, Ch. 9 s.2(*d*).

[86] See *infra*, Ch. 4 s.4(*b*).

[87] Ibid.

[88] *Infra*, Ch. 5 s.4(*b*).

[89] *Infra*, Ch. 9 s.2(*a*)–(*c*).

[90] See *infra*, Ch. 8 n. 114.

[91] Ibid.

[92] Ibid.

[93] See *infra*, Ch. 4, text at n. 220.

[94] *Infra*, Ch. 4, text at n. 221.

[95] *Infra*, Ch. 4, text at n. 219. In this Agreement reference is made to a particular tidal datum. Art. 6 states that in the area where the 12-mile territorial waters of the islands of Al-Arabiyah (Saudi Arabia) and Farsi (Iran) overlap, a boundary line 'shall be drawn so as to be equidistant throughout its length from the lowest low-water lines in each Island'.

[96] *Infra*, Ch. 4, text at n. 226.

apparently to include an area of proven hydrocarbon resources under the jurisdiction of Saudi Arabia.[97] Under the Colombia–Ecuador Boundary Agreement, a fishing 'buffer zone' 10 miles in width on each side of the boundary is established beyond 12 miles measured 'from the coast'.[98] Part of the buffer zone on the side of Ecuador is inside that State's straight baseline system which closes the waters landward as 'internal waters'.[99]

A departure from the two States' baselines may also be made in order to simplify the equidistance line for plotting or administrative purposes. The Anglo-Norwegian Agreement of 1965 provides that the shelf boundary is based, with certain minor divergencies for administrative convenience, on equidistance.[100] Similarly, the India–Sri Lanka Agreement of 1976 provides for a single maritime boundary drawn from selective basepoints to diminish the number of turning-points that would have resulted from the strict application of the equidistance method.[101] The same principle is applied in the Agreements of India with the Maldives, Indonesia, and Thailand,[102] as well as in the Agreements between Colombia and Panama,[103] Mexico and the United States,[104] and, more recently, the United Kingdom and Ireland.[105]

Finally, departure from the established baselines may prove a 'good policy' in order to avoid an actual or potential dispute on those baselines. The 1977 Maritime Boundary Agreement between Cuba and Haiti is based on a modified median line in the narrow Windward Passage, giving no effect to the straight baseline system of Haiti which seems to close, *inter alia*, the Gulf of Gonava.[106] In the 1986 Shelf Boundary Agreement between Libya and Malta, the parties negotiated a modified median line (based on the ICJ's decision of 1985)[107] which obviously disregarded

[97] *Infra*, Ch. 4, text at n. 225.
[98] Agreement of 23 Aug. 1975, Art. 2, *LIS* 69.
[99] See comment ibid. 5.
[100] See *infra*, Ch. 4, text at n. 229.
[101] See *infra*, Ch. 5, text at n. 185.
[102] *Infra*, Ch. 5, text at n. 186.
[103] *Infra*, Ch. 5, text at n. 202.
[104] *Infra*, Ch. 5, text at n. 181.
[105] According to D. H. Anderson (Deputy Legal Adviser, Foreign and Commonwealth Office), 'bay closing lines and straight baselines . . . were not taken into account in those areas where the method of equidistance was relevant' (his comment (unpublished document) on the Agreement which was signed on 7 Nov. 1988). Text of the Agreement is reproduced in *LSB* 13 (1989), 48.
[106] The closing line for the Gulf of Gonava flanks the southern area of the Windward Passage. The construction of a median line from it would shift that line about 13 miles towards the coast of Cuba. The two States, however, ignored each others' straight baselines and established a modified median line, apparently to avoid controversies relating to those baselines. See Prescott, *The Maritime Political Boundaries of the World*, 341.
[107] See *infra*, Ch. 9 s.2(*f*).

Libya's challenge to the legality of the straight baselines claimed by Malta around, *inter alia*, the Filfla Rock.[108]

In this respect, four interesting agreements reveal that, alternatively, the State which claimed hitherto no straight baselines may use straight baselines comparable to the other's for the purpose of arriving at 'an equitable' equidistance line. Thus, before the conclusion of the Indonesia–Malaysia Territorial Sea Agreement of 1970, Malaysia had never officially promulgated straight baselines.[109] For the purpose of that agreement, Malaysia constructed straight baselines, apparently to put itself on an equal footing in the delimitation of its boundary with Indonesia, which had previously drawn archipelagic baselines. The agreement specified that the territorial sea boundary between the two States was a median line measured from their respective straight 'baselines'.[110] This practice was also followed in the Agreement between the two States delimiting their continental shelves in the Strait of Malacca and the South China Sea.[111] In the Shelf Boundary Agreement between Canada and Greenland, the southern segment was developed on the equidistance principle, drawn from straight baselines on both coasts. The straight baselines were drawn specially for delimitation purpose by Canada to be of comparable length to the existing (since 1963) Danish straight baselines.[112] The Maritime Boundary Agreement between Kenya and Tanzania establishes baselines for both States. Some of those baselines, however, do not seem to influence the course of the boundary line, based on modified equidistance.[113]

These extremely varied examples indicate that although 'equidistance' is frequently applied in order to reconcile respective interests, baselines as established by States do not form an essential element in the appreciation of the appropriateness of that rule. States feel free to choose—and indeed they very often do so—certain basepoints for the purpose of effecting an equidistance line. As a consequence of this freedom, however, States do not (nor do they feel obliged to) point out their legal perception that equidistance allows such a departure from the strict formulae of the 1958 Conventions. In fact, a party's agreement could well be based on what is politically feasible at the time of the agreement, having nothing to do with the party's perception of the rules of law. It is, partly, because of this probability that the ICJ has, as will be seen later,[114] expressed doubt regarding the formulation in law of the 'equidistance–special circumstances' rule.

[108] See *supra*, text at n. 69.
[109] See comment in *LIS* 35.
[110] *LIS* 50.
[111] Agreement of 27 Oct. 1969, *LIS* 1.
[112] Agreement of 17 Dec. 1974, *LIS* 72: 7.
[113] Agreement of 9 July 1976, *LIS* 92.
[114] *Infra*, Ch. 4 s.4(*c*).

4. CONCLUSIONS

The above analysis indicates that 'equidistance' does not involve an imperative reference to 'the nearest point of the baselines from which the breadth of the territorial sea of the States is measured'. While this understanding is manifested in the *application* of 'equidistance' in codification, adjudication, and State practice alike, the same sources do not reveal clear understanding with regard to the *status* given to that rule.

The Geneva Conventions' linguistic construction disguises to a great extent the real intention of the drafters to codify a general rule of delimitation embodied in the combined 'equidistance–special circumstances' rule. This rule assumes a liberty to depart from strict equidistance from the baselines. The drafting history reveals, however, the reasons behind this inaccuracy in drafting. The apparent association between 'equidistance' and the baselines, for instance, resulted from the drafters' intention to adopt a 'preference' among other solutions which could equally apply.

State practice indicates the same conclusion. Admittedly, though, because treaty delimitations are not necessarily or solely inspired by a sentiment of legal obligation, and are often viewed *ex aequo et bono*, the relevant value of that source in illustrating the content of 'equidistance–special circumstances' loses much of its legal weight. Yet a careful analysis of that practice shows that a maritime boundary agreement assumes a margin of appreciation of various geographical/economico-political factors. A mutually accepted decision on these factors makes up the 'content' of a maritime boundary agreement. In this respect, the extensive use of 'equidistance', as that practice shows, cannot be understood unless, *inter alia*, a departure from strict conformity with the baselines is assumed in the spirit of that rule.

Adjudication, however, seems to construe 'equidistance' as a mere geometrical method which is applicable in principle by reference to the established baselines. The departure from this strict method is made by the ICJ by reference to 'equitable principles', and not by virtue of the elasticity inherent in a rule which appears to relate, nevertheless, to what the Court expresses as the only 'primordial' rule of delimitation: namely, the attainment of an equitable result. This is a contention that will be discussed further in later chapters.

It remains to clarify the meaning of the contention that 'equidistance–special circumstances' is a 'rule of law'. A consideration of a rule of delimitation should guide the decision-maker to an equitable decision as to where to draw the boundary. In this respect, the quality 'rule of law' is attributed to a principle which serves to create parameters within which an equitable line may be drawn. The actual drawing of the boundary line will be effected according to one or several methods of delimitation. This is a very important distinction in relation to the proper identification of the rules of law on delimitation; and it will be seen that international adjudication makes this distinction very clearly.[115] In this latter sense it is not difficult to agree with the ICJ that 'equidistance', whether drawn from respective baselines or modified, is *a method*.

However, 'equidistance' is submitted in this study to have attained the quality of a delimitatory rule as explained above. It is a submission based, firstly, on the drafting history of the rule in the Geneva Conventions and its long-standing application by States. The analysis of those sources reveals that 'equidistance'—or rather 'equidistance–special circumstances'—connotes a legal operation, whereby (1) the basic characteristic of the law on delimitation (that is contractual law) and (2) the ultimate objective of that law (that is the construction of a boundary line in the light of the special or relevant circumstances of the area considered) can be crystallized. In other words, the presumption in favour of the rule lies in its ability to provide significant and reliable evidence of a substantive standard of equity. It is this conclusion which, secondly, supports the elevation of 'equidistance–special circumstances' to a positive, contemporary rule of law. Unhappily, the ICJ fails, as will be seen, to distinguish between these two different meanings of 'equidistance', although it is generally agreed that the extensive use of the rule in State practice provides 'impressive evidence that the equidistance method can in many different situations yield an equitable result'.[116]

[115] See Ch. 4 s.3(*a*). [116] *Libya–Malta* case, para. 44.

2

Delimitation of the Territopial Sea: The Origin and Development of the Equidistance–Special Circumstances Rule

1. INTRODUCTION

The aim of the law on delimitation has always been to establish principles and rules that ensure an equitable delimitation in accordance with the geography of the area concerned. Upon the establishment of a definite concept of the territorial sea in the early nineteenth century[1] attention began to focus on developing principles and rules which ensured equitable division of the territorial sea between States. The long-standing application of these principles and rules on territorial sea boundaries constitutes the principal *sedes materiae* for the contemporary law on maritime boundaries in general. A clear example of this association is the recent incorporation in the law of the sea of the 'equidistance–special circumstances' rule which, as will be established, was already operating in earlier territorial sea boundaries between States.

During the discussions held in the International Law Commission in preparation for the 1958 Conference, the question of which method or methods should be employed for the delimitation of the territorial sea was referred to a Committee of Experts called upon to assist the Commission in its work. The Committee advised in 1953 that:[2]

> An international boundary between countries the coasts of which are opposite each other at a distance of less than ? miles should as a general rule be the median line, every point of which is equidistant from the baselines of the States concerned . . . There may, however, be special reasons, such as navigation and fishing rights, which may divert the boundary from the median line.
>
> . . . the (lateral) boundary through the territorial sea—if not fixed otherwise—should be drawn according to the principle of equidistance from the respective coastlines.

[1] On the historical development of the concept of the territorial sea see, e.g. S. Swarztrauber, *The Three-Mile Limit of Territorial Seas* (Naval Institute Press, Annapolis, Md., 1972), Ch. 6; D. O'Connell, *The International Law of the Sea* (Clarendon Press, Oxford, 1982), i. 60–147.

[2] See Report of the Committee of Experts, UN Doc. A/CN. 4/51/Add.1.

Moreover, the report of the Committee remarked that

> it considered it important to find a formula for drawing the international boundaries in the territorial waters of States, which could also be used for the delimitation of the continental shelves of two States bordering the same continental shelf.[3]

Endorsing the Committee's report, the Commission formulated its final draft articles in 1954 and stated that 'the boundary of the continental shelf should be drawn according to the same principles as those to be adopted for the delimitation of the territorial sea'.[4]

As will be seen, the 1958 Conference basically adopted the Commission's draft articles, and 'equidistance' became the delimitatory rule for all the maritime zones which were under consideration, namely the territorial sea, the contiguous zone,[5] and the continental shelf. The Convention on Fishing and Conservation of Living Resources of the High Seas also refers, in Article 7, back to Article 12 of the Territorial Sea Convention, thus also adopting 'equidistance'. A relevant point is that, in the deliberations following the Committee's report, 'equidistance' was never regarded as an absolute rule. It was always associated with the key element 'special circumstances'; and, as a result of this association, the rule—combining the two elements—raised almost no opposition, and no alternative solution was ever formally proposed.[6]

This complicated formulation of 'equidistance' was mainly attributed to State practice. Equidistance was already applicable in the delimitation of various zones in diverse situations to an extent which can scarcely be claimed for any other rule of delimitation. Section 2 of this chapter attempts to explore the real content of this rule as reflected in early and contemporary agreements on territorial sea delimitation. Section 3 will continue this enquiry by reflecting on the history of the codification of the rule leading to the 1958 Conference and its reaffirmation in the 1982 Convention. In section 4 some concluding remarks will be given.

[3] Ibid.

[4] Report of the Commission to the General Assembly, *Yearbook of the ILC* (1954), ii. 157.

[5] The Commission's draft articles of 1954 did not give any provision on the delimitation of the contiguous zone, and it was later at the 1958 Conference's First Committee that the 'median line' was also adopted for the delimitation of the contiguous zone.

[6] The only official proposal to the contrary which was put to discussion was made by Venezuela. It provided that 'the boundary of the continental shelf . . . should be determined by agreement . . . or by other means in accordance with international law' (Doc. A/CONF. 13/C.4/L.42). However, this proposal was criticized by most delegates and was eventually rejected by 32 votes to 3, with 19 abstentions. See UNCLOS I, *Official Records*, vi. 94–7.

2. THE CONTENT OF 'EQUIDISTANCE' IN STATE PRACTICE

By the time the concept of the territorial sea was firmly established, there already existed a limited number of agreements which provided for equal division of 'water frontiers' in lakes and rivers.[7] This contributed to the formulation of the general belief among nineteenth-century scholars that the appurtenance of equal sovereigns should reach out equally and abut in the middle of the waters. But it was also clear that the rigid application of this view did not always guarantee an equitable delimitation owing to the presence of local conditions which varied from one area to another. The solution was to develop a rule which would provide a specific method of delimitation and at the same time was capable of adjustment in order to meet the peculiarities of the area to be delimited. Thus, in 1872 D. Field suggested a combined principle of delimitation in an international code he proposed to the International Law Association:

> The limits of national territory, bounded by a river or other stream, or by a strait, sound, or arm of the sea, the other shore of which is the territory of another nation, extend outward to a point equidistant from the territory of the nation occupying the opposite shore, or if there be a stream, or navigable channel . . . to the mid-channel; or, if there be several channels, to the middle of the principal one.[8]

Field's solution was endorsed by other writers,[9] for it upheld a practical understanding which was to be reflected in State practice, that is, the meaning of 'median' or 'equidistance' line was not confined to the strict geographical median/equidistant line drawn between two shores. It was extended to mean, *inter alia*, the line of the 'middle of the channel', i.e. the thalweg line.[10] In fact, there was some evidence which suggested that the use of both the lines together was not an unknown practice. For instance, in 1892 the United States Supreme Court interpreted the phrase 'a line drawn along the middle of the river Mississippi from its source to the river Iberville', included in the 1763 Treaty between France, Spain, and the United Kingdom, as meaning 'along the middle of the channel of

[7] See e.g. L. Bouchez, 'The Fixing of Boundaries in International Boundary Rivers', 12 *ICLQ* (1963), 85.

[8] D. Field, *Outlines of an International Code* (Baker, Voorhis & Co., New York, 1872), 16.

[9] e.g. T. Twiss (*The Law of Nations*, i (Clarendon Press, Oxford, 1861), 251) recommended the median line as a general rule and the central deep-water line as a supplementary rule.

[10] The thalweg has been the general method for the delimitation of navigable rivers. See W. Boggs, *International Boundaries* (Columbia University Press, New York, 1940), 178–84. On recent State practice see D. Bowett, *The Legal Regime of Islands in International Law* (Oceana Publications, New York, 1979), 61–72.

the river Mississippi'.[11] In 1906, the same Court referred to the complementary use of the two lines in clear terms: 'where there is no necessary tract of navigation, the line of demarcation is drawn in the middle . . . but whenever a deep water sailing channel [is found] therein . . . the rule of thalweg applies.'[12]

However, following the development of shipborne commerce, mutual preservation of equal rights of access in deep-water channels became an important factor in the delimitation of adjacent territorial sea boundaries. For this practical reason, the thalweg line soon combined with the geographical median line, and the two became one combined rule, which allowed for the special circumstance of navigation. States have widely incorporated this understanding in their treaty practice. For instance, Article 1 of the 1846 Treaty between the United Kingdom and the United States provided that the water boundary in the Pacific border region should pass 'through the middle of [the channel which separates the continent from Vancouver's Island, i.e. the thalweg line], and of Fuca's Strait, to the Pacific Ocean [i.e. the median line]'.[13]

According to the Protocol of 1873 between the two parties, the line was ultimately fixed by connecting median points with many straight lines and one curved line. The parties applied the median line principle flexibly by ignoring rocks and taking navigational routes into account.[14] In fact, this combined use of the median/mid-channel line seems to have been the common practice in the late nineteenth century.[15]

The early twentieth century experienced ever increasing use of the geographic/thalweg lines as ones that might supplement each other. For instance, in the 1912 Declarations between France and Germany relating to the delimitation of the boundary between the French possessions of

[11] *Iowa* v. *Illinois*, 147 US 1, 8 (1892), quoted in *Louisiana* v. *Mississippi*, 202 US 1, 50 (1906).

[12] *Louisiana* v. *Mississippi*, 50.

[13] Treaty between Great Britain and the USA for the Settlement of the Oregon Boundary, 15 June 1846, 34 *BFSP* (1846), 14. The parties differed over the choice of the channel in the coastal archipelagos. They submitted the dispute to the German emperor, who ruled in 1871 in favour of the US position that the boundary should be the middle of the Haro Channel. *San Juan Water Boundary* case, B. Moore, *A History and Digest of International Arbitrations to which the United States has been a Party*, i (Government Printing Office, Washington, DC, 1898), at 230. A similar dispute arose between the same parties over an unclear description of the boundary in the Portland Channel areas between Alaska and British Columbia. Following the submission of the dispute to an *ad hoc* tribunal in 1903, the line was fixed by using simple geographic median lines and mid-channel lines. See O'Connell, *The International Law of the Sea*, ii (1984), 664–5.

[14] Protocol of 10 Mar. 1873, 1: Papers Relating to the Foreign Relations of the United States (1873), map facing p. xxv.

[15] e.g. British Honduras (Belize)–Mexico, 8 July 1893, 85 *BFSP* (1893), 58–60; France (Mauritania)–Spain, establishing the territorial sea boundary in the Western Bay, 27 June 1900, 32 *MNRG*, 59.

Dahomey and the Sudan, and the German territory of Togo, the thalweg line was drawn from the designated terminus of the river boundary up to a point in the lagoon where the thalweg was identifiable, from which point the boundary followed the true median line.[16] Several other agreements adopted the same solution.[17]

At this stage it was possible for Boggs to develop the equidistance concept in 1937. Drawing upon the median concept of land boundaries across rivers and lakes and a limited number of territorial sea boundaries, he identified three variations to this concept, namely a line (1) bisecting construction lines drawn between adjacent salient points on opposite coasts; or (2) equally dividing the surface areas; or (3) equally dividing the main navigable channel.[18]

This general understanding continued to form the basis for more recent territorial sea boundary agreements. The 1974 Treaty between the Federal Republic of Germany and the Democratic Republic of Germany established a line of 7.99 miles in the Bay of Ldieresisbeck. The line initially followed an equidistance line, then deviated in a north-easterly direction to coincide with the south-eastern edge of a shipping route ordered by British occupation authorities in 1945. The line was claimed by the FRG so as to preserve access to the ports of Ldieresisbeck and Travemula.[19]

An interesting example in this connection is the Judgment of the *ad hoc* Tribunal in the *Beagle Channel* case.[20] The case was primarily concerned with a dispute over treaty interpretation between Argentina and Chile relating to the maritime boundary and sovereignty of certain islands in the Beagle Channel.[21] With regard to the boundary issue the Tribunal adopted in 1977 a combined median/special circumstances line:

[16] Declarations of 28 Sept. 1912, 106 *BFSP* (1912), 1001.

[17] See e.g. Agreement between the same parties for the delimitation of the boundary between French Equatorial Africa and Cameroon (ibid. 974); Norway–Sweden, 15 Mar. 1904, establishing the boundary in the Gulf of Christiana (see the *Grisbadarna* case, 4 AJIL (1910), 227–8); UK–USA, 21 May 1910, dividing the boundary in the Passamaquoddy Bay and the 1925 extension of the boundary through the middle of the Grand Mannan Channel. See 'Information and Observations Submitted by Governments Regarding the Question of the Territorial Sea', *Yearbook of the ILC* (1953), ii. 86.

[18] W. Boggs, 'Problems of Water-Boundary Definition: Median Line and International Boundaries through Territorial Waters', 27 *Geographical Review* (1937), 445.

[19] Treaty of 29 June 1974. See *LIS* 74. See also the following earlier agreements: the Aaland Islands Convention, 20 Oct. 1921, establishing, *inter alia*, the territorial sea boundary between Finland and Sweden, 114 *BFSP* (1921), 421; Finland–USSR, 28 Oct. 1922, 19 *League of Nations Treaty Series*, 183, 194; Singapore–Johore, 19 Oct. 1927, establishing the territorial sea boundary in the Johore Strait, M. Whiteman, *Digest of International Law* (Department of State Publication, Washington, DC, 1963–70), iv. 311.

[20] *Beagle Channel* Arbitration (1977), repr. in 17 ILM (1978), 634.

[21] In 1881 Argentina and Chile concluded a treaty which had described the eastern mouth of the channel so vaguely that both the maritime boundary and jurisdiction over the islands of Picton, Lennox, and Nueva were disputed in the late 19th century and it was only in 1971 that the parties agreed to go to arbitration. For details see D. Greig, 'The Beagle Channel

In drawing its own line . . . the Court has been guided by . . . mixed factors of appurtenance, coastal configuration, equidistance, and also of convenience, navigability, and the desirability of enabling each party so far as possible to navigate in its own waters. None of this has resulted in much deviation from the strict median line except . . . near Gable Island where the habitually used navigable track has been followed.[22]

In some cases the boundary line is described as perpendicular to the general direction of the coasts, but, owing to the straight coastal formation in most of these instances, the resultant boundaries are technically median or equidistant lines.[23] For instance, in a reply to the Preparatory Committee of the Hague Conference, the Belgian Government stated its policy to be that the boundary 'between the contiguous territorial seas of two neighbouring countries should lie along a line perpendicular to the coast, at the end of the land frontier'.[24] Later, the French Government affirmed this view and explained that 'No interests are injured by this delimitation, since the adjacent French and Belgian coasts are straight and the frontier is perpendicular to the coast.'[25]

For the same reason, the French authorities in Tunisia had tacitly accepted the perpendicular line proposed in 1913 by the Italian authorities

Arbitration', 7 *AYBIL* (1976–7), 332–85. Following the announcement of the Judgment in 1977, which gave the three islands to Chile, Argentina repudiated the Award, as it realized the dramatic incursion of the Chilean straight baselines system which enclosed the three islands, from which a claim to a 200-mile epicontinental sea was made (see the Declaration of Nullity by Argentina, repr. in 17 ILM (1978), 739–50). After six years of negotiations, the two States reached in 1984 an agreement resolving their dispute. The main provisions of this agreement were (1) that the three islands be confirmed as Chilean territory; and (2) that Chile's territorial waters around the islands should be confined to a belt of 3 miles, while beyond this area a line drawn south from Cape Horn would be the boundary between the Chilean and Argentine claims of 200-mile EEZs. See Treaty of Peace and Friendship between Argentina and Chile, text in *LSB* 4 (Feb. 1985), 50.

[22] In 17 ILM (1978), 634, para. 10.

[23] The only known example of a 'perpendicular' line drawn between complicated coasts is that awarded in 1909 by the Court of Arbitration in the dispute between Norway and Sweden over their territorial sea boundary in the Fjord area, otherwise known as the *Grisbadarna* case. The reasons announced by the Court for implementing a perpendicular line were (1) that this line was 'much more in accord with the ideas of the seventeenth century' (when the parties first initiated their boundary in the area); (2) that it took account of the two parties' established fishing activities; and (3) the necessity of indicating the boundary in a clear and unmistakable manner (see the Award's English version in 4 AJIL (1910), 226, esp. 232, 234). However, the Court's line was correctly described as 'a compromise solution' which aimed at dividing the 'losses' and 'gains' between the two parties. The issue at stake was the rich lobster grounds of the Grisbadarna and Skjottegrunden banks. And the Court found it 'just and lawful' to apply a line which allocated Grisbadarna to Sweden and Skjottegrunden to Norway. See analysis of the case by Sang-Myon Rhee, 'Sea Boundary Delimitation between States before World War II', 76 *AJIL* (1982), 555, at 566–71.

[24] Belgian letter of 19 Dec. 1928, in 'General Observations Submitted by Certain Governments, Basis of Discussion', LN Doc. C.74.M.39.1929, v. 120.

[25] Doc. A/CN. 4/71/Add.2, 89.

in Libya for delimiting the fishing jurisdiction off Ras Ajdir.[26] On their metropolitan lands the same States had agreed in 1892 on a perpendicular line for the delimitation of their fishery jurisdiction in the Bay of Menton.[27] Albania and Yugoslavia agreed in 1926 on the same method for delimiting their territorial sea in the Adriatic.[28] The convenience of the perpendicular line where the coast is generally straight is confirmed in more recent territorial sea agreements. Examples are found in the Poland–Soviet Union Treaty of 1958[29] and the Brazil–Uruguay Treaty of 1975.[30]

Since most coastlines are irregular with indented coastal formations and scattered islands, the perpendicular line was found to be of limited assistance. On the other hand, the development of technical means of delimitation induced States to determine 'every segment' of their territorial waters. The equidistance rule appealed to these moves toward precise delimitation, for no other rule had proved to have such certainty in application.[31] States, therefore, sought to provide a detailed description of their territorial sea boundaries, either by producing a certain meridian[32] or an azimuth,[33] or by plotting a line connecting geodetic co-ordinates of latitude and longitude.[34]

[26] See *infra*, Ch. 8 s.4(*b*)(ii)(2).
[27] See *infra*, Ch. 8 n. 42.
[28] See *infra*, Ch. 8 s.3(*a*).
[29] Treaty of 18 Mar. 1958, Art. 1 provides that the territorial sea boundary between the two States 'shall follow a line perpendicular to the shoreline at the terminal point of the Polish–Soviet State Frontier . . . and running to the point of intersection with the outer limit of the territorial waters of [Poland]. An extension of this line in the same direction to the point of intersection with the outer limit of the territorial waters of [the Soviet Union] shall constitute the boundary of the territorial waters of the Soviet Union.' *LIS* 55.
[30] According to the Treaty of 21 July 1972, the maritime boundary is defined by a line which runs 'on a bearing of 128 sexagesimal degrees . . . to the outside limit of the territorial sea of both countries'. The Treaty describes the line as 'running in a direction nearly perpendicular to the general line of the coast'. This line is considered a precise demarcation of the 'equidistance principle' which the parties had previously agreed upon to govern their territorial sea boundary. See *LIS* 73.
[31] The equidistance rule has gained great impetus especially since Boggs's precise definition and description of plotting it on a chart. See Boggs, 'Problems of Water-Boundary Definition'. See also his article 'Delimitation of Seaward Areas under National Jurisdiction', 45 *AJIL* (1951), 240.
[32] See e.g. France–Spain, 30 Mar. 1879, establishing the territorial sea boundary in the Bay of Figuier by a meridian passing through the middle of a line connecting two points on the coast, 70 *BFSP* (1879), 176–8. The same method was reaffirmed in the Territorial Sea Boundary Agreement concluded between the two States in 1974. See *LIS* 83. The meridian line, reflecting an equidistance line, was also established in the 1885 Treaty between Portugal and Spain for the delimitation of their territorial waters in the estuary of the Río Guadiana, 14 *MNRG* 2nd, 77.
[33] e.g. Turkey–Soviet Union, 17 Apr. 1973, establishing the territorial sea boundary in the Black Sea. The line was drawn for 12 miles along an azimuth of 290°. See *LIS* 59.
[34] The determination of the boundary by this method, without spelling out the principles of law employed, is the most common practice in contemporary negotiated boundary agreements. See e.g. Canada–France, 27 Mar. 1972, determining the limits of the territorial sea of Canada and the fishery zone of St Pierre and Miquelon, *LIS* 57; Indonesia–Singapore,

In all these and other precisely delimited boundaries 'equidistance' was applied with a flexibility necessary to reflect various relevant factors, whether related to the geography of the coast, such as the presence of islands,[35] or the instability of the coast at the international boundary terminus.[36] Equidistance has also been applied with the aim of respecting legitimate claims such as historic or other vested rights.[37] It was this practical utilization of Boggs's equidistance line that must have attracted the attention of the Committee of Experts and led them to recommend it to the International Law Commission.[38]

25 May 1973, establishing their maritime boundary in the Strait of Singapore, *LIS* 6; India–Sri Lanka, 8 July 1974, establishing their respective historic waters boundary in Palk Bay, *LIS* 66; Australia–Papua New Guinea, Dec. 1978, delimiting, *inter alia*, the territorial waters of the Australian off-lying island groups Abusi and Saibi, see *infra*, Ch. 10, text at nn. 81–2.

[35] When islands are sufficiently close to their own mainland coast, they are in principle given full effect for determining territorial sea boundaries by the equidistance method. For example, Art. 1 of the 1980 Boundary Agreement between Burma and Thailand provides that the boundary, including that dividing their 12-mile territorial waters, is an equidistance line from the respective coasts, including islands. See *LIS* 102. But on grounds of 'special circumstances', small islands in most of the cases are either ignored or given reduced effect. See e.g. *supra*, Ch. 1 s.3(*b*)(iii).

[36] e.g. following the Treaty of 1885 between Spain and Portugal, delimiting their territorial sea in the Río Guadiana estuary by a meridian line, a dispute arose over the location of the line because of the fluctuations of the delta. On 27 Mar. 1893 Spain and Portugal settled the problem by placing the boundary midway between the two claimed meridians (14 MNRG 2d, 414). According to the 1970 Mexico–USA Agreement, the boundary line in the Gulf of Mexico is determined by two straight baselines. The first begins at the centre of the mouth of the Río Grande and runs in a straight line to a fixed co-ordinated point (irrespective of any future change in the mouth of the river). From there the boundary continues seaward in a straight line which represents 'a practical simplification' of an equidistance line drawn in accordance with 'Arts. 12 and 24 of the Territorial Sea Convention'. See *LIS* 45.

[37] e.g. in the 1910 Boundary Agreement between the UK and USA regarding the Passamaquoddy Bay (*supra*, n. 17), the boundary was a negotiated line that deviated from a strict median line because historical fishing rights and mutual convenience of navigation were taken into account. On the other hand, historic claims can become a main obstacle to establishing a territorial sea boundary. Good examples are the 'Wielingen' controversy between Belgium and The Netherlands, and the 'Ems–Dollard' affair between Germany and The Netherlands. Wielingen is the principal channel of navigation in the River Scheldt, where the territorial sea boundary between Belgium and The Netherlands is partly delimited. The boundary is not complete due to a Netherlands claim to the whole channel on historic grounds. The claim is denied by Belgium. Ironically, The Netherlands denied a historic claim to the mouth of the River Ems by Germany, where the water boundary between Germany and The Netherlands is partly delimited by the Treaty of Meppen of 1824. However, with regard to the Ems–Dollard dispute the two States reached an agreement in 1984 which concentrated on economical and environmental aspects, but left the boundary question almost unresolved. See G. Tanja, 'A New Treaty Regime for the Ems–Dollard Region', 2 *IJECL* (1987), 123.

[38] One finds it difficult to assume otherwise since Boggs himself was a member of that Committee. His understanding of a complicated rule of 'equidistance–special circumstances' was established as early as 1937. See *supra*, text at n. 18.

3. THE CODIFICATION OF THE 'EQUIDISTANCE–SPECIAL CIRCUMSTANCES' RULE

(*a*) *The codification of the rule in the Territorial Sea Convention*

As indicated above, following the recommendation of the Committee of Experts, the Commission adopted the equidistance rule for the delimitation, *inter alia*, of the territorial sea.[39] Thus in the final draft articles of 1956 the Commission formulated articles 12 and 14 on the delimitation of the territorial sea which were modelled on the Commission's draft article on the delimitation of the continental shelf.[40] Article 12 provided in part:

> 1. The boundary of the territorial sea between two States the coasts of which are opposite each other . . . shall be fixed by agreement between those States. Failing such agreement and unless another boundary line is justified by special circumstances, the boundary is the median line.

Similarly, article 14 read:

> 1. The boundary of the territorial sea between two adjacent States shall be determined by agreement between them. In the absence of such agreement, and unless another boundary is justified by special circumstances, the boundary is drawn by application of the principle of equidistance.[41]

At the Geneva Conference, these two texts were reworded and unified. The Commission's texts were criticized as being incapable of dealing with the difficult situation where two States claimed different limits of territorial waters.[42] The Commission's prescriptive formula clearly presupposed final agreement on a uniform breadth for the territorial sea and therefore only applied if the claims of two States were of equal breadth. To amend this situation, Norway proposed a prohibitive formula which provided an express rule against the unilateral extension by one State of its territorial sea boundary beyond the median line. The Norwegian proposal also combined draft articles 12 and 14 in one article, since the problems dealt with in the two articles 'were so closely interrelated in some cases as to be practically indistinguishable'.[43] Articles 12 and 14

[39] See *supra*, n. 3 and accompanying text.

[40] The Commission had found it best to deal first with the delimitation of the continental shelf, and then to draft on the same lines the corresponding delimitation provision in respect of the territorial sea. See the Commission's comment in *Yearbook of the ILC* (1954), ii. 157; and Art. 72 (delimitation of the continental shelf) ibid. (1956), ii. 300.

[41] Ibid. (1956), ii. 271–2.

[42] See comment by the Norwegian delegate, UNCLOS I, *Official Records*, iii. 187–8.

[43] Ibid.

were consequently amended and amalgamated in one article, the first sentence of which reads

> 1. Where the coasts of two States are opposite or adjacent to each other, neither of the two States is entitled, failing agreement between them to the contrary, to extend its territorial sea beyond the median line every point of which is equidistant from the nearest points on the baselines from which the breadth of the territorial sea of each of the two States is measured.

As a result of this unification, together with the insertion of the specific circumstance of 'historic title',[44] the provision on special circumstances proved too long to be maintained in the same sentence providing for the equidistance principle. Thus, a separate sentence was finally drafted as the second sentence of Article 12(1):

> The provisions of this paragraph shall not apply, however, where it is necessary by reason of historic title or other special circumstances to delimit the territorial seas of the two States in a way which is at variance with this provision.

The above formulation raises two questions of immediate concern. First, there is the question of the status which Article 12 tends to attach to the 'special circumstances' clause: is this clause meant to be part of, or an exception to, the equidistance rule? Second, there is the problem of determining which factors may be regarded as 'special circumstances'.

(*b*) *The status of the 'special circumstances' clause*

Recalling that there is considerable evidence in the nineteenth and early twentieth centuries of State practice suggesting a broad understanding of the role of equidistance, it is unfortunate that the early discussions of the ILC reveal a stricter view on the content of this rule. In these discussions, the Commission viewed equidistance as a mere technical method which might not be capable of adaptation to adverse situations.[45] This was perhaps a natural consequence of the unrealistic target which the Commission was initially aiming at, that is, the formulation of a clear-cut technical method of delimitation, which, nevertheless, had to be capable of general application. Having this approach, it was only natural that 'equidistance' was initially proposed as the 'general rule', but envisaging 'exceptions'.[46] The Commission's initial strict approach is evident in the language used in the Special Rapporteur's Questionnaire to the Committee of Experts. The Committee was ostensibly asked to determine *which of*

[44] See *infra*, text at n. 70.

[45] See the views of members of the Commission in the 1951 and 1952 Sessions, *Yearbook of the ILC* (1951), i. 282, and (1952), i. 180–6.

[46] See ibid. (1953), i. 128–32.

the following methods should be used for the territorial sea delimitation of adjacent States:[47]

A. by continuing the Land Frontier? B. by a perpendicular line . . . ? C. by a line drawn vertically on the general direction of the coastline? D. by a median line? If so, how should this line be drawn? To what extent should islands, shallow waters and navigation channels be accounted for?

Question D in particular indicates the Commission's (or its Rapporteur's) complete disregard for established State practice whereby the combined use of the equidistance/thalweg line provided a clear indication on the development of the equidistance–special circumstances rule.[48]

However, following further discussions, the strict view on the role of equidistance began to change. A much broader reconstruction of 'equidistance–special circumstances' in one combined rule began to emerge. In its 1953 report on the delimitation of the continental shelf the Commission stated:

Having regard to the conclusions of the committee of experts . . . *the Commission now felt in the position to formulate a general rule, based on the principle of equidistance* . . . The rule thus proposed is subject to such *modifications* as may be agreed upon by the parties. Moreover, while in the case of both kinds of boundaries the rule of equidistance is the general rule, it is subject to modification [not 'exception' as was previously announced] in cases in which another boundary line is justified by special circumstances. As in the case of the boundaries of coastal waters, provisions must be made for departures necessitated by any exceptional configuration of the coast, as well as the presence of islands, or navigable channels. *To that extent the rule adopted partakes some elasticity.*[49]

Considering the applicability of the 'equidistance–special circumstances' rule by third-party settlement, the Commission emphasized its nature as a rule of law:

no special provision was considered necessary for submitting any resulting disputes to arbitration. Such arbitration, while expected to take into account the special circumstances calling for modification of the major principle of equidistance, is not contemplated as arbitration *ex aequo et bono. The major principle must constitute the basis of the arbitration, conceived as settlement on the basis of law, subject to reasonable modifications necessitated by the special circumstances of the case.*[50]

At the 1958 Conference, the debate on the delimitation provisions indicated the conference's intention to formulate 'equidistance–special

[47] UN Doc. A/CN. 4/61/Add.1.
[48] See *supra*, s.2.
[49] *Yearbook of the ILC* (1953), ii. 216 (emphasis added).
[50] Ibid. (emphasis added).

circumstances' as one combined rule. Attempts by certain delegates to delete reference to 'special circumstances'[51] were overwhelmingly defeated in both the First and Fourth Committees.[52] The situation was properly explained by the United Kingdom delegate:

> the median line would always provide the basis for delimitation. If both the States involved were satisfied with the boundary provided by the median line, no further negotiation would be necessary; if a divergence from the median line appeared to be indicated by special circumstances, another boundary could be established by negotiation, but *the median line would still serve as the starting point.*[53]

That this was the general understanding of the Conference appears, almost invariably, in the resultant delimitation provisions. According to these provisions, 'special circumstances' would 'necessitate' (Article 12) or 'justify' (Article 6 of the Shelf Convention) a deviation from a strict equidistance line.

Irrespective of any significance attributed to the difference in terminology,[54] what becomes evident is that the mere existence of factors likely to fall within the category of 'special circumstances' is insufficient *per se* to invoke the application of the clause and permit a deviation from the equidistance line. It is necessary therefore to prove, as in the case of Article 12, that those special circumstances 'necessitate' a deviation. This is not equivalent to saying that 'special circumstances' are confined to exceptional cases requiring an extraordinary burden of proof. It merely indicates that the proof of the 'necessity' of, or 'justification' for, qualified equidistance is an essential operation for applying the rule itself. The provision's aim is to place an obligation to consider what factors can be regarded as 'special' circumstances (and consequently justify modification) but not to place restriction on the initial right to recourse to 'special circumstances' as one element of a combined equidistance–special circumstances rule. This obligation can therefore be regarded as introducing an essential quotient of those considerations of equity and justice which were foremost in the minds of the members of the Commission.

However, the uncertainty over the relation between 'equidistance' and 'special circumstances' was reduced when the Court of Arbitration explained in 1977 the 'equidistance–special circumstances' rule. This Award will be considered elsewhere in the context of the continental

[51] See the proposals of Yugoslavia (A/CONF. 13/C. 1/L.60) and Portugal (A/CONF. 13/C. 1/L.101).

[52] These proposals were rejected in the Fourth Committee (dealing, *inter alia*, with the delimitation of the shelf) by 39 votes to 9, with 8 abstentions, UNCLOS I, *Official Records*, vi. 97. With regard to territorial sea delimitation, the First Committee (territorial sea) adopted the special circumstance clause by 38 votes to 15, with 19 abstentions, ibid. iii. 193.

[53] Ibid. vi. 92 (emphasis added).

[54] See *infra*, s.3(*c*).

shelf, but it is certain that the same considerations will have a bearing on the case of territorial waters.

(*c*) *The content of the 'special circumstances' clause*

Article 12 includes no definition of the concept of 'special circumstances'. The experts referred in their Report only to navigation and fishing interests, by way of example.[55] Also among the geographical factors which are generally accepted as capable of qualifying as special circumstances are islands and exceptional coastal configurations.[56] The experts did not include islands in their reference to special circumstances, but the *travaux* leave little reason to doubt that certain islands could be regarded as special circumstances. It was only at the 1958 Conference that Norway had secured the incorporation of a specific reference to historic title. These latter categories of special circumstances are considered below in some detail, but two points of immediate concern are mentioned now.

First, these examples of special circumstances may also apply in the delimitation of shelf boundaries under Article 6 of the Shelf Convention. As seen earlier, the legislative history of the two provisions demonstrates their common origin in the Report of the Experts, on which the Commission had relied in drafting the two provisions. Under Article 12 a departure from strict equidistance line is only possible if special circumstances 'necessitate' it. Under Article 6 strict equidistance is already excluded if special circumstances 'justify' another boundary line. This slight difference in wording does not indicate, as some imply,[57] that 'special circumstances' bear in Article 12 a connotation different from that which they bear in Article 6. The difference in wording is explicable partly by the fact that the drafting of the two Conventions was done by different Committees.[58]

Secondly, the enquiry *in abstracto*, into which factors may qualify

[55] See *supra*, text accompanying n. 2. See also comment by François, *Yearbook of the ILC* (1954), i. 100.

[56] See *Yearbook of the ILC* (1954), ii. 158

[57] See H. Jayewardene, 'The Regime of Islands in International Law', Ph.D. thesis (Cambridge, 1981), 395–7. In the *North Sea* cases Germany contended that the difference in wording implied a wider scope of application in the Shelf Convention. Denmark and The Netherlands rejected this contention, pointing out that there was no indication in the *travaux* that the conference considered that there would be any essential differences in the impact of 'special circumstances' in the delimitation of territorial sea boundaries from their impact on the delimitation of continental shelf boundaries (*North Sea Continental Shelf* cases, *Pleadings*, i. (1968), 63, 68–9, 420, 485). The Court did not elaborate on this point. It referred, however, in para. 59 to the logical proposition that the distorting effect attributed to coastal configurations in the delimitation of the lateral boundaries of the territorial sea is far more limited than in the case of the continental shelf. This difference comes naturally as a result of the very much greater distances involved in the delimitation of the continental shelf.

[58] See *supra*, n. 52. See also *infra*, s.3(*c*)(ii).

under the category of 'special circumstances', is believed to be an unnecessary investigation for both practical and juridical reasons. From the operative point of view, the term 'special circumstances' refers to the variety of situations which may be classified as 'special'. What is admitted as a 'special circumstance' in a specific case cannot automatically be regarded as such in another. When considered as a juridical concept, the 'scope' given to it will determine its ultimate impact on the law on delimitation. In this respect, it can safely be said that both State practice and adjudication envisage, as will be established, 'special circumstances' as a concept which embodies a general appeal to equitable considerations rather than as a notion of defined or limited categories of circumstances with an exceptional effect.[59]

(i) Islands as a special circumstance

The Report of the Experts did not explicitly list islands with the special reasons which, in their opinion, may divert the boundary from the strict median line. The Report however stated that, unless otherwise agreed between the parties concerned, 'all islands should be taken into consideration in drawing the median line'.[60] This should not be interpreted to mean that islands have significance as special circumstances only within the narrow, subjective context of agreement. The aim in providing a delimitation provision was to secure a legal solution when the parties could not agree. The special circumstances proviso is a necessary element thereto and was formulated to temper the rigidity of the equidistance method and secure an equitable result. There is no justification for regarding islands as falling outside the clause in situations where the parties are not agreed on their inclusion. Moreover, the *travaux* clearly indicated that islands had consistently been recognized as the typical example of 'special circumstances' in both the delimitation of the territorial sea and the continental shelf. The Report indicated that certain rocks and shoals could be 'disregarded in laying down the median line'.[61] During the 1952 Session of the International Law Commission, Hudson referred to several difficult situations occasioned by islands. Particular reference was made to an adjacent State situation where the frontier reached the sea.[62] In his comment on what might constitute 'special circumstances', François mentioned the case where a small island opposite one State belonged to another.[63] Similarly, reflecting on the Experts' Report,

[59] *Infra*, Ch. 4 ss. 3, 4.
[60] *Supra*, n. 2.
[61] Ibid.
[62] *Yearbook of the ILC* (1952), i. 180–1.
[63] Ibid. (1953), i. 128.

François referred to the case of islands not taken into account by agreement between the parties and gave, as an example, the case where an island was further away from the State to which it belonged than from the other State.[64]

At UNCLOS I, Fitzmaurice, the United Kingdom delegate, gave the example of small islands as a case likely to cause complications.[65] In the context of continental shelf delimitation he referred specifically to islands in these terms:

> For the purpose of drawing a boundary, islands should be treated on their merits, very small islands or sand cays on a continuous continental shelf and outside the belts of territorial sea being neglected as base-points for measurement and having only their own appropriate territorial sea.[66]

In the First Committee, the Burmese delegate sought to add to the Commission's general proposition that 'every island has its own territorial sea' the following: 'provided that when an island belonging to one State is within the territorial sea of another State, that island shall not have its own territorial sea.'[67] He criticized the failure of the Commission to specify the position of a foreign island situated within the territorial sea of another State. As he said, 'if such an island were to have its own territorial sea, it would also be entitled to its own contiguous zone and continental shelf and the result would only be an intolerable conflict of jurisdiction', and the 'median line principle' could at the very best only provide a solution on the landward side of the island.[68] In the Fourth Committee a proposal specific to islands was made by Italy which in effect ignored islands for the purposes of a median line boundary between opposite States.[69]

These statements and proposals are of some significance, not only in the context of indicating the self-evident proposition that islands may well qualify as constituting 'special circumstances', but also in that they seem to indicate that not all islands can have equal impact. The smaller and/or the further the island from the mainland of the owning State the less effect on the boundary it would have. It will be seen that this approach was reinforced in UNCLOS III.

[64] Ibid. (1954), i. 100, para. 55.

[65] UNCLOS I, *Official Records*, iii. 189.

[66] A/CONF. 13/42, p. 93.

[67] A/CONF. 13/C. 1/L.3.

[68] UNCLOS I, *Official Records*, iii. 161. See also the Burmese statement (ibid. iii. 4) that Art. 10 of the Convention 'called for some clarification, since insurmountable difficulties would arise if any island belonging to one State was situated in the territorial sea of another State'.

[69] A/CONF. 13/C. 4/L.25.

(ii) 'Historic title' as a special circumstance

In the final draft articles prepared by the Commission in 1956, there was no reference to historic titles. In the course of UNCLOS I, Norway suggested that the equidistance line 'Shall not apply, however, where one of the States concerned through prescriptive usage has acquired the right to delimit its territorial sea in a way which is at variance with the provision'.[70]

The words 'through prescriptive usage' were replaced by the words 'by reason of historic title', and thus Article 12(1) was completed. The *travaux* therefore shed little light on the impact of this specific category of 'special circumstances'. It has been assumed, however, that strict equidistance would not apply in the case of *longi possessio*, or historic rights recognized in international law.[71] This would seem to suggest that 'historic title' as a special circumstance ought to enjoy 'preferential treatment' in the process of securing the balance necessary for employing the 'equidistance–special circumstances' rule.

It is mere speculation that the *travaux* uphold this view. As pointed out above, the peculiar formulation of Article 12, when compared with the corresponding shelf provision, was partly attributable to the insertion of 'historic title' in the special circumstances clause. Although now used to cover both 'historic title' and 'other special circumstances', the words 'necessary by reason of' were originally designed to fit a clearly defined exception, that is 'prescriptive use', which was finally replaced by the phrase 'historic title'. In this connection, if there is the slightest logic in the contention that the word 'necessary' bears a connotation stronger than that attributed to the word 'justified' in Article 6, this will only be true for the special case of 'historic title'. This could mean that a 'historic title', unlike 'any other special circumstances', must be given full recognition in the delimitation at variance with the equidistance line. This interpretation may attract some support from the general understanding that historic titles represent established rights of a legal value equal to that attributed to rights acquired by an explicit agreement.[72]

However, within the context of Article 12, the impact of 'historic rights' on a strict median line may take one of two forms:

1. A historic title may take the form of a historic claim to a wider marginal belt of territorial sea. In fact, the specific reference

[70] A/CONF. 13/C. 1/L.97.

[71] UNCLOS I, *Official Records*, iii. 187.

[72] See ibid. In the *Grisbadarna* case the Permanent Court of International Justice asserted that 'It is a settled principle of the law of nations that a state of things which actually exists and has existed for a long time should be changed as little as possible . . . this rule is specially applicable in a case of private interests.' *Supra*, n. 23, p. 233.

to historic title in Article 12 appears to have been inserted to cover the situation where a State 'might be entitled by historical title or prescriptive usage to go further in delimiting its territorial sea from that of neighbouring or opposite States than the median line would allow'.[73]

2. The more common case is where a historic title takes the form of a right to construct straight baselines either on the basis of the principle of historic bays or by invoking long-standing socio-economic considerations which justify liberal interpretations of the requirements on straight baselines.[74] These baselines increase the outer limit of the territorial sea by enclosing large areas of sea within it. The juridical status of such areas is that of internal waters. Ideally, such claims can be considered a 'special circumstance' for the purpose of Article 12.

(*d*) *The reaffirmation of Article 12 in the 1982 Convention*

Following the submission of several proposals on the delimitation of the territorial sea,[75] the Second Committee of UNCLOS III, which dealt, *inter alia*, with the regime of the territorial sea, was able at the Caracas Session to identify the 'Main Trends' and to present in a systematic way the formulae which best reflected those trends.[76] With regard to the delimitation of the territorial waters,[77] the Second Committee presented

[73] Comment by the Norwegian delegate, UNCLOS I, *Official Records*, iii. 190.

[74] See *supra*, Ch. 1 s.2(*b*).

[75] See e.g. the proposals of The Netherlands, Doc. A/CONF. 62/C. 2/L.14; Romania, L.18; Bulgaria, GDR, Poland, USSR, L.26; Turkey, L.9.

[76] UNCLOS III, *Official Records*, vi. 136. For the text of the document 'Main Trends', see ibid. iii. 107–42.

[77] At the Caracas Session there was a proposed article for the delimitation of the contiguous zone, identical to Art. 24(3) of the Territorial Sea Convention. The article was included in the 'Main Trends' but never repeated in the following Negotiating Texts of the Conference. There has never been a convincing explanation for that omission. Although they are not too frequent, situations may occur in practice when the delimitation of contiguous zones is required, or of a contiguous zone, on the one hand, and an economic zone on the other. During the discussion of the Consultative Group of the Second Committee at the 1977 Session, Cyprus demanded an explanation of this omission. Explaining why the delimitation provision had been omitted from the RSNT, the Chairman said that several delegations opposed this provision. Vukas, a member of the Yugoslav delegation, referring to his own notes at the informal meetings of the Second Committee, disputed the correctness of the Chairman's statement. He explained that only The Netherlands 'expressly denied the usefulness' of the insertion of such a provision. The omission of a provision on the contiguous zone delimitation was specifically opposed by Cyprus, Peru, and Pakistan. Vukas's explanation of this omission is that, in view of the rarity of contiguous zone delimitations, the delegates at the Conference felt reluctant to create additional problems for the already complex negotiations on the exclusive zone/shelf delimitation. See B. Vukas, 'The LOS Convention and Sea Boundary Delimitation', in Vukas (ed.), *Essays on the Law of the Sea* (Zagreb, 1985), 147, at 153–64.

certain formulae, including Formula A,[78] which merely recited the provision embodied in Article 12 of the Territorial Sea Convention. Right from the beginning, it was obvious that Article 12 would be supported on the basis of international law theory and also as being in keeping with traditional practice, having been widely used in bilateral treaties, especially with regard to narrow seas.[79]

The retention of Article 12 was also favoured because it was seen as providing a balanced solution, since it contained the objective criterion of equidistance, while being sufficiently flexible to assimilate special circumstances.[80] Thus, when the issue of the delimitation of territorial waters was referred to Negotiating Group 7 (one of several groups appointed by the 1978 Session of the Conference to deal with outstanding hard-core problems),[81] the retention of Article 12 was made without difficulty. The formula was regarded as being likely to command a wide support, even amongst States which were reluctant to accept the extension of the territorial sea to 12 miles.[82] The retention of Article 12 was also explained by the lack of support for the alternative formula, that is, the delimitation of the boundary 'according to equitable principles'. Primarily, objection was made on the basis that this latter formula introduced an element of subjectivity and ambiguity,[83] and did not deal with the problem adequately.[84] Consequently, the 1958 formula was reaffirmed in all the negotiating texts made out by the conference and entered finally, with minor linguistic changes,[85] into the 1982 Convention as Article 15.

However, quite central to these moves to reformulate, *inter alia*, the territorial sea delimitation provisions was the issue of islands, which had not been dealt with comprehensively in UNCLOS I and consequently was the source of difficulty in many boundary delimitations. The historical development of this issue as well as the outcome in UNCLOS III is discussed in literature quite extensively.[86] What is proposed here is to

78 UNCLOS III, *Official Records*, iii. 111.
79 Ibid. ii. 101, 111.
80 Ibid. ii. 119.
81 Doc. A/CONF. 62/62, para. 11(5).
82 Save for the question of passage through straits which was not of course directly relevant to the delimitation problems. See e.g. 'Statement of the Representative of the USA in Response to the Statement by the Vice-Chairman of the Group of Coastal States', UNCLOS III, *Official Documents*, xii. 111.
83 UNCLOS III, *Official Records*, ii. 111.
84 Ibid. ii. 119.
85 On the basis of a suggestion by the Yugoslav delegation, the words 'provisions of this paragraph' at the beginning, and 'with this paragraph' at the end of the second sentence, were replaced by 'The above provisions' and 'therewith', respectively. See Vukas, 'The LOS Convention', 151.
86 The legal regime of islands is discussed thoroughly in the following studies: Bowett, *The Legal Regime of Islands in International Law*; C. Symmons, *The Maritime Zones of Islands in International Law* (Martinus Nijhoff, The Hague, 1979); H. Dipla, *Le Régime*

recall the interesting phenomenon which might indeed be considered an exception to the basic trend at UNCLOS III, which was aimed at expanding, rather than restricting, the breadth of maritime zones, as was exemplified by the generally approved expanded territorial sea of 12 miles and the 200-mile EEZ. With regard to islands, the trend was generally aimed at curtailing the traditional view that every island, properly so called, generates, *inter alia*, a territorial sea. In fact, the origin of this trend started even before the Conference,[87] but was clearly demonstrated in many proposals submitted at the Caracas Session.[88]

Certain proposals stipulated a specific size criterion for the definition of islands, so as to exclude small formations, even if constantly above sea level. The Maltese proposal, for instance, defined an island as 'a naturally formed area of land, more than one square kilometre in area, surrounded by water, which is above water at high tide'.[89] Even in the case of islands qualified under this definition, Malta did not propose a uniform entitlement to 'ocean space' (defined as comprising waters and seabed areas beyond the internal waters). According to Article 11(3) of the proposal, when islands are less than 10 sq.km. in area, 'the jurisdiction of the State exercising sovereignty or control may extend only to a belt' not exceeding 12 miles from the applicable baseline. Some proposals aimed at giving smaller features such as 'rocks' and 'islets' more restricted zones, or indeed none at all. An interesting proposal was that of Romania, which defined an islet as 'a naturally formed elevation of land . . . less than one kilometre in area'.[90] Another concept, an 'island similar to an islet', was defined as a naturally formed area of land which 'is more than one square kilometre but less than . . . kilometres in area which is not or cannot be inhabited (permanently) or which does not or cannot have its own economic life'. Under these provisions the possibility that such 'smaller' formations might generate a territorial sea was not completely ruled out, but, considering the general aim of these provisions, it was also clear that such formations were not meant to have an automatic right to a territorial sea.

juridique des îles dans le droit international de la mer (Presses Universitaires de France, Paris, 1984).

[87] See e.g. the Maltese Draft Treaty (1971) and the OAU'S Declaration of Principles (1971), both submitted to the Seabed Committee in 1973, and resubmitted at the Caracas Session. See *infra*, nn. 89, 91, respectively.

[88] There were, however, a number of countering proposals, submitted by island States and States possessing a large number of islands, which all aimed at treating islands on a par with continents. See e.g. the Greek draft articles on National Maritime Zones, submitted at the Caracas Session (A/CONF. 62/C. 2/L.22); draft articles jointly submitted by New Zealand, Fiji, Tonga, and Western Samoa (A/CONF. 32/C. 2/L.30).

[89] The Maltese Draft Treaty as amended in 1973 (UN Doc. A/AC. 138/SC.II/L.28).

[90] Art. 11 of the Romanian draft articles on the Regime of Islands (UN Doc. A/CONF. 62/C. 2/L.53). See also a similar definition in Art. 1 of the Maltese submission.

Other proposals at the Conference introduced criteria based on general conceptions such as the habitability, economic viability, and proximity of an island to the mainland. According to the joint African draft articles,[91] even an island (defined as 'a vast naturally formed area of land') did not enjoy full maritime jurisdiction if it was 'non-adjacent' to the owning State's coastline. If islands were considered 'non-adjacent' (i.e. 'not situated in proximity of the coasts of States to which they belong'), their 'maritime space' was to be delimited on the basis of relevant factors taking into account equitable criteria, including the 'size of the islands', 'their geographical configuration', their adjacency to 'the maritime zone of another State', and their influence on 'the equity of the delimitation'. Also formulated in general terms was the Turkish submission, which was indeed more restrictive, inasmuch as islands might not have any 'marine space' if they were 'without economic life and situated outside the territorial sea of a State'. Furthermore, 'rocks' were equated with 'low-tide elevations' as not having any 'marine space' of their own.[92]

The direct impact of these moves on the resulting 1982 Convention has been limited. It is only in relation to the economic zone or continental shelf entitlement that Article 121(3) provides that rocks 'which cannot sustain human habitation or economic life of their own' are excluded from any entitlement. But these moves certainly indicate that the traditional view that all islands or insular territories of whatever significance should have identical maritime zones to those of mainlands is no longer valid. The present view seems to suggest that an island's entitlement to a maritime zone is ultimately determined in the light of factors peculiar to that island, e.g. its economic importance, political status, size, location, etc. These factors are also relevant in determining the interrelated, albeit distinct, question of how much territorial sea the island is entitled to in a territorial sea boundary with another adjacent or opposite State. A typical case of outstanding peculiarity is the 'Spanish enclaves' on the Mediterranean coast of Morocco. As will be seen elsewhere, four of these enclaves are 'isolated' insular features.[93]

4. CONCLUSIONS

The early development of the law on maritime boundaries through territorial sea boundary agreements indicates that States utilize 'modified

[91] Joint African draft articles (A/CONF. 62/C. 2/L.62). This African proposal reflected the spirit of the OAU's Declaration on Issues of the Law of the Sea which was submitted to the Seabed Committee in 1973. In spelling out the 'Main Trends', the Second Committee inserted the Joint African and Romanian draft articles as Formulae B and C, respectively, in pt. XIII (regime of islands).

[92] Draft articles of Turkey, submitted at the Caracas Session (A/CONF. 62/C. 2/L.55).

[93] See *Infra*, Ch. 8 s.4(*b*)(iii).

equidistance' as the central rule of delimitation. This modification, however, has been regarded as an essential element for the formulation of a general rule of law, since equidistance, as these agreements show, is not envisaged in the narrow geographical sense of the word.

This understanding continued to develop in recent State practice. In the interest of attaining the goal of acceptable 'division', special circumstances have been closely and inherently associated with the idea of a median/equidistant line. The thalweg line, for instance, has been applied in association with the median line so as to guarantee equal access to navigable waterways. Despite its limited applicability, the perpendicular line has been found ideal for dividing the territorial sea into equal parts when the coast is generally straight. When it can be constructed, the perpendicular line is useful not only in testing the equity of an equidistant line but also in simplifying such a line. From the outset, the presence of islands or historic rights has demanded special treatment in the construction of a median line. These factors were therefore perceived as complementary elements for the application of the equidistance rule. It is this perception which enriched and enhanced the general applicability of this rule; so much so that the International Law Commission was able to identify its basic characteristics after a careful recommendation by the Committee of Experts. The decision at UNCLOS III to retain the formula of Article 12 confirms the flexibility and thus the adequacy of that rule, at least for the delimitation of the extended, 12-mile territorial sea. Quite contemporary to this latter adoption was the trend to treat islands restrictively, so that the question of their entitlement to, *inter alia*, territorial waters as well as their impact on boundary delimitation would be considered in the light of the subjective circumstances of each case. There is little doubt that such a trend offers some clarity as to the 'scope' of the element of 'special circumstances'. This indicates at the same time the capacity of 'equidistance–special circumstances' to absorb the developments directed by the sources of the law. It is because of this capacity to evolve in accordance with the sources of the law that 'equidistance–special circumstances' is submitted to reflect a 'rule of law'. The existence of a substantive formula within that rule would nevertheless prevent such evolution from reaching the point of whatever might be subjectively felt to be equitable in a particular case. This is a major value for the certainty of the law which the judicially announced principle of equity fails to attain, but this is a central argument that will be discussed extensively in the next two chapters.

3

Delimitation of the Continental Shelf: The Applicable Law Following the Adoption of the 1958 Convention

1. INTRODUCTION

Article 6 of the 1958 Continental Shelf Convention deals in two paragraphs with inter-State delimitation in situations where the States involved are either opposite or adjacent to each other. It reads as follows:

1. Where the same continental shelf is adjacent to the territories of two or more States whose coasts are opposite each other, the boundary of the continental shelf appertaining to such States shall be determined by agreement between them. In the absence of agreement, and unless another boundary line is justified by special circumstances, the boundary is the median line, every point of which is equidistant from the nearest point of the baselines from which the breadth of the territorial sea of each State is measured.

2. Where the same continental shelf is adjacent to the territories of two adjacent States, the boundary of the continental shelf shall be determined by agreement between them. In the absence of agreement, and unless another boundary line is justified by special circumstances, the boundary shall be determined by application of the principle of equidistance from the nearest point of the baselines from which the breadth of the territorial sea of each State is measured.

As stated earlier, following the Report of the Committee of Experts in 1953, the discussions in the International Law Commission revealed a new approach towards the role and purpose of a delimitatory rule.[1] The discussions which took place in 1953 (when the essence of Article 6 was finally drafted) tended toward formulating a rule of general application, namely, the 'equidistance–special circumstances' rule. This replaced the hitherto strict view which aimed at codifying a geometrical method which might be subject to exceptions.[2]

[1] See *supra*, Ch. 2 s.3(*b*).

[2] When the Commission took up in 1950 the question of codifying rules for the legal regime of the continental shelf, discussions concerning the nature and extent of the rights of a coastal State over the continental shelf were still somewhat tentative and exploratory. (See e.g. François's First Report, which contained only a few remarks on the apportionment *or internationalization* of the continental shelf in the case of several States being interested. *Yearbook of the ILC* (1950), i. 215f., 305). It is therefore scarcely surprising that the

The purpose of the following section is to reflect upon the drafting history of Article 6 in order further to examine the credibility of this conclusion. In this respect, it is also relevant to recall that after the 1958 Shelf Convention came into force, two decisions dealt with Article 6 extensively, namely the *North Sea* and the *Anglo-French* Decisions. They revealed almost opposite views as to the content of that Article. This was mainly due to their different readings of the role of 'equidistance' as embodied therein. In a third section, it is proposed to identify these differing interpretations and point out their impact on the application of Article 6. The position of the subsequent decisions as well as State practice regarding Article 6 will be considered in the next chapter.

2. THE FORMULATION OF THE 'EQUIDISTANCE–SPECIAL CIRCUMSTANCES' RULE

The linguistic construction of Article 6 may suggest an apparently hierarchical (and therefore non-complementary) relationship between the concepts of 'equidistance' and 'special circumstances' which the phrase 'unless another boundary line is justified' would seem to indicate. Although this literal interpretation is inconsistent, as will be established, with the intention of the drafters,[3] one immediate suggestion is that the reason why the rule included in Article 6 has been often misunderstood, especially in UNCLOS III,[4] lies in the way it was drafted, encouraging confusion as to the proper meaning of the provision as a whole.[5]

Commission at this early stage should not have had clear ideas about the criteria for delimiting the continental shelf boundary; or that some members could not have envisaged whether there could be a combined 'equidistance–special circumstances' rule and should have simply fallen back upon 'arbitration' or 'agreement'. (See the Report of the Commission on its 1950 Session, ibid. (1950), ii. 384, 'The Continental Shelf'; the Commission's discussion on the subject in the 1951 Session, ibid. (1951), i. 285–8).

[3] Pt. III of the Vienna Convention on the Law of Treaties deals with rules regarding application and interpretation of treaties. Art. 32 provides that recourse may be had to the preparatory work of the treaty and the circumstances of its conclusion, in order to determine its meaning when the literal interpretation of the terms of the treaty 'leaves the meaning ambiguous' or 'leads to a result which is manifestly absurd or unreasonable'.

[4] See *infra*, Ch. 4 s.1.

[5] One result of this imprecise drafting is the contention that Art. 6 is 'so framed as to put second the obligation to make use of the equidistance method, causing it to come after a primary obligation to effect delimitations by agreement' (see *infra*, text accompanying n. 45). There is considerable evidence in the *travaux* to indicate that this was never the intention of the drafters of Art. 6. Very briefly, following François's draft Art. 7 in 1953 which provided that 'the boundary of the continental shelf . . . should be drawn according to the principle of equidistance', Cordova remarked that there 'was no reason . . . why the system laid down in Art. 7 should be mandatory'. It was therefore suggested, on the basis of a proposal by Pal, that the rule of equidistance would apply only 'in the absence of agreement'. As explained by its author, the clause was introduced so as 'to give States the

However, the discussions which took place after the 1953 Session should dispel all doubts as to the intention of the framers to formulate a general rule of law, enshrined in the 'equidistance–special circumstances' rule. Endorsing the recommendation of the Committee of Experts that 'equidistance' be adopted for the delimitation of the continental shelf, François submitted in 1953 a new draft article (Article 7) providing, *inter alia*, that:

1. in the case of opposite States, the boundary should be 'the median line every point of which is equidistant from two opposite coasts';
2. in the case of adjacent States, the boundary 'should be drawn according to the principle of equidistance from the respective baselines'.[6]

This unqualified adoption of 'equidistance', despite the Committee's remark that 'in a number of cases [equidistance] may not lead to an equitable solution which should then be arrived at by negotiations',[7] was criticized by members of the Commission.[8] Specifically, it raised the concern that 'there might be special reasons, such as navigation and fishery rights, which might divert the boundary from the median line . . . The Commission should perhaps consider whether rules should be laid down for such special cases.'[9] François suggested that the rule of equidistance should be recognized only 'as a general rule'.[10] But this latter proposal also met with opposition. Lauterpacht, for instance, contended that it was at least arguable that the words 'as a general rule' would deprive the rule of its legal character. He suggested that reference to exceptions should be spelt out, but it would be better to specify the cases rather than to open the door to difficulties of interpretation.[11] There were therefore two divergent views, one which considered that the words 'as a general rule' 'deprived the text of its juridical significance', and François's

opportunity of settling differences by agreement'. On the basis of this understanding, François considered the clause 'to be superfluous'. 'Obviously', he said, 'when two parties agreed to follow the prescribed rules and arrived at a settlement which did not conflict with the interests of a third party, no possible objection could be raised' (see *Yearbook of the ILC* (1953), i. 125–8). It was therefore in accordance with this perception that the clause gained the general approval of the Commission and was finally redrafted in 1956. Judge Morelli correctly contended that if the clause was meant to indicate some chronological priority, 'several questions would be raised, to which it would not be easy to reply. At what moment would it be necessary to establish that the condition of absence of agreement, to which the functioning of the equidistance rule is subordinated, is fulfilled? What is the legal situation either before that moment or before the conclusion of an agreement if any? Is it community? What would be the extent of the continental shelf subject to such a community?' (His dissent in the *North Sea Continental Shelf* cases, ICJ Reports (1969), 205).

[6] *Yearbook of the ILC* (1953), i. 106.

[7] Report of the Committee of Experts, A/CN. 4/51/Add.1.

[8] See comments by members of the Commission, *Yearbook of the ILC* (1953), i. 127–8.

[9] Comment by Sandstrom, ibid. i. 126.

[10] Ibid. i. 128.

[11] Ibid. i. 126, 130–1. Hsu also agreed with Lauterpacht, i. 131.

view which considered the words necessary 'in order to ensure the proper application of the principle of law'.[12] It was at this point that Sandstrom proposed a formula which had earlier been suggested by Spiropoulos, to the effect that the words 'unless another boundary line is justified by special circumstances' be substituted for François's proposal.[13]

This is the origin of the formulation of 'special circumstances' in the text of Article 6. It is therefore clear that the introduction of this clause was not meant to break the delimitation provision into two: a delimitatory rule and exception therefrom. This is what Lauterpacht had wanted and what the Commission objected to.[14] Rather, the clause was inserted so that the rule adopted 'partakes of some elasticity'.[15] The elevation of 'equidistance' into an acceptable rule of law, rather than its subjection to a set of exceptions, was the purpose of this insertion.

The drafters of Article 6 were not unaware of the fact that the strict equidistance method had only a limited scope of application. It would be absurd to suggest that their prolonged discussions on the subject were concerned only with the drafting of a strict technical method which could hardly be applied in concrete situations. On the other hand, the Commission was certainly aware of the fact that the inclusion of a clause on 'special circumstances' meant not so much a limited list of exceptions to a generally applicable rule, but more an element of complementary function within a general rule. In its 1956 comment on its final draft articles, the Commission pointed out that 'provision must be made for departure [from the strict equidistance line] and that this case may *arise fairly often* so that *the rule adopted* is fairly elastic'.[16] A formula which is made to cater for such departure cannot be perceived as an 'exception', since it deals with situations which 'arise fairly often'.

The discussion in the Fourth Committee of UNCLOS I focused on the 'special circumstances' element rather than on the 'equidistance' clause. Although three States, namely Greece, Portugal, and Yugoslavia, opposed the inclusion of this element,[17] most delegations defended the Commission's draft. They envisaged the 'special circumstances' as a necessary element for the effective application of the whole delimitation provision. The delegate of Tunisia felt 'that delimitation . . . should take account of the geographical configuration of the region, and that considerable flexibility would have to be used in applying that Article'.[18] The delegate of

[12] See comment by Chairman, ibid. i. 131.

[13] Ibid. para. 16. For Spiropoulos's proposal see ibid. i. 130.

[14] *Supra*, n. 11. See also his proposal in *Yearbook of the ILC* (1953), i. 132 and the comments by members of the Commission that followed it.

[15] See the Commission's 'comments on the draft articles', ibid. ii. 213, at 216.

[16] *Yearbook of the ILC* (1956), ii. 300 (emphasis added).

[17] A/CONF. 13/39, paras. 8, 12, 35.

[18] UNCLOS I, *Official Records*, vi. 22, para. 35.

Indonesia found that the 'International Law Commission's text was sufficiently flexible to provide all States whatever their geographical situation with the necessary safeguard'.[19] Several other States expressed similar views.[20]

It is therefore clear that the framers of the Convention intended to attribute to the substantive rule in Article 6 a broad construction by combining the clauses 'equidistance' and 'special circumstances' in one comprehensive delimitatory rule. The contention that 'special circumstances' is envisaged as an exceptional clause was never expressed, either by the ILC or at the Conference. Such an interpretation would contradict the clear fact that 'special circumstances' as a practical implementation of equity will exist in almost every case of delimitation.[21] The 'exception' will in reality turn into a general rule, according to this interpretation, and thus deny the applicability of 'equidistance' in almost every case. Certainly, this cannot be claimed as the result which this immense drafting history of Article 6 was intended to lead to.

This drafting history might suggest that 'equidistance' was initially conceived as a mere technical method of a geometrical nature,[22] but in the course of its later legal evaluation both in the International Law Commission and in the jurisprudence, it became transformed into a 'naturalized' rule the full effect of which may appear when envisaged within the context of the 'equidistance–special circumstances' rule. This transformation is not a phenomenon unique to the case under consideration. As Judge Tanaka pertinently observed:

> The incorporation of the equidistance rule as a geometrical technique into a legal norm exemplifies an extremely widespread phenomenon which can be observed in regard to several kinds of extra-legal-social and cultural norms and such fields as usage, ethics and technique.[23]

This proper interpretation of the rule of delimitation embodied in Article 6 is shared even by those who argue that such an interpretation has been recognized not so much in the codification but through the application of the rule in State practice.[24] As O'Connell observed, this 'unitary rule' interpretation 'is warranted only if the rule is seen, not as an

[19] Ibid. vi. 98, para. 39.

[20] See e.g. statements by delegates of Iran, the UK, Italy, Sweden, the USA, France, ibid. vi. 92–6.

[21] See analysis of judicial cases and State practice *infra*, Ch. 4 ss. 3, 4, respectively.

[22] It is evident, for instance, that the Committee of Experts adopted 'equidistance', *inter alia*, for 'technical' conveniences, 'bearing in mind in particular the practical difficulties of the navigator'. *Supra*, n. 7.

[23] Judge Tanaka, his dissent to the *North Sea* cases, 183.

[24] See *infra*, Ch. 4 s.4 and Ch. 5 s.4(*b*).

autonomous one conventionally agreed upon, but as an expression of an unstated, more basic rule, of equitable delimitation.'[25]

Whatever the view taken, what seems to be valid is that the treatment of the rule embodied in Article 6 as a purely geometrical method excluded by special circumstances is not warranted. An interesting development which clearly indicates that the literal, exceptional connotation of the word 'unless' in Article 6 was never intended is that, during the drafting of the 1982 Convention, the text which contained reference to equidistance (later superseded by what is now Article 83(1)) qualified that reference by indicating that equidistance should be used 'where appropriate, and *taking account of* [not unless] all circumstances prevailing in the area concerned'.[26] Obviously, a strict equidistance line would be a less crucial determining factor under this amendment and primary consideration would be paid to the equidistance–special circumstances as a whole.

3. THE INTERPRETATION OF ARTICLE 6 BY THE *NORTH SEA* AND THE *ANGLO-FRENCH* DECISIONS

(*a*) *The* North Sea *Cases*

The *North Sea* cases concerned the delimitation of the continental shelf situated off the North Sea coasts of the Federal Republic of Germany, Denmark, and The Netherlands. It was owing to the fact that both Denmark and The Netherlands, as Applicants, were in the same interest that the Court joined the proceedings in the two cases.[27]

Conforming with the parties' request, the Court attempted to identify principles of general equity applicable to the delimitation of the area under consideration. Within this attempt the Court was compelled to elaborate on the content and significance of the rule of delimitation embodied in Article 6 of the Shelf Convention. Denmark and The Netherlands were both parties to the Convention. The Federal Republic had signed but not ratified the Convention.

Denmark and The Netherlands contended that the delimitation should be governed by Article 6, either by reason of estoppel or as embodying a rule of customary law.[28] They claimed that, since no special circumstances

[25] D. O'Connell, *The International Law of the Sea*, ii (Clarendon Press, Oxford, 1984), 700.

[26] A/CONF. 62/WP.10/Rev.2, Arts. 74(1), 83(1). See also the submission of the 'Equidistance Group' at Negotiating Group 7, *infra*, Ch. 4 n. 4.

[27] *North Sea* cases, 8.

[28] Ibid. paras. 37 ff.

existed in the North Sea, their boundaries with the Federal Republic were to be determined by application of the equidistance method.[29]

The Federal Republic contended that the equidistance rule as laid down in Article 6 was not applicable to it because (1) it had not ratified the Convention or otherwise become bound by it; (2) the rule of Article 6 had not become customary international law; and (3), in any event, the rule could not be used where it did not result in a just and equitable apportionment of the shelf.[30] The Federal Republic further contended that, even if Article 6 had become customary law, special circumstances existing in the North Sea would exclude its application.[31]

The effect of the strict application of the equidistance line on the area to be delimited was described by the Court in the following terms:

> in the case of a concave or recessing coast such as that of the Federal Republic on the North Sea, the effect of the use of the equidistance method is to pull the line of the boundary inwards, in the direction of the concavity. Consequently, where two such lines are drawn at different points on a concave coast, they will, if the curvature is pronounced, inevitably meet at a relatively short distance from the coast, thus causing the continental Shelf area they enclose, to take the shape of an apex to seaward . . . The effect of concavity would of course equally be produced for a country with a straight coastline if the coast of adjacent countries protruded immediately on either side of it.[32]

It was this factual situation which formed a sound basis for the Court to attack the delimitatory rule of Article 6. For, although the Court found the application of Article 6 was not obligatory between the parties, and it was, therefore, unnecessary for it 'to go into certain questions relating to the interpretation . . . of that provision',[33] the Court viewed 'equidistance' as a strict method of delimitation.

Pointing out that there existed an 'essential difference' in the delimitation of the continental shelf areas between opposite and adjacent States,[34] the Court asserted that in certain geographical circumstances

[29] See Submissions No. 3, Counter-Memorials of Denmark and The Netherlands, in *Pleadings* (1968), i, respectively, 221, 375.

[30] See Submissions No. 3 in the Memorial of the Federal Republic of Germany, ibid. i. 91.

[31] Ibid. i. 422 ff.

[32] Para. 8.

[33] Para. 34. See also paras. 69, 81.

[34] See paras. 57–8. Highlighting the difference between 'opposite' and 'adjacent' States, the Court sought support from the misleading linguistic construction of Art. 6 which talks of the 'median line' for the delimitation of the continental shelf between opposite States and 'the principle of equidistance' for the delimitation of the continental shelf between adjacent States (see para. 47). This distinction, however, was not meant to reflect a difference of substance. As seen *supra*, Ch. 2 s.3(*a*), this distinction was criticized during the First Committee's discussion on the delimitation of the territorial sea. The fact that it was maintained in Art. 6 is explained by accidents of drafting the two Articles. in different

'which are *quite frequently met with*, the equidistance method, despite its known advantages, leads *unquestionably* to inequity in the following sense':

The slightest irregularity in a coastline is automatically magnified by the equidistance line as regards the consequences for the delimitation of the continental shelf. Thus it has been seen in the case of concave or convex coastlines that if the equidistance method is employed, then the greater the irregularity and the further from the coastline the area to be delimited, the more unreasonable are the results produced.[35]

This comment obviously rests on a mistaken conception of the rule embodied in Article 6; for the Court seemed to be unduly concerned with the 'equidistance method' in isolation from the companion element of 'special circumstances'. The Court did not attempt, in its exposition of the rule of equity, to establish whether the application of this complementary element would rectify this 'unquestionable inequity' resulting from the geography of the area.

After all, this was a very controversial issue between the parties in the dispute. It was argued on behalf of Denmark and The Netherlands that the application of 'special circumstances' admitted the possibility of a modification of the equidistance method on the basis of geographical configuration 'only in cases where a particular coastline, by reason of some exceptional feature, gave the State concerned an extent of continental shelf abnormally large in relation to the general configuration of its coast'.[36] Contrary to this argument, the Federal Republic contended that the special circumstances clause did not constitute an exception to the rule of equidistance, but that these two elements were valid on an equal footing, so that the equidistance element had no priority over the special circumstances element. Germany, minimizing the significance of the former element, advocated the 'equidistance–special circumstances' approach to cover the case where the concavity of its 'coastline front' would give rise, on the basis of Article 6, to 'an unjust and inequitable apportionment'.[37]

Furthermore, it was even proposed on behalf of the Federal Republic that a certain construction of the baselines from which the equidistance

Committees, which also resulted, as pointed out *supra*, Ch. 2 s.3(*a*), in drafting Arts. 6 and 12 in somewhat different terms. However, there is one important distinction which reflects, as the Court of Arbitration in the *Anglo-French* Arbitration pointed out (18 *RIAA*, para. 95), a factual but not legal difference between adjacent and opposite coasts. This distinction is considered *infra*, text at n. 69. See also *infra*, Ch. 4 s.3(*b*)(iii)(1).

[35] Para. 89 (emphasis added).

[36] See the Danish and Netherlands Counter-Memorials, *Pleadings*, i. 214, 367.

[37] See Argument of Professor Jaenicke, Agent of the Federal Republic, in *Pleadings*, ii. 43 ff.

line could be drawn would rectify this inequity. The proposal put forward was that this baseline should not be based upon 'the angled inward-curving North Sea Coast'[38] of the Federal Republic. Rather, it should be represented by a straight baseline connecting the island of Sylt and the island of Borkum.[39] This submission attracted the attention of at least one member of the Court,[40] and, consequently, further explanations were submitted, by both Counsel and Agent of the Federal Republic.[41] The Court, however, failed to examine the plausibility of this proposal, designed to illustrate the adaptability inherent in the rule of Article 6. Article 6 could be expected to lead to an equitable result on the understanding, *inter alia*, that the baseline to be employed for the purpose of the geometrical construction will vary from case to case, i.e. from the strict version used in measuring the limit of the territorial waters to certain modified baselines employed because of special circumstances in the geography of the region.[42]

It is true that the Court pointed in the *dispositif* to the 'general configuration of the coasts of the parties' as a factor to be taken into account between the parties involved.[43] Yet this was in connection with the Court's exposition of the factors meant to reflect its general approach that delimitation was 'to be effected by agreement in accordance with equitable principles'. But in relation to Article 6, the Court simply expressed its doubt as to the proper 'part played by the notion of special circumstances' and 'to the exact meaning and scope' thereof.[44] The Court appeared to have overlooked the proposition that what it cited as 'relevant factors to be taken into account' are in fact what might be considered as 'special circumstances' under Article 6.

This negative approach by the Court emanated from its failure to give due weight to the evidence in the *travaux* and State practice of the application of the combined 'equidistance–special circumstances' rule embodied in Article 6. Envisaging the rule as a mere geometrical method, not applicable between the parties to the dispute, the Court sought evidence from the linquistic construction of Article 6. The Court especially gave significance to the hierarchical order that might be misleadingly deduced from such a construction.

[38] See Argument of Professor Oda, Counsel for the Federal Republic, ibid. ii. 62.

[39] Ibid. ii. 63.

[40] See Questions by Judge Fitzmaurice, ibid. ii. 63, 64.

[41] See Replies by Professors Jaenicke and Oda, ibid. ii. 64–5, 193–201.

[42] See *supra*, Ch. 1 s.3(*b*). The Court, however, without admitting that this is an element of adaptability in the rule of Art. 6, recommended the Federal Republic's suggested straight line: 'Where the parties wish to employ in particular the equidistance method of delimitation, the establishment of one or more baselines of this kind can play a useful part in eliminating or diminishing the distortions that might result from the use of that method.' Para. 98.

[43] Para. 101.

[44] Para. 72.

Article 6 is so framed as to put second the obligation to make use of the equidistance method, causing it to come after a primary obligation to effect delimitation by agreement. Such a primary obligation constitutes an unusual preface to what is claimed to be a potential general rule of law.[45]

This is of course an approach which not only ignores the proper meaning of the clause intended by the framers of Article 6,[46] but also confuses a procedural rule, not solely related to delimitation issues,[47] with the necessity of defining substantive rules of delimitation. The inclusion of the agreement clause is meant to reflect the general principle that States are entitled to reach agreement on what they consider equitable and fair. The Court itself referred to this fact in the sentence following the one quoted above: 'it is well understood that, in practice, rules of international law can, by agreement, be derogated from in particular cases, or as between particular parties.' But the Court insisted on giving some weight to the explicit inclusion of this clause in Article 6. This weight, however, was left undefined: 'but this is not normally the subject of any express provision, as it is in Article 6.'

Neither is there anything specific to delimitation in the requirement to reach such agreement 'in accordance with equitable principles', and the conclusion that 'equidistance' should be denied the status of a rule of law solely on account of incidents of incompatibility with this requirement could only be justified if it had been proven that the line reflecting equitable principles could not be, or could only by coincidence be, an equidistance line. On the contrary, had the Court allotted to Article 6 a proper interpretation, it would have discovered little difference between this rule and its decision that the delimitation was to be 'effected by agreement in accordance with equitable principles, and taking account of all the relevant circumstances'.[48]

This short-sighted view of the Court is pointed out by Judge Ammoun. Agreeing with the majority judgment on the application of the legal concept of equity, he argued that equity itself, in the true sense of the term, constituted a subsidiary source of international law, *praeter legem*, its role being to fill any vacuum in the law. Since the current state of the law provided no positive legal rule of delimitation, equity was necessary, in Judge Ammoun's view, to fill the vacuum. But he did not think that the Court went far enough: '[t]he Court should, therefore, after having first excluded the application of the [strict] equidistance line as a rule of law', he said, 'state the rule which is capable of being adopted by application of the principle of equity.'[49] In Judge Ammoun's opinion this rule would have been the 'equidistance–special circumstances' rule.

This combined rule is appropriate because it provides the juridical

[45] Ibid. [46] See *supra*, n. 5. [47] See *infra*, Ch. 4 s.2(*b*).
[48] Para. 101. [49] Judge Ammoun, his Sep. Op., the *North Sea* cases, 145.

framework for the evaluation of a given factual situation, in other words, in reflecting equity. He said:

a special circumstance affecting the equidistance method may be the effect of a particular legal situation: a treaty, or historic waters. It may also be the consequence of geographical considerations. On the basis of the map and measurements already mentioned, the configuration of the coast of the Federal Republic of Germany constitutes such a circumstance, which should be taken into account to avoid the inequitable application of the equidistance line pure and simple.[50]

However, in one respect the Judgment does assist in a proper understanding of the rule embodied in Article 6. The Court seemed in effect to confer upon the 'special circumstances' clause a broad construction. In its exposition of the rules of customary law on delimitation, it was apparent that the role of equity, as conceived by the Court, was similar to that of the concept of special circumstances. The Court explicitly stated that 'the exception in favour of special circumstances' was introduced 'in pursuance' of the belief that the determination of the boundaries should be effected on 'equitable principles'.[51] The identification of 'special circumstances' with equity is implied in the *dispositif*:

delimitation is to be effected by agreement in accordance with equitable principles, and taking account of all the relevant circumstances, in such a way as to leave as much as possible to each Party all those parts of the continental shelf that constitute a natural prolongation of its land territory into and under the sea, without encroachment on the natural prolongation of the land territory of the other.[52]

In the previous passages the Court dealt with the nature of these 'equitable principles' and 'relevant circumstances' at some length. First, it said:

there is no legal limit to the considerations which States may take account of for the purpose of making sure that they apply equitable procedures, and more often than not it is the balancing-up of all such considerations that will produce this result rather than reliance on one to the exclusion of all others. The problem of the relative weight to be accorded to different considerations naturally varies with the circumstances of the case.[53]

The Court went on to add that '[i]n balancing the factors in question it would appear that various aspects must be taken into account. Some are related to the geological, others to the geographical aspect of the situation, others again to the idea of the unity of any hydrocarbon deposit.'[54] Finally, the Court directed attention to the additional factor of a reasonable degree of proportionality which a delimitation effected according to equitable principles ought to reflect.[55]

[50] Ibid. 149. [51] Para. 55. [52] Para. 101.
[53] Para. 93. [54] Para. 94. [55] Para. 98.

This by implication means that the equitable factors which could justify a boundary, other than a strict equidistance line, under international customary law are the same as the factors which would constitute special circumstances under Article 6. This, of course, implies a broader construction to the element of special circumstances; for, according to the Court's view, the valid test to establish whether a given factor is regarded as a special or relevant circumstance is whether its application would lead to an equitable delimitation in the concrete situation.

But, as already pointed out, this positive feature in the Court's analysis does not come from what should have been a proper interpretation of the whole rule of Article 6. Rather, the Court develops this view as part of its construction of the general principle of equity. It is true that the essential relationship between 'equidistance' and 'special circumstances', as co-ordinante principles, may not be readily apparent from the language of Article 6; and it was for this reason that the Court denied Article 6 'the fundamentally norm-creating character', although it was prepared to identify this character with the rule of equidistance '*in abstracto*'.[56] Nevertheless, in attempting to ascertain the content of that Article, there was no necessity to be bound by the precise language of the Article itself. A more creative, and legitimate,[57] approach might have been for the Court to extract the unified rule of 'equidistance–special circumstances' from the intention of the drafters and the actual practice of States, both treated by the Court very casually. This approach might have proven ultimately to be more productive for future settlements and would indeed have helped the successive efforts, both in UNCLOS III and adjudication, to discern some positive rules on delimitation.

(*b*) *The* Anglo-French *Arbitration*

(i) Introduction

The Court of Arbitration was charged in 1977 with drawing the continental shelf boundary between the United Kingdom and France in the English Channel area. Both the United Kingdom and France had ratified the 1958 Shelf Convention, and France had made a number of reservations to Article 6[58] (which the Court in this case adjudged to be valid).[59] The United Kingdom had regarded them as unacceptable. The Court thus found it necessary to judge not only the validity of the reservations but

[56] Para. 72.
[57] See *supra*, n. 3.
[58] Text in 'Multilateral Treaties in Respect of which the Secretary-General Performs Depositary Functions' (ST/LEG/SER.D/7), Ch. 21 s.4.
[59] See paras. 71–4.

also the effect of objections to reservations. Without going into the substantive law on which the Court based its conclusion on these points,[60] the Court found that the effect of the reservations and objections was to render some parts of the delimitation area subject to the regime of Article 6 and others to the principles of customary law. The practical effect of this finding was regarded by the Court as minimal. It stated that:

in the circumstances of the case, the rules of customary law lead to much the same result as the provisions of Article 6 . . . the effect of applying or not applying the provisions of the Convention, and in particular of Article 6, will not make much practical difference, if any, to the actual course of the boundary.[61]

The reasoning behind that conclusion is of extreme importance. The Court construed Article 6 as a single, combined equidistance and special circumstances rule, with neither element dominating the other, whose aim is to bring about an equitable delimitation. It is the purpose of the following analysis to examine the holding of the Court only as it bears upon this explanation and the principles that consequently followed it.

(ii) The Court's interpretation of Article 6

Perhaps the most important dictum of the Court was its reading of Article 6 as embracing one single rule, 'a combined equidistance–special circumstances' rule. It was this interpretation of the rule in Article 6 which enabled the Court to lay down certain principles which certainly reflect general statements of the law on maritime boundaries.

It had been argued by the United Kingdom that Article 6 placed an onus of proof upon France to show the existence of any special circumstances which would justify displacing a median line boundary.[62] In other words, the equidistance or median line was the general rule, and 'special circumstances' the exception.

The Court considered that this view did not place Article 6 in its true perspective. In its view, 'Article 6 . . . does not formulate the equidistance principle and "special circumstances" as two separate rules. The rule . . . is a single rule, a combined equidistance–special circumstances rule.' The Court then stated:

This being so, it may be doubted whether, strictly speaking, there is any burden of proof in regard to the existence of special circumstances. The fact that the rule

[60] A concise study of the Court's ruling on these points may be found in A. Boyle, 'The Law of Treaties and the Anglo-French Continental Shelf Arbitration', 29 *ICLQ* (1980), 498; D. Colson, 'The United Kingdom–France Continental Shelf Arbitration', 72 *AJIL* (1978), 95 at 100–2.

[61] Para. 65.

[62] Para. 67.

is a single rule means that the question whether 'another boundary is justified by special circumstances' is an integral part of the rule providing for application of the equidistance principle. As such, although involving matters of fact, that question is always one of law which, in case of submission to arbitration, the tribunal must itself, *proprio motu*, take cognisance when applying Article 6.[63]

This integration of the two elements underlies two principal features of Article 6, both referred to by the Court, namely (1) predictability in the sense that the relevance of special circumstances in the application of Article 6 'does not depend on a claim to invoke special circumstances having been advanced by the interested State when ratifying or acceding to the Convention',[64] and (2) elastcity, in the sense that Article 6 allows the decision-maker the necessary discretion for appreciating 'the geographical and other circumstances relevant to the delimitation'.[65] Consequently, the possibility of any difference in the appreciation of these circumstances under Article 6 and under customary law almost disappears.

(iii) The relationship between Article 6 and customary law

As seen earlier, the Court in the *North Sea* case made only a casual connection between Article 6 and customary law. But the Court of Arbitration took the view that customary law and Article 6 were virtually the same.

Having combined equidistance and special circumstances in one rule, the Court noted that the interrelationship of Article 6 and customary law was self-evident. For, although aware of the fact that 'equidistance' as a conventional rule 'ultimately possesses an obligatory force which it does not have in the same measure under the rules of customary law', the Court maintained that the combined character of the equidistance–special circumstances rule 'means that the obligation to apply the equidistance principle is always one qualified by the condition "unless another boundary is justified by special circumstances"'.[66] Such an approach has the effect of assimilating the conventional and customary rules of delimitation, and on this the Court was unequivocal:

In short, the role of the 'special circumstances' condition in Article 6 is to ensure an equitable delimitation; and the combined 'equidistance–special circumstances' rule, in effect, gives particular expression to a general norm that, failing agreement, the boundary between States abutting on the same continental shelf is to be determined on equitable principles.[67]

[63] Para. 68.
[64] Para. 69.
[65] Ibid. However, as the Court later stated, this discretion is envisaged as 'lying not outside but within the rules of the law'. See *infra*, text at n. 81.
[66] Para. 70.
[67] Ibid.

Having found that both Article 6 and the rules of customary law (as explained in the *North Sea* cases) led to the same result—that is, a boundary of an equitable character—the Court was bound to determine that the role of equidistance as *a strict method of delimitation* is always relative to the particular geographic situation. It held that it is the

geographical and other circumstances of any given case which indicate and justify the use of the equidistance method as the means of achieving an equitable solution rather than the inherent quality of the method as a legal norm of delimitation.[68]

In this regard the Court reviewed the distinction made in the *North Sea* cases regarding opposite and adjacent States and found that the distinction was 'one derived not from any legal theory, but from the very substance' of the two geographical situations. In other words, equidistance used in the geographical situation encountered in the 'opposite' relationship is less likely to produce inequity than in the 'adjacent' geographical relationship.[69]

Upon further review of the principles of the *North Sea* cases, the Court found that the fundamental concepts of those cases—namely (1) that a coastal State has an inherent right to the continental shelf which constitutes the natural prolongation of its land territory and (2) that, failing agreement, the boundary must be determined in accordance with equitable principles—are of 'general character and applicable to a delimitation under Article 6 no less than under customary law'.[70] While upholding the second concept and identifying it with the appreciation of the geographical and other circumstances in a concrete case, thus giving 'equity' a substantive meaning,[71] the Court reduced the legal effect of the concept of natural prolongation.[72] Commenting on the International Court's con-

[68] Ibid.

[69] The Court explained that 'whereas in the case of opposite States a median line will normally effect a broadly equitable delimitation, a lateral equidistance line extending outwards from the coasts of adjacent States for a long distance may not infrequently result in an inequitable delimitation by reason of the distorting effect of individual geographical features. In short, it is the combined effect of the side-by-side relationship of the two States and the prolongation of the lateral boundary for great distances to seawards which may be productive of inequity and is the essence of the distinction between "adjacent" and "opposite" coasts situation' (para. 95).

[70] Para. 77.

[71] In a clear indication of the role of equity in maritime boundaries, the Court declared in para. 70 that 'the question whether the use of the equidistance principle or some other method is appropriate for achieving an equitable delimitation is very much a matter of appreciation in the light of the geographical and other circumstances'. In other words, 'equity' is identified with the determination of the factors which ought to define the course of the boundary in a given case. See *infra*, Ch. 4 s.3.

[72] While rendering it the central point in its reasoning, the Court in the *North Sea* cases made no attempt to explain the meaning of this 'fundamental concept of natural prolongation'. However, when used in a geophysical sense the notion of natural prolongation is

clusion that 'the continental shelf of any State must be the natural prolongation of its land territory and must not encroach upon what is the natural prolongation of the territory of another State', the Court observed that

> this conclusion states the problem rather than solves it. The problem of delimitation arises precisely because in situations where the territories of two or more States abut on a single continuous area of continental shelf, it may be said geographically to constitute a natural prolongation of the territory of each of the two States concerned.[73]

This was the factual situation that the Court encountered in its *ratio decidendi* in relation to the delimitation of the Channel Islands region. Having found that the English Channel constituted a single physical prolongation from both of the territories abutting upon it,[74] the Court declared that the natural prolongation of territory 'cannot be said to require that the continental shelf to the north and north-west of the Channel Islands should be considered as automatically and necessarily appurtenant to them rather than to the French Republic'.[75]

Instead, the Court decided that the presence of the Channel Islands close to the French coast must be considered, prima facie, as a circumstance creative of inequity and 'a special circumstance' within the meaning of Article 6. This prima-facie finding was then tested against the other 'relevant circumstances' or 'equitable considerations' related to the Islands. Considerations of equity were therefore identified with the 'relevant' or 'special' circumstances. It is natural that the latter should have a broad construction.

highly questionable; for although a major geological or geomorphological discontinuity may not often exist between opposite States (the Timor Trough between Australia and Indonesia is the only known example), it is extremely unusual to find such a major discontinuity extending laterally across the shelf in the vicinity of the land boundary of adjacent States. Furthermore, even in instances where such a discontinuity is said to exist, the question may arise as to which of the two geophysical elements, morphology or geology, can identify the natural prolongation; and, in case of discrepancy, which one is to prevail. In the *Jan Mayen* Conciliation (20 ILM (1981), 797) the Commission found that, according to the scientific data available, the northern part of Jan Mayen Ridge could morphologically be considered a southward extension from the shelf of Jan Mayen and could not be regarded as an extension from the Icelandic Shelf. But, geologically, the Ridge was found to be a microcontinental one that pre-dated both Jan Mayen and Iceland. The Conciliation Commission, therefore, concluded that the concept of natural prolongation 'would not form a suitable basis for the solution' of the boundary issue between Jan Mayen and Iceland. Ibid. 822, 824.

[73] Para. 79.

[74] Para. 191. The UK had argued that the depression called the Hurd Deep formed a limit or boundary to the 'natural prolongation' of any French territory. The Court rejected this argument.

[75] Para. 192.

(iv) Broad interpretation of the 'special circumstances' element

In support of its identification of Article 6 with customary law, the Court stated that

> the provisions of Article 6 do not define the condition for the application of the equidistance–special circumstances rule, moreover, the equidistance–special circumstances rule and the rules of customary law have the same object—the delimitation of the boundary in accordance with equitable principles. In the view of the Court, therefore, the rules of customary law are relevant and even essential means both for interpreting and completing the provisions of Article 6.[76]

The determination of factors that may be regarded as 'special circumstances' would therefore be governed by the ultimate purpose of achieving an equitable delimitation. Such an approach would ensure that all factors held to be creative of inequity in accordance with the rules of customary law would necessarily fall within the 'special circumstances' clause of Article 6. Thus when the Court concluded that the existence of the Channel Islands close to the French coast, if permitted to divert the course of the mid-channel line, would effect a radical distortion of the boundary, creative of inequity, it went on to test this conclusion by applying the equidistance–special circumstances rule of Article 6. It was concerned, in this connection, to see whether a departure from the strict median line was 'justified' in relation to both parties and in the light of the 'relevant circumstances'.[77] Among the 'relevant circumstances' invoked by the United Kingdom and accepted by the Court were the particular character of the Islands as populous islands of a certain political and economic importance, their close ties with the United Kingdom, the latter's responsibility for their defence and security, and the impossibility of the Islands having any appreciable area of continental shelf except in the open waters to their west and north.[78] Following the same line of enquiry, the Court also acknowledged that defence and security interests might in principle be accepted as special circumstances which might have some effect.[79]

These circumstances were allowed by the Court to have very little effect on the course of the boundary. But the Court's reasoning does indicate a wide interpretation of the concept of 'special circumstances'. This could be sustained only by a proper interpretation of Article 6 as comprising one combined equidistance–special circumstances rule.[80]

[76] Para. 75.
[77] Para. 197.
[78] Ibid.
[79] Para. 188.
[80] Envisaging Art. 6 as prima facie containing a principal rule (equidistance), subject to exception (special circumstances), Judge Briggs argued that the Court's interpretation of

The Court, however, was concerned about any misunderstanding that may flow from this wide interpretation of the 'special circumstances' concept. Having decided that the position of the Scilly Isles constituted a 'special circumstance' justifying a boundary other than the strict median line, it warned that this cannot be interpreted as giving it 'carte blanche to employ any method that it chooses in order to effect equitable delimitation'. The Court stated that the dictum advanced by the 1969 Judgment that there is 'no legal limit to the considerations which States may take account of for the purposes of making sure that they apply equitable principles' was made in connection with cases where 'the Parties had retained the actual delimitation of the boundary in their own hands for further negotiation'. Furthermore, this dictum should be read 'in the light of certain other observations of the International Court in the same Judgment. In these other observations, it was stressed that any recourse to equitable considerations must be to considerations "lying not outside but within the rules" of law, and that there is no question of any decision *ex aequo et bono* (para. 88); and . . . it was also stressed that "there can never be any question of completely refashioning nature (para. 91)"'.[81]

(v) Further identification of Article 6 with customary law

The Court's elucidation of the relationship between Article 6 and the corresponding rules of customary law was not confined to the former's identification with the rules of customary law discussed in the *North Sea* cases. The Court further pointed out the relation between Article 6 and the existing efforts at the time to formulate delimitatory rules within UNCLOS III. Rejecting the French argument that the evolution of customary law stimulated by the proceedings of that Conference had rendered the 1958 Convention obsolete,[82] the Court emphasized that this finding 'does not mean that it regards itself debarred from taking any account in these proceedings of recent developments in customary law. On the contrary, the Court has no doubt that it should take due account of the evolution of the law of the sea in so far as this may be relevant in the context of the present case.'[83]

Art. 6 'constitutes some threat that the rule of positive law expressed in Art. 6 will be eroded by its identification with subjective equitable principles, permitting attempts by the Court to redress the inequities of geography'. His Sep. Declaration, 18 *RIAA*, 126.

[81] Para. 245. The Court, however, was criticized as having employed elements of subjective evaluation of the equities and of choice of means of observing them, especially with regard to the Channel and Scilly Islands. See, e.g. D. Bowett, *The Legal Regime of Islands in International Law* (Oceana Publications, New York, 1979), 221–2, 230–1; E. Lauterpacht, 'Equity, Evasion, Equivocation and Evolution in International Law', *American Branch of International Law Association: Proceedings and Committee Reports* (1977–8), 1, at 41–3.

[82] Para. 45.

[83] Para. 48.

Accordingly, the Court did examine Article 71(1) of the Revised Single Negotiating Text, the then current forerunner of what is now Article 83(1) of the 1982 Convention.[84] Reflecting on that provision, the Court found

> no reason to suppose that, if [the provisions of Article 71(1)] were applicable, they would make any difference to the determination of the boundary in the present case. Those texts speak of delimitations between 'adjacent' or 'opposite' States in accordance with equitable principles as distinct cases; and they envisage that, where appropriate, the equidistance or median line shall be employed, taking account of all the relevant circumstances. Since it is the geographical circumstances which primarily determine the appropriateness of the equidistance or any other method of delimitation in any given case, [Article 71(1)] would not appear to visualise the solution of cases like the present one on principles materially different from those applicable under the 1958 Convention or under general international law.[85]

This clarification of the relationship between the Geneva Convention and UNCLOS III may now be said to have lost validity, since the text of Article 71(1) has been replaced by the much altered Article 83(1) of the 1982 Convention. None the less, the Court's clarification remains of interest as a possible pointer to the way in which Article 83(1) of the 1982 Convention may be interpreted, especially given the fact that Article 83, as will be seen below, provides no more than a starting-point for the determination of substantive rules on delimitation.

4. CONCLUSIONS

The drafting history of Article 6 reveals that the Article was not meant to introduce an a priori method of delimitation the negation of which can be 'justified' only by proving the presence of 'special circumstances'. Rather, the whole proviso—'equidistance–special circumstances'—was meant to formulate a single rule of two complementary elements.

The language in which this combined rule was framed in the Geneva Conventions certainly was not empty of inaccuracy. It conveyed hierarchical and conditional connotations that gave the Court in the 1969 Judgment good reason to demean Article 6 to a purely geometrical method of one restricted form of application. The Court could, however, be criticized for its failure to seek to identify, irrespective of the language

[84] Art. 71(1) provided that: '1. The delimitation of the continental shelf between adjacent or opposite States shall be effected by agreement in accordance with equitable principles, employing, where appropriate, the median or equidistance line, and taking account of all the relevant circumstances.'

[85] Para. 96.

employed, the real implications of the complementary element of 'special circumstances' in bringing a degree of flexibility to the rule in Article 6. Equidistance was stripped of its 'naturalizing' element[86] and was thus exposed simply as a strict, geometrical method, unable to produce an equitable result.

In the *Anglo-French* Arbitration, on the other hand, the Court of Arbitration was able to identify the two elements in Article 6 as constituting one combined rule; and, consequently, the Court attempted to develop a comprehensive statement of the law: that is, Article 6 is a substantive expression of the established law on delimitation. In other words, the Court was able to reach the conclusion that the two elements in Article 6 formed a substantive rule which was capable of explaining in concrete terms the content of the principle of equity. This is a positive statement by the Court which obviously aimed at ensuring that its reasoning should be seen as a development from the 1969 *North Sea* decisions; and consequently, implying an invitation to subsequent decisions to adopt or at least reflect upon its reasoning. This is submitted to be a significant indicium by the Court of Arbitration which subsequent adjudication chose, unfortunately, to ignore. This will be discussed in the following chapter.

[86] See *supra*, text at nn. 22–3.

4

Delimitation of the Continental Shelf/EEZ under the Influence of the 1982 Convention

1. THE DELIMITATION PROVISIONS OF THE 1982 CONVENTION

The drafting history of the final delimitation provisions of the 1982 Convention has been reviewed extensively.[1] These provisions, however, cannot be evaluated without appreciating the circumstances which led to the adoption of what now become Articles 74(1) (delimitation of the exclusive economic zone) and 83(1) (delimitation of the continental shelf) of the Convention.[2]

As has been noted, the delimitation of ocean boundaries between States with opposite or adjacent coasts was one of the hard-core issues facing UNCLOS III, and was eventually referred for resolution to Negotiating Group 7.[3] While there was early agreement on the rules for delimiting the territorial sea, the provisions for delimiting the continental shelf and the exclusive economic zone were among the last to be drafted. This was because two main proposals already existed (when Group 7 embarked upon its mandate), representing the 'two main interest groups', namely the 'Median Line Group' and the 'Equitable Principles Group'.

One major shortcoming in the discussions of Group 7 was that no attempt was made to question this apparent opposition between the two proposals, whilst both groups accepted that either principle should be qualified by 'special circumstances' (the Median Line Group proposal), or the need to reflect all 'relevant circumstances' (the Equitable Principles Group proposal).[4] The failure to realize the fallacy of this opposition is

[1] See e.g. a series of review articles by Oxman and Stevenson in the *AJIL*, respectively volumes: 68 (1974), 1; 69 (1975), 1; 69 (1975), 763; 71 (1977), 247; 73 (1980), 1; 74 (1980), 1. See also a concise review of the work of Group 7 by Judge Oda in his Dissent in the *Tunisia–Libya* case, ICJ Reports (1982), 234–64.

[2] Because the resultant delimitation provisions for the continental shelf and the exclusive economic zone in the Convention are identical, and because their difficult negotiating histories are essentially the same, they are treated together.

[3] See *supra*, Ch. 2 s.3(*d*).

[4] The so-called proponents of the equidistance principle supported the proposal that the delimitation of the EEZ/continental shelf 'shall be effected by agreement employing, as a general principle, the median or equidistance line, *taking into account any special circumstances* where this is justified' (informal proposal of 24 countries participating in Group 7,

perhaps explained by two facts. First, many delegates at the Conference expressed scepticism regarding the exact meaning and scope of the rule embodied in Article 6 of the Shelf Convention. This mainly reflected the 1969 Judgment's misinterpretation of and, consequently, attack on Article 6. The explanation offered by the 1977 Award of the 'equidistance–special circumstances rule' and the Award's identification of this rule with customary international law coincided with the creation of Group 7. The debates in the Group hence persisted without the chance to evaluate fully the reasoning behind this important dictum. Secondly, a large number of States participating in UNCLOS III had unresolved boundary problems with neighbouring States. Having envisaged a sharp difference between the two proposals, the delegates of those States had to ensure that the resultant provisions would not jeopardize their own interests *vis-à-vis* other States.[5]

Unfortunately, this unwarranted conflict between the two groups was to persist throughout the negotiations of Group 7. About half of the States involved decided that their interests would best be served by the 'equitable principles' approach. The rest thought that the delimitation formula of Article 6 had been fair and sound. The task of Group 7 was, consequently, framed with the purpose of developing a formula integrating both positions. During session after session, submissions and counter-submissions were advanced by the two groups. The process by which the final text of paragraph 1 of Articles 74 and 83 was settled represented an unusually exposed example of a compromise-orientated approach. As the negotiations continued after the 1978 Session (when Group 7 first met), it became clear that, if provisions were ever to be drafted, a 'neutral formula', containing no substantive criteria, might have to be the solution.[6] All efforts, especially by the Chairman of the Group, to gain consensus on a minimum substantive rule proved fruitless.[7] The negotiations of Group 7 in full had virtually ended and the

Doc. NG7/2). Whereas the advocates of the equitable principles submitted that such a delimitation 'shall be effected by agreement, in accordance with equitable principles *taking into account all relevant circumstances*' (informal proposal of 30 participants in Group 7, Doc. NG7/10).

[5] See e.g. the counter-proposals of Greece and Turkey which were specifically related to their dispute over the Aegean continental shelf, *infra*, Ch. 10 s.3(*a*), (*b*).

[6] Reporting on the work of Group 7 in 1978, the Chairman of the Group remarked that 'Like before . . . no approach or formulation received such widespread and substantial support that would offer a substantially improved prospect of a consensus', Doc. NG7/21. See also his Reports on the work of the Group in the 1979 and 1980 Sessions, respectively text at n. 11 and next footnote.

[7] During the last meeting of Group 7 (9th Session), the Chairman made the following proposal: 'The delimitation of the exclusive economic zone/the continental shelf between States with opposite or adjacent coasts shall be effected by agreement in conformity with international law. Such an agreement shall be in accordance with equitable principles,

possibility that the Conference would fail to adopt any rule on the subject in the Convention was imminent. However, on the final day of the resumed Tenth Session, the President of the Conference, under increased pressure to resolve the issue and complete the Convention, introduced his proposal, which was to be incorporated in the 1982 Convention as Articles 74(1) and 83(1) respectively. 'Judged only by the standard of alchemy', Allott comments, 'could it be said to be anything but bizarre':[8]

> The delimitation of [the exclusive economic zone/the continental shelf] between States with opposite or adjacent coasts shall be effected by agreement on the basis of international law as referred to in Article 38 of the Statute of the International Court of Justice, in order to achieve an equitable solution.

Introducing this proposal, the President reported that 'during his consultations, he had gained the impression that the proposal enjoyed widespread and substantial support in the two most interested groups of delegations, and in the Conference as a whole'.[9] It will, however, be demonstrated that this 'impression' lost sight of the real situation, in both Group 7 and Plenary.

2. THE CONTENT OF THE CONVENTION FORMULA

(a) *The formula's alleged value of consensus*

As a product of the 'consensus' technique employed throughout the Conference, paragraph 1 of Articles 74 and 83 is said to have provided consensus on whatever rules it holds 'and that is indeed its merit'.[10] But even this assumed quality is debatable, for the record shows that at no stage was there consensus in Group 7. In his report to the Plenary on the work of Group 7 at the Eighth Session, the Chairman stated that even a precise formula providing merely for 'delimitation by agreement' in accordance with international law 'did not receive any particular sympathy from either side'.[11] Similarly, in his final conclusions on the

employing the median or equidistance line, where appropriate, and taking account of all circumstances prevailing in the areas concerned.' But, having suggested this text, the Chairman made it clear that during the negotiations 'no agreement could be reached in the delimitation of the exclusive economic zone or the continental shelf . . . Owing to the obvious difficulties in agreeing upon a more detailed definition, it has been indicated by some delegations that the final solution might be found in a concise formulation merely identifying the two most fundamental elements of delimitation, that is, that it shall be effected by an agreement and based on international law'. Doc. A/CONF. 62/L.47.

[8] P. Allott, 'Power Sharing in the Law of the Sea', 77 *AJIL* (1983), 1, at 20.
[9] A/CONF. 62/SR. 154, p. 2.
[10] Oda, *Tunisia–Libya* case, 246.
[11] Doc. NG7/44.

work of Group 7, the Chairman advised that the formula suggested by him during the last session[12]

> has been described as quite unacceptable by the members of the group supporting the median line approach. *Because of this firm refusal by a notable part of the members of the group to adopt the present formula of paragraph 1 of Articles 74 and 83 it is clear that it cannot be considered a text which could provide consensus on this issue.*[13]

Although this formula was incorporated in the second and, afterwards, the third revision of the ICNT, the 'Equitable Principles Group' firmly rejected it. At the Plenary Meeting on 28 July 1980 the delegate of Ireland introduced a letter addressed to the President of UNCLOS III by the countries sponsoring NG7/10.[14] It submitted *inter alia* that the

> new formulations as they appear [in the ICNT Revision 2] *'did not emerge from negotiations themselves' nor did those formulations receive 'the widespread and substantial support' required in plenary to offer a substantially improved prospect of consensus.*[15]

The countries which signed this letter considered that 'the new formulations will not be helpful for further negotiation'. Nevertheless, this formula remained unchanged until, as already mentioned, replaced by the Conference President's formula on the final day of the Tenth Session.

The President's formula was more haphazard than any of the previous proposals. One delegate to the Conference reveals that the formula 'was worked out in secret by two States representing their respective groups and the President, and was available even in private only for a few days'. And its 'hasty insertion in the draft convention was questioned publicly by many members of the Conference, such as the United States, China, Portugal, Iran, Egypt, Libya and Kuwait'.[16]

It is clear therefore that the Conference's resultant formula does not, as generally claimed, reflect general consensus. The formula was suggested officially only one day before the close and was incorporated, again on the same day, when delegates must have felt, after almost ten years of negotiation, that the Conference should give up and improvise a text 'that delegations on both sides, privately [looked] upon with embarrassment'.[17]

[12] See *supra*, n. 7.

[13] Doc. A/CONF. 62/L.47 (emphasis added).

[14] See *supra*, n. 4.

[15] A/CONF. 62/SR.130, p. 18 (emphasis added).

[16] See B. Oxman, 'The Third United Nations Conference on the Law of the Sea: The Tenth Session (1981)', 76 *AJIL* (1982), 1, at 14–15.

[17] Statement by the head of the US delegation to the Conference, 28 Aug. 1981, quoted ibid. 15. Allott, 'Power Sharing in the Law of the Sea', 20, describes the inevitable outcome that had been 'imposed' on the 'two opposite groups' in the following metaphor: 'the growing impatience of the rest of the Conference and the threat of a third-party intrusion

Having said this, one should not feel that UNCLOS III failed completely to achieve consensus. If one goes beyond this formula, some *de facto* consensus on a *substantive* rule seems to emerge. The technique of treaty drafting employed by the Conference within its pattern of preparation for provisions representing the 'Main Trends', followed by debate, review, and revision of these provisions, with commentaries by the Chairman explaining both the areas of consensus and difference between the two groups, is probably more valuable for identifying the law on delimitation envisaged by UNCLOS III than the resultant provisions.

This process seems to indicate consensus on one specific rule, to which paragraph 1 of Articles 74 and 83 fails to refer. Both the parties in Group 7 agreed on the role of the 'special/relevant circumstances' in determining the criteria or methods of delimitation. The proposals submitted by either side,[18] by the Chairman,[19] or by the so-called neutral States,[20] included reference to this rule. Reflecting on the areas of consensus between the 'two groups', the Chairman reported on the work of the Group that

> there appears to be general agreement as regards two of the various elements of delimitation: first, consensus seems to prevail to the effect that any measure of delimitation should be effected by agreement, and second, all the proposals presented refer to relevant or special circumstances as factors to be taken into account in the process of delimitation.[21]

In this connection, one would discover little difference between a delimitation based on equity, by taking account of all relevant circumstances, and one which is effected in accordance with the 'equidistance–special circumstances' rule, since, following the correct application of Article 6,[22] strict equidistance would have to be adjusted, or even abandoned, to reflect the special circumstances. The reasons behind the failure of the Conference to recognize this factor were mentioned earlier. However, this false assumption of a difference was maintained, until it finally resulted in the adoption of a non-committal formula, which 'says nothing of significance while worse still, trying to give a contrary impression by introducing unnecessary language and avoiding recognized terminology associated with the jurisprudence and scholarship on the subject'.[23]

into their relationship made the groups feel even less assured as masters of their own fate. In the later stages they suffered like Damocles as one person or another—the independent chairman of the relevant negotiating group, the President of the Conference, an influential delegation not a member of either group—threatened to personify the otherwise disembodied international legislator and provide a solution.'

[18] See e.g. proposals at n. 4 *supra*.

[19] See e.g. his proposal at n. 7 *supra*.

[20] See e.g. proposals presented at the 8th Session by Mexico and Peru, Doc. NG7/29/Rev.1.

[21] Doc. NG7/22.

[22] *Supra*, Ch. 3 s.2.

[23] Oxman, 'The Third UN Conference: The Tenth Session', 15.

(*b*) *The principle of agreement*

The indication that the boundary shall be effected by agreement epitomizes clearly the approach of the Decisions. In the *North Sea* cases the Court stated that 'delimitation must be the subject of agreement between the States concerned'.[24] Unilateral delimitation is, therefore, rejected; and thus in the *Guinea–Guinea Bissau* Arbitration the Tribunal refused to 'take into consideration a delimitation which did not result from negotiations or an equivalent act in accordance with international law'.[25] Within this context, the rule embodied in Articles 74(1) and 83(1) simply reiterates a general rule of conduct in all inter-State relations.[26] Another aspect of the rule is that States are under a duty to negotiate in 'good faith' toward an agreement, and to use the methods available for the peaceful settlement of disputes. But this aspect is vague and open-ended. In the *North Sea* cases the Court, in defining the 'obligation to negotiate meaningfully', referred to the cases of the *Free Zones of Upper Savoy and the District of Gex*[27] and *Railway Traffic between Lithuania and Poland*,[28] and merely quoted from the latter that negotiations must be pursued 'as fast as possible with a view to concluding agreements.'[29] In the *Fisheries Jurisdiction* case the Court remarked that the most appropriate method for resolving the dispute would be negotiation, and that to direct the parties to negotiate 'is therefore a proper exercise of the *Judicial Function* of the Court'.[30] However, the Court never attempted to explain the substance, if any, of this obligation.

The duty to negotiate may refer here to the specific obligation to enter into negotiations with a view to achieving agreement, with the corollary that each party must be prepared to modify its position.[31] Even accepting this moral-orientated interpretation, it is submitted that the duty to negotiate in good faith is of little utility for the purpose of Articles 74(1) and 83(1) for at least two obvious reasons. First, the stipulation that States must be prepared to modify their positions can make sense only if certain directions guiding the parties to some substantive rules are available.

[24] ICJ Reports (1969), para. 85.

[25] 25 ILM (1986), para. 94.

[26] As the Chamber in the *Gulf of Maine* case indicated, the principle of delimitation by agreement 'enunciate[s] . . . something which is a principle of international law'. ICJ Reports (1984), para. 86.

[27] PCIJ, Series A, 22 (1929), 13.

[28] PCIJ, Series A/B, 42 (1931), 116.

[29] *North Sea* cases, para. 87.

[30] *Fisheries Jurisdiction* case (*UK* v. *Iceland*), Merits, ICJ Reports (1974), 32.

[31] In Doc. 7/3 of 21 April 1979, submitted to Negotiating Group 7, Morocco had explained that the parties should not regard formal negotiations as a *sui generis* prerequisite of the automatic application of a specific method, resulting from the impossibility of reaching an agreement. No sensible negotiations, it stressed, can be conducted if one party sticks to its viewpoint without considering any modification.

These directions are non-existent in the formula under consideration. Secondly, the 'obligation to negotiate' has little practical use, since a third-party decision-maker would be most unlikely to demand proof of complete fulfilment of this condition before allowing States to appear before it to solve their boundary dispute, given that the provisions define neither the manner of conducting negotiations, nor their duration, except by a loose reference in paragraph 2 to a 'reasonable period of time'.[32]

(*c*) *The principle of conformity with international law*

It is submitted that a treaty text which merely provides that delimitation shall proceed on the basis of international law is superfluous. The superfluity is apparent when such a directive is contained within an instrument that is to represent the most ambitious scheme of codification and progressive development of international law ever attempted.[33] Moreover, taking the development of this 'international law' by international adjudication, the hallmark of which has been to subject delimitation to the fundamental principle of equity, Articles 74(1) and 83(1) are in fact telling States to apply 'equitable principles' to achieve, as the last directive therein requests, an 'equitable solution'. The formula stops here and provides no further guidelines as to how to distinguish between 'equity' as an ultimate end to a dispute and 'equitable principles' as a set of legal delimitatory rules. The fear thus arises that the vagueness already inherent in the notion of equity is exacerbated by this circularity of equitable means and ends with no conceptual tool to indicate how such a circle is to be broken other than by further reference to adjudication.[34]

[32] This is separate from the case when one of the two States (both parties to the compulsory procedures provisions of pt. xv of the 1982 Convention) refuses the jurisdiction of a third-party decision-maker on grounds of the failure of the other claimant State to 'exchange views regarding [the settlement of the boundary dispute] by negotiation or other peaceful means' (Art. 283); or to discharge its duties 'in a manner which would not constitute an abuse of rights' (Art. 300). In this case, the decision-maker will have to decide upon this preliminary issue before dealing with the substantive dispute. This, however, is related to the application of the provisions on disputes settlement, and not to the delimitation provisions.

[33] The explicit invocation of 'international law' was apparently made in order to satisfy the States favouring the median line who stressed, during the second part of the Conference's 9th Session, that they attached great importance to the insertion of a reference to international law into the text. The 'Equitable Principles Group' indicated that such a reference would add nothing of substance and was, therefore, superfluous. For a concise review of the meetings between representatives of the two 'Groups' during the 9th Session, see J. Symonides, 'Delimitation of Maritime Areas between States with Opposite or Adjacent Coasts', 13 *PYIL* (1984), 19, at 30–2.

[34] In both the *Tunisia–Libya* and *Libya–Malta* cases the Court expressed its concern that the quality which usually characterizes the term 'equitable principles' 'is not entirely satisfactory because it employs the term equitable to characterize both the result to be achieved and the means to be applied to reach this result'. See *infra*, text at n. 61.

The reference to international law and Article 38 of the Statute of the ICJ[35] merely sets general guidelines which could be extended to other dimensions, including not only fully formed international law but also those vehicles showing a developing pattern of behaviour (such as judicial decisions and negotiated agreements). Inviting States to solve their boundary disputes in accordance with this unwoven fabric of materials is, it is submitted, practically asking them to 'find' the law themselves. If an international dilemma is infrequent and there is a low risk of conflict, imprecisely determined rules of law may permit individualized settlements without exacerbating that risk; but when many disputes are possible, highly flexible, individualized rules of law may be counter-productive and inefficient. Article 6 of the Shelf Convention may be considered for this reason a more realistic approach. The references to median or equidistance lines and special circumstances not only constrain the behaviour of States concerned in relation to their unresolved boundary disputes, but also provide them with a 'point of departure' for the 'discussion' of the law applicable. In contrast, there is a dangerous lacuna in the 1982 Convention formula. Since no substantive rules are provided, the process of clarification by judicial and arbitral decision will always be the most likely solution that States should expect when seeking to resolve their shelf/EEZ boundary disputes on the basis of the 1982 Convention.

(*d*) *The principle of equitable solution*

One issue settled in the doctrine is the distinction between the use of equity as a legal norm and determinations *ex aequo et bono*, implying 'decisions based on practical considerations and expediency, disregarding, if necessary, existing law and recognized rights'.[36] International adjudication concurs with this view.[37] Behind this basic unanimity, however, considerable differences of opinion exist as to the proper meaning, function, and basis of this elusive concept of 'equitable principles'. In doctrine, there is disagreement over its introduction into international law.[38] The sources from which the concept derives its validity as a prin-

[35] Reflecting on the discussions of Group 7, Symonides mentions that this provision was made under the insistence of the 'Equidistance Principle Group', for they assumed that the significance of Art. 6 of the Shelf Convention had been stressed by the said Art. 38, since it had first mentioned conventional law. 'Delimitation of Maritime Areas', 37 n. 44.

[36] B. Cheng, 'Justice and Equity in International Law', 8 *CLP* (1955), 185, at 204.

[37] See e.g. the *North Sea* cases, para. 88; the *Anglo-French* Arbitration, 18 *RIAA*, para. 245; the *Tunisia–Libya* case, para. 71.

[38] Cf. e.g. G.P. Schwarzenberger, 'The Concept of Equity in International Law', *New Commonwealth Research Material*, 1 (1934), 17; E. McWhinney, 'Equity in International Law', in R. Newman (ed.), *Equity in the World's Legal System* (Bruylant, Brussels, 1973), 581; M. Akehurst, 'Equity and General Principles of Law', 25 *ICLQ* (1976), 801.

ciple of law are unsettled.[39] Nor is the proper function of the concept in international law agreed.[40]

However, leaving this confusion in the doctrine aside, all the Decisions affirm, as will be seen, the requirement in law that maritime boundaries are to be effected according to equitable principles. While these cases offer some explanation of the notion of equity—in a broad sense—they never attribute to it any identifiable content that may help to explain what it means in boundary delimitation. An attempt will be made later to 'distil' from these cases some recognizable framework for the notion of equitable principles. What it is intended to emphasize here is that the analysis of the concept of 'equitable principles' in these cases has been inconsistent. The particular formulation of this concept in Articles 74(1) and 83(1) appears to relate to this inconsistency, pointing the concept of 'equitable delimitation' in an uncertain and rather unsafe direction.

Equity was given its judicial blessing in the 1969 cases. It must be recalled, however, that the Court in those cases was faced with the very powerful submissions of Denmark and The Netherlands which argued for the elevation into a rule of law of 'equidistance' in its strict, literal interpretation.[41] Considering that the Court was primarily concerned with the unreasonableness of this rigid application of equidistance, its emphasis on equity would seem understandable in that context.[42] In that Judgment, however, the Court never maintained that in all circumstances, whatever the situation, the application of 'equitable principles' justifies unrestricted discretionary power to employ any method in order to achieve an 'equitable solution'.

[39] For a concise review of literature on this issue see Akehurst, 'Equity and General Principles of Law', 807 ff.

[40] Equity is described by Fitzmaurice as playing that part attributed to it in English law, that is, to mitigate the severity of legal rules. See his Sep. Op. in the *Barcelona Traction* case, ICJ Reports (1970), 85. Whereas in the *Tunisia–Libya* case the Court held that, in contrast with domestic law, 'this role of equity has no parallel in the development of international law; the legal concept of equity is a general principle directly applicable as law' (para. 70). In his Sep. Op. in the *Tunisia–Libya* case, Judge Aréchaga states that to resort to equity 'means, in effect, to appreciate and balance the relevant circumstances of the case, so as to render justice, not through the rigid application of general rules and principles and of formal legal concepts, but through an adaptation and adjustment of such principles, rules and concepts to the facts, realities and circumstances of each case', para. 24. O'Connell suggests that equity 'plays a role in the international judicial process: in the correction of over-rigorous law, in the filling of gaps and in the abrogation of law' (his *International Law* (Stevens & Sons, London, 1970), i. 14). Whereas Akehurst contends that equity *contra legem* is inadmissible in international law in the absence of express authorization to do so. 'Equity and General Principles of Law', 807.

[41] *North Sea* cases, paras. 37–40.

[42] See *Supra*, Ch. 3 s.3(*a*). The Court, nevertheless, observed that it was 'a truism to say that the delimitation must be equitable; rather is the problem above all one of defining the means whereby the delimitation can be carried out in such a way as to be recognized as equitable'. Para. 92.

On the contrary, the Judgment contained various indications as to the applicable substantive rules and also the factors to be taken into consideration.[43] More precisely, the Court affirmed that

it is not a question of applying equity simply as a matter of abstract justice, but of applying a rule of law which itself requires the application of equitable principles, *in accordance with the ideas which have always underlain the development of the legal regime of the continental shelf in this field*, namely: . . . (*b*) the parties are under an obligation to act in such a way that, in the particular case, and taking all the circumstances into account, equitable principles are applied,—for this purpose the *equidistance method can be used but other methods exist and may be employed*, alone or in combination, according to the areas involved.[44]

Recognizing the dangers of running too far with a concept as abstract as 'equitable principles', the Court of Arbitration in the *Anglo-French* Arbitration took great pains to ensure that the legal process of maritime boundary delimitation remained subject to concrete legal rules. Applying the 'equitable principles' idea of the 1969 cases to Article 6 of the Shelf Convention, the Court held that

the appropriateness of the equidistance method or any other method for the purpose of effecting an equitable delimitation is a function or reflection of the geographical and other relevant circumstances of each particular case. The choice of the method or methods of delimitation in any given case, whether under the 1958 Convention or customary law, has therefore to be determined in the light of those circumstances and of the fundamental norm that the delimitation must be in accordance with equitable principles.[45]

The above statements are significant in that they stress the a priori importance in boundary delimitation of a *method*. The particular method to be used, as a next step in the analysis, depends primarily upon geography. It is neither a matter of complete freedom of choice, nor of rigid application of a particular method such as equidistance. Of course other factors could affect the choice or modification of the method employed, but, throughout its decision, the Court stressed that equity is achieved through a method which is primarily a function of geography and the other factors underlying the concept of the continental shelf. Thus when France argued that, in the Atlantic Region, the Court was free to select any method, or a variety of methods, which in its view ensured an equitable delimitation,[46] the Court's reaction was firm and unambiguous. It said it did not have 'carte blanche to employ any method that it chooses in order to effect an equitable delimitation'.[47]

Unfortunately, in the *Tunisia–Libya* case the Court appears to have

[43] See paras. 83–101. [44] Ibid. para. 85 (emphasis added).
[45] Para. 97. [46] Paras. 207, 218. [47] Para. 245.

gone beyond the circumscribed definition of equity enunciated by the Court in 1969 and applied by the Court of Arbitration in the 1977 Award. In a statement surprising for its scope, the Court said:

It is . . . the result which is predominant; *the principles are subordinate to the goal.* The equitableness of a principle must be assessed in the light of its usefulness for the purpose of arriving at an equitable result.[48]

Equity according to this dictum seems to relate to establishing a compromise between different claims. It becomes, in Judge Gros's words, 'an appraisal of the expediency of a result, which is the very definition of the arbitrary'.[49] This 'arbitrariness' is further enhanced by the Court's refusal to identify equitable principles:

It is not every such principle which is in itself equitable; it may acquire this equity by reference to the equitableness of the solution. The principles to be indicated by the Court have to be *selected* according to their appropriateness for reaching an *equitable result.*[50]

The influence of the language of paragraph 1 of Articles 74 and 83 on this dictum is clear.[51] The problem, however, is that it is very doubtful whether the drafters of the Convention intended the interpretation given by the Court. As seen above, the *travaux* suggest that the phrase 'equitable solution' was merely chosen as a compromise, so that, in the interest of satisfying both of the 'two groups' at Negotiating Group 7, reference to both 'equidistance' and 'equitable principles' could be omitted from the final text. It is submitted, however, that this obscure interpretation by the Court seems to concede that the choice of method is a matter of indifference to the law and delimitation is governed solely by the equity of the result. This, it is submitted, is falling into that operation of *justitia distributiva* which the Court itself has so roundly rejected.[52]

Fortunately, in the *Gulf of Maine* case, the Chamber's analysis of the principles of equity reiterates that of the 1969 and 1977 Decisions, though in a more detailed and precise manner. The Chamber mentioned that its final aim was to produce 'an equitable overall result'.[53] But, throughout its Judgment, the Chamber made it clear that the subject-matter which

[48] Para. 70 (emphasis added).
[49] His dissent to the *Gulf of Maine* case, 383.
[50] Para. 70 (emphasis added).
[51] Just before quoting the above statement, the Court referred to the change in the final text of Art. 83(1) when 'equitable principles' were replaced by 'equitable solution' and thus the Court stated that in the final text '[e]mphasis is placed on the equitable solution which has to be achieved. The principles and rules applicable to the delimitation of continental shelf areas are those which are appropriate to bring about an equitable result'. Ibid. para. 50.
[52] Para. 71.
[53] Para. 241.

should reflect equity is not the solution. Nor could equity be found in the 'principles' or 'rules' of international law governing the matter; for the peculiarity of 'each particular case' precludes 'the possibility of those conditions arising which are necessary for the formation of principles and rules of customary law giving specific provisions for'[54] delimitation. Rather, equity is introduced as an 'identifier' of those 'criteria and particular methods that may be used to ensure *in concreto* that a particular situation is dealt with in accordance with the principles and rules in question'.[55] More positively, the 'equity' of a certain criterion or method is not so much judged by the result achieved as by recognition of 'the geographic configuration of the area and other relevant circumstances'.[56] A fact-evaluation process was suggested by the Chamber to identify equity.

> The essential fact to bear in mind is . . . that the criteria in question are not themselves rules of law and therefore mandatory in the different situations, but 'equitable', or even 'reasonable', criteria, and that what international law requires is that recourse be had in each case to the criterion, or the balance of different criteria, appearing to be most appropriate to the concrete situation.[57]

It is submitted that this is a positive step toward clarifying the meaning of equity;[58] and, thus, for the first time in international adjudication, the Chamber did feel able to identify certain 'equitable criteria' that may be taken into consideration for any international maritime delimitation.[59]

In *Libya–Malta*, however, the Court seemed capricious in its analysis of the principle of equity. On the one hand, it seemed unimpressed by the Chamber's attempt to allot to equity some identifiable content. The Court

[54] Para. 81. In para. 80 the Chamber stressed 'the distinction to be drawn between what are principles and rules . . . and what could be better described as the various equitable criteria and practical methods'.

[55] Para. 80.

[56] Para. 112.

[57] Para. 158.

[58] Judge Gros, however, maintains that the Chamber's approach in this regard does not differ from that of the 1982 Judgment: 'Like that of the Court in 1982, the Judgment of the Chamber has attempted to construct, in support of unsuccessful codification of maritime delimitation, a doctrine of equitable result, demonstrated by the progression of the reasoning through the contradictions it seeks to efface.' His Dissent to the *Gulf of Maine* case, 388.

[59] In para. 157 the Chamber identified some 'equitable criteria' such as 'the criterion expressed by the classical formula that the land dominates the sea; the criterion advocating, in cases where no special circumstances require correction thereof, the equal division of the areas of overlap of the maritime and submarine zones appertaining to the respective coasts of neighbouring States; the criterion that, whenever possible, the seaward extension of a State's coast should not encroach upon areas that are too close to the coast of another State; the criterion of preventing, as far as possible, any cut-off of the seaward projection of the coast or of part of the coast of either of the States concerned; and the criterion whereby, in certain circumstances, the appropriate consequences may be drawn from any inequalities in the extent of the coasts of two States into the same area of delimitation.'

merely quoted with approval its statement in the 1982 Decision, which, in its view, confirmed the result-orientated approach in Article 83(1). In this Article, the Court recalled, 'Emphasis is placed on the equitable solution which has to be achieved. The principles and rules applicable to the delimitation of continental shelf areas are those which are appropriate to bring about an equitable result.'[60] And, like the 1982 Judgment, the Court also in this case derived some support for its view on equity from the 1982 Convention. For, although acknowledging that the term 'is not entirely satisfactory because it employs the term equitable to characterize both the result to be achieved and the means to be applied to reach this result',[61] the Court nevertheless resolved that '[i]t is however the goal—the equitable result—and not the means used to achieve it, that must be the primary element in this duality of characterization'.[62]

On the other hand, in the subsequent paragraphs of the Judgment, the Court seemed to take a more fact-intensive approach to identifying equitable principles. The Court explained that the justice of which equity is an emanation is not abstract justice. It is equity according to the rule of law, and 'its application should display consistency and a degree of predictability; even though it looks with particularity to the peculiar circumstances of an instant case, it also looks beyond it to principles of more general application'.[63]

The Court then went on to identify some of those principles of general application. These are very similar to the 'criteria' of general application enumerated by the Chamber.[64] The most illuminating part of the Judgment with regard to equity is its emphasis that these general principles, identified with equity, are linked to a geographical situation. 'In interpreting [these principles], it must be borne in mind that the geography which is not to be refashioned means those aspects of a geographical situation most germane to the legal institution of the continental shelf.'[65]

[60] ICJ Reports (1985), para. 28.
[61] Para. 45.
[62] Ibid.
[63] Para. 45.
[64] See *supra*, n. 59. The Court identified certain 'equitable principles [which] are expressed in terms of general application . . . the principle that there is to be no question of refashioning geography, or compensating for the inequalities of nature; the related principle of non-encroachment by one party on the natural prolongation of the other, which is no more than the negative expression of the positive rule that the coastal State enjoys sovereign rights over the continental shelf off its coasts to the full extent authorized by international law in the relevant circumstances; the principle that although all States are equal before the law and are entitled to equal treatment, equity does not necessarily imply equality . . . nor does it seek to make equal what nature has made unequal; and the principle that there can be no question of distributive justice.' Para. 46.
[65] Para. 47.

Thus, in adopting the 'criterion of distance', the Court emphasized the importance of the fact-evaluation analysis for the assessment of the equitableness of that criterion.

In this assessment, account must be taken of the fact that, according to the 'fundamental norm' of the law of delimitation, an equitable result must be achieved on the basis of the application of equitable principles to the relevant circumstances.[66]

However, despite this uncertainty in identifying 'equity' in the abstract, the Decisions seem to follow a similar analytical approach regarding its application to the concrete cases. This approach, however, does not succeed in eliminating the subjective evaluations which would naturally follow from applying this elusive concept of equity. The following section discusses this approach and demonstrates the element of subjectivity inherent therein.

3. THE ELEMENTS OF EQUITY AS IDENTIFIED BY ADJUDICATION

(*a*) *Introduction*

The jurisprudence of the Decisions relates the principle of equity to the factors that are deemed relevant to the delimitation. As stressed forcefully by the Court in *Tunisia–Libya*, 'There can be no doubt that it is virtually impossible to achieve an equitable solution in any delimitation without taking into account the particular relevant circumstances of the area.'[67]

The identification of the relevant circumstances characterizing an area is therefore the first step in applying equity. The next step, as the Decisions reveal, is to consider the weight each of these circumstances may have in effecting the boundary line. The last step is to find the appropriate method or methods best reflecting the previous two-step process. Simple as this process may sound in the abstract, its application by the Decisions illustrates how difficult it is to predict an equitable result by any process of logical deduction from this decision-making process.

Before discussing this issue, one important factor must be stressed. As the EEZ regime lacks the jurisprudence and precedent enjoyed by the shelf regime,[68] it is relevant to ask to what extent the 'elements of equity'

[66] Para. 62.
[67] Para. 72.
[68] It is sufficient to say that four of the six contemporary cases on maritime boundaries dealt with the delimitation of the continental shelf exclusively.

relating to the delimitation of shelf are also applicable to the delimitation of the EEZ. It is obvious that certain factors might be more relevant to an equitable delimitation of the shelf than to the delimitation of the EEZ and vice versa. The peculiarity of those factors to one regime rather than the other will be pointed out, whenever applicable. It would, however, appear safe to state that the jurisprudence of the Decisions indicates that all the factors of 'the first order'[69] are, due to their essential as well as 'neutral' nature, germane to the delimitation of the shelf and the superjacent waters. The following enumeration of the factors considered by the Decisions is therefore submitted to be relevant for both shelf and EEZ delimitation.

(*b*) *Identification of the relevant factors*

(i) The primacy of the factor of geography

Geography in the context of maritime boundary delimitation is identified according to a political factor, that is, a coastal State's territorial jurisdiction. The primacy of this factor in effecting an equitable delimitation is self-evident. Coastal geography is the legal source of the power which a State may exercise over maritime extensions seaward. Thus in the *Anglo-French* Arbitration the Court translated the notion of natural prolongation of a land territory, which had been central to the 1969 *North Sea* Judgment, into a politico-geographical correlation between a State's coast and the offshore shelf area: 'In international law . . . the concept of the continental shelf is a juridical concept which connotes the natural prolongation under the sea not of a continent or geographical land mass but of the land territory of each State.'[70]

This approach logically asserts that geography should be accepted as it is, and the Decisions are unanimous on this point. They all declare that the facts of geography 'are not the product of human action amenable to positive or negative judgment, but the result of a natural phenomenon, so that they can only be taken as they are',[71] or, as is often put, there is 'no question of refashioning geography, or compensating for the inequities of nature'.[72]

This, in turn, leads to another logical proposition, which is that 'equity does not necessarily imply [absolute] equality',[73] but broad equality, if

[69] See analysis of the Chamber in the *Gulf of Maine* case, *infra*, text at nn. 75–6.
[70] Para. 191.
[71] *Gulf of Maine* case, para. 73.
[72] *Libya–Malta* case, para. 46.
[73] *North Sea* cases, para. 91; *Libya–Malta*, para. 46.

the States involved have broadly comparable coasts; and, conversely, equity does not require equality between coasts which reflect no equality. Thus, in the *Guinea–Guinea Bissau* Arbitration, the broadly equal coasts of the two adjacent States dictated a broadly equal division of the area of maritime jurisdiction between them by virtue of a line perpendicular to the general direction of the coast (with special provision being made for the first segment close to the coast).[74] Conversely, in *Libya–Malta* and *Gulf of Maine*, where, as will be seen later, there was a considerable difference in coastal lengths, the provisional median line adopted by the Court and the Chamber was 'transposed' to reflect the difference in coastal lengths.

Finally, this emphasis on geography suggests that factors emanating from other considerations are of a secondary order in determining the equity of the delimitation line. In the *Gulf of Maine* case, after reviewing the arguments of the parties regarding the socio-politico-economic aspects of the area under consideration, the Chamber stressed that 'the crux of the matter lies elsewhere':

> The Chamber is convinced that for the purpose of [effecting a delimitation on the basis of law], international law . . . does no more than lay down in general that equitable criteria are to be applied, criteria which are not spelled out but which are essentially to be determined in relation to what may be properly called the geographical features of the area.[75]

Only when, on the basis of these geographical criteria, the drawing of a delimitation line has been envisaged, may the other factors be brought in and taken into account 'in order to be sure of reaching an equitable result'.[76]

Going along with this basic premiss of geography, the Decisions also reveal that various other political, social, and economic factors, reflecting world social justice, could not lead to an equitable delimitation, because ideas of the way in which they should be taken into account are bound to vary widely, and, most importantly, they have little comnection with the decisive factor of coastal geography. Also, for the latter reason, some geophysical factors—not intimately related to coastal geography—are treated as irrelevant. The rationale for the Decisions' exclusion of such 'factors' illuminates their specific approach to the principle of equity. Therefore, it is perhaps reasonable first to elaborate on this rationale in connection with some of these irrelevant 'factors'.

[74] *Infra*, Ch. 5, s.3(*c*).
[75] Para. 59.
[76] Ibid. See also para. 110.

(ii) Factors regarded as irrelevant

1. *Geophysical discontinuity within 200 miles*

The relevance of the factor of geology or geomorphology, as initiated by the *North Sea* cases,[77] is certainly not applicable in the delimitation of the exclusive economic (or fishery) zones. Irrespective of the geophysical characteristics of the seabed, a coastal State is entitled to an EEZ up to 200 miles from its territorial sea's baseline.[78]

With regard to the delimitation of the continental shelf within 200 miles (or 400 miles if the two States are opposite), the Decisions' approach has been somewhat cautious. The introduction in the 1982 Convention of the distance criterion of 200 miles for, *inter alia*, a State's jurisdiction over its continental shelf has certainly accelerated the relative demise of the natural prolongation concept, first postulated in the *Anglo-French* Arbitration.[79] As observed by the Court in the *Tunisia–Libya* case, the legal concept of the continental shelf 'as based on the "species of platform" has thus been modified by this criterion'.[80]

Nevertheless, the Decisions (*Libya–Malta* apart) appear to have made their exclusion of geophysical factors case-specific and reserved the possibility of their consideration in other cases. The Decisions have tended to stress the fact that the relevant area was one of 'uniformity', so that no features likely to distort the continuity of the shelf area could be identified.

In *Tunisia–Libya*, the Court decided that it was only 'in the particular geographical circumstances of the present case' that 'the physical structure of the continental shelf areas is not such as to determine an equitable line of delimitation'.[81] However, even if they cannot be counted as such, according to the Court, mere geophysical discontinuities can always be taken into account 'as one of several circumstances relevant for an equitable solution'.[82]

Even in the case of a single boundary, the Decisions do not seem to depart from this approach. In the *Gulf of Maine* case, the Chamber examined the possible existence of a natural boundary in, *inter alia*, the seabed.[83] This was undertaken to identify whether any such boundary could serve 'as the basis for carrying out a delimitation of the kind requested of the Chamber'.[84] The Chamber felt that it did not have to decide on the matter before it as there were 'no geological, geomorphological or other factors sufficiently important, evident and conclusive to represent a single, incontrovertible natural boundary'.[85] Implicitly,

[77] Para. 101(D)(2). [78] Art. 57 of the 1982 Convention.
[79] *Supra*, Ch. 3 s.3(*b*)(iii). [80] Para. 47. [81] Para. 133.
[82] Paras. 68, 80. [83] Paras. 43 ff. [84] Para. 55. [85] Para. 56.

therefore, the Chamber did not exclude the relevance of the natural prolongation factor to the law on delimitation. The Chamber, however, was quick to qualify this implicit conclusion by stating that any delimitation was a 'legal-political operation',[86] and, consequently, it was not necessarily true that where 'a natural boundary is discernible, the political delimitation necessarily has to follow the same line'.[87]

A little more certain was the Chamber's refusal to extend the natural prolongation principle to the water column. The United States represented the North-East Channel as not only marking the physical boundary between its seabed and that of Canada but also forming a natural boundary between distinct oceanic and ecological regimes in the water column.[88] This was forcefully rejected by the Chamber. It held that it was not 'convinced of the possibility of discerning any genuine, sure and stable "natural boundaries" in so fluctuating an environment as the waters of the ocean, their flora and fauna'.[89]

In *Guinea–Guinea Bissau* the Tribunal once again observed the possibility when '[t]he characteristics of a continental shelf may serve to demonstrate the existence of a break in the continuity of the shelf or in the prolongation of territories of the States which are parties to a delimitation'.[90]

However, the judicial review of the geophysical factor reflected some change in the *Libya–Malta* case. The Court first referred to the development of the law regarding coastal State entitlement to an area of continental shelf up to a distance of 200 miles, irrespective of the physical characteristics of the seabed. This development, the Court maintained, reflected a kind of intrinsic relationship between the distance criterion and a State's title to the continental shelf, which in turn rendered the concept of natural prolongation in the physical, i.e. geological/geomorphological, sense obsolete.[91] Nevertheless, the Court attempted to keep

[86] Ibid.
[87] Ibid.
[88] Paras. 51–3.
[89] Para. 54.
[90] Para. 117.
[91] After referring to the 1969 and 1982 Judgments whereby, to quote from the latter, identification of natural prolongation was considered, 'where the geographical circumstances are appropriate, to have an important role in defining equitable delimitation', the Court commented that 'to rely on this jurisprudence would be to overlook the fact that where such jurisprudence appears to ascribe a role to geophysical or geological factors in delimitation, it finds warrant for doing so in a regime of the title itself *which used to allot those factors a place which now belongs to the past, in so far as sea-bed areas less than 200 miles from the coast are concerned*'. Para. 40 (emphasis added). It should be noted, however, that in 1982, when the Court decided *Tunisia–Libya*, the distance principle was already established in the Negotiated Text.

'natural prolongation' alive by attributing to it a role complementary to the distance principle. This will be discussed elsewhere.[92]

2. *The size of the land mass*

The reason for excluding this factor is obvious. The capacity to engender continental shelf/exclusive zone rights derives from sovereignty over the coast of a land mass; the land mass *per se* cannot provide the factual basis and legal justification for a State's maritime jurisdiction. In *Guinea–Guinea Bissau* both parties, and in *Libya–Malta* the former, argued that considerations should be given to proportionality in relation to the land mass of each State. The Tribunal ruled in the former case that this claim

> does not constitute a relevant factor . . . The rights which a State may claim to have over the sea are not related to the extent of the territory behind its coasts, but to the coasts themselves and to the manner in which they border this territory.[93]

In *Libya–Malta*, the Court stressed that 'landmass has never been regarded as a basis of entitlement to continental shelf rights'. It found no support 'in the practice of States, in jurisprudence, in doctrine, or indeed in the work of the Third United Nations Conference on the Law of the Sea'.[94]

However, whilst saying so the Court identified, as one of the 'relevant circumstances' justifying modification of its initial median line, the existence of the Maltese islands as '*a minor feature* of the northern seaboard of the region in question',[95] and as '*a relatively small feature* in a semi-enclosed sea'.[96] This obviously indicates that when islands are involved, the size of the land mass might be considered *de facto* a relevant factor. It is this factor which ultimately determines an island's coastal extension, a primary factor in the delimitation process.[97]

3. *The relative economic position of the parties*

A State's comparative wealth is determined by factors such as population, industrial activity, personal income, per capita income, tax burden, proved recoverable hydrocarbon and other resources, etc. These factors are generally regarded as variables of unpredictable nature and hence unhelpful to boundary delimitation designed to achieve permanency. In the *Tunisia–Libya* case, the former invoked economic considerations by drawing attention to its relative poverty *vis-à-vis* Libya in terms of absence of natural resources, especially hydrocarbons, compared with the relative wealth of Libya.[98] The Court ruled that these economic considerations cannot be taken into account for the delimitation of the continental

[92] See *infra*, Ch. 9 s.2(*f*). [93] Para. 119. [94] Para. 49. [95] Para. 69.
[96] Para. 73. [97] See *infra*, s.3(*b*)(iii)(1). [98] Para. 106.

shelf area between the parties because 'They are virtually extraneous factors since they are variables which an unpredictable national fortune or calamity . . . might at any time cause to tilt the scale one way or the other.'[99]

In the *Gulf of Maine* case, both of the parties, in support of their cases, submitted a comparative analysis of the respective importance of the fisheries resources for what was called the one-dimensional economy of Lunenburg County and for the diversified, urbanized economy of Massachusetts.[100] The Chamber refused to take account of these and other similar aspects. It observed that such activities 'may require an examination of valid considerations of a political and economic character' resulting possibly in a decision *ex aequo et bono*.[101] More interesting was Canada's reliance on the relative wealth factor in a local, socio-economic context. According to Canada, the delimitation line affecting Georges Bank 'should ensure the maintenance of the existing fishing patterns that are . . . vital to the coastal communities of the region in question'.[102] This consideration was so strong that the Chamber alluded to its 'obvious socio-economic impact on the communities inhabiting certain counties of Nova Scotia'.[103] However, even within this context the Chamber maintained that there was no reason 'to consider *de jure*' that the delimitation line carried out in the region 'must result in each Party enjoying an access to the regional fishing resources which will be equal to the access it previously enjoyed *de facto*'.[104]

The Chamber, however, observed that the overall result, achieved through the application of equitable criteria and appropriate methods, should not lead to 'catastrophic repercussions for the livelihood and economic well-being of the population of the countries concerned'.[105] Implicitly, should it lead to such consequences, an appropriate deviation from the line based on geographical criteria would be permissible. In view of this ambiguous notion of 'catastrophic repercussions',[106] it may be said that the Chamber's observation simply pays lip-service to mass data

[99] Para. 107. In the *Libya–Malta* case the same argument was brought by Malta. In response to this, the Court said that it did not 'consider that a delimitation should be influenced by the relative economic position of the two States in question . . . neither the rules determining the validity of legal entitlement to the continental shelf, nor those concerning delimitations . . . leave room for any considerations of economic development of the States in question'. Para. 50.

[100] Para. 58.

[101] Para. 59.

[102] Para. 234.

[103] Para. 236.

[104] Ibid.

[105] Para. 237.

[106] Professor Bowett interprets the Chamber's statement as setting out 'extremely stringent conditions' under which the Chamber might be able to accept the relevance of

concerning human and economic geography presented by the parties in support of their respective claims on this issue.[107]

None the less, the Chamber's overall observations seem to indicate that the relative wealth might best be envisaged in a 'local' context, reflecting the well-being or livelihood of a particular coastal community. Viewed as such, it is submitted that the interests of a certain community may call for recognizing an urgent 'localized' need for exclusive exploitation. This condition could be used as an auxiliary factor. Maritime boundary disputes usually involve geographic features such as promontories and islands. Deciding how much weight should be given to such features has been one of the most intricate and subjective issues in boundary settlements. There is of course a simple indicium that the livelihood of the inhabitants of those features is more related to sea activity and thus more affected by the boundary settlement than the inhabitants of the hinterland. A reasonable means of effecting an equitable boundary involving these features might be reached through the consideration of any *local* demographic and socio-economic patterns associated with them. In the context of EEZ/FZ delimitation in particular, it is pertinent to recall that the concept of the EEZ, as the very name suggests, is a reflection of the socio-economic relationship between the coastal State and its offshore areas.[108] It is submitted that economic considerations, especially those connected with a coastal community's exceptional dependence on fisheries (the same is applicable to the sedentary resources of the continental shelf), are bound to be treated as factors assisting the creation of an equitable maritime boundary.[109] There is considerable evidence in State practice indicating that States are prepared to give weight to local socio-economic factors in drawing their maritime boundaries.[110]

(iii) The relevant factors

1. *The factor of geography*

This factor consists of three elements, all of which are considered essential by the Decisions for effecting an equitable delimitation. These

'economic factors'. His 'Economic Factors in Maritime Delimitation Cases', in *International Law at the Time of Its Codification: Essays in Honour of Robert Ago*, i (Giuffrè Editore, Milan, 1987), 45, at 54–5.

[107] See e.g. the *Gulf of Maine* case: *Supplementary Evidence and Miscellaneous Documents, Annex to the Reply submitted by Canada*, vol. ii, 12 Dec. 1983; *Analytical Annex to the Reply, Submitted by the United States of America*, vol. ii, 12 Dec. 1983.

[108] See e.g. D. Attard, *The Exclusive Economic Zone in International Law* (Clarendon Press, Oxford, 1987), ch. 1.

[109] Bowett, 'Economic factors', 62.

[110] e.g. Argentina–Uruguay (1973); Kenya–Tanzania (1976); Australia–Papua New Guinea (1975), all discussed *infra*, Ch. 5 s.4(*b*).

are the general configuration of the coasts, including the presence of any special or unusual features, the relationship of one coast to another, and the overall extension of the shelf/exclusive zone area abutting on these coasts.

The invocation of this three-dimensional factor as the most decisive aspect of equity is inherent in the straightforward idea that as each coastal State has equal entitlement to maritime jurisdiction, its coasts are presumed to possess an equal capacity to generate an area of maritime jurisdiction. But the actual capacity for generating exclusive rights depends *in concreto* upon physical factors with which States are not equally endowed. As the Court said in the *Tunisia–Libya* case, it is the 'geographical correlation between coastal submerged areas off the coasts [which is] the basis of the coastal State's legal title'.[111]

The role of this factor in deciding the criteria and methods applicable to delimitation is probably best illustrated in the Court's *dispositif* in *Libya–Malta*. The Court found that the factors to be taken into account in achieving an equitable delimitation were the following:

(1) the general configuration of the coasts of the Parties, their oppositeness, and their relationship to each other within the overall geographical context;
(2) the disparity in the lengths of the relevant coasts of the Parties and the distance between them;
(3) the need to avoid in the delimitation excessive disproportion between the extent of the continental shelf areas appertaining to the coastal State and the length of the relevant parts of its coast, measured in the general direction of the coastline.[112]

The determination of this factor, however, gives rise to many problems and it is indeed the decision-maker's reflection on these problems which ultimately determines which method of delimitation is appropriate. As far as the problem of determining the lengths of the 'relevant coasts' is concerned, all the Decisions submit that the appropriate method is to 'follow the general direction of the coastline'.[113]

In respect of off-lying features, minor features, such as small islands, rocks, and minute coastal configurations, are to be ignored.[114] The Decisions, however, reflect considerable uncertainty as to the weight given to offshore islands. Giving full effect to Ushant Island in the *Anglo-French* Arbitration[115] (10 miles off the coast of Brittany) and ignoring completely Jerba (screening the Tunisian coast) and granting half-effect to the

[111] Para. 73.
[112] Para. 79(B).
[113] See the *North Sea* cases, para. 101; the *Tunisia–Libya* case, para. 131; the *Gulf of Maine* case, para. 221; the *Guinea–Guinea Bissau* Arbitration, para. 97.
[114] *North Sea* cases, para. 57; the *Gulf of Maine* case, para. 201.
[115] Para. 253.

Kerkennahs (11–14 miles off coast) in the *Tunisia–Libya* case[116] are obvious examples of the Decisions' subjective approach to the question of islands. The Decisions, however, reveal that the weight islands receive depends ultimately on their own characteristics and other prevailing circumstances. The multitude of factors upon which the Tribunal in the *Anglo-French* Arbitration built its decision with regard to the Channel Islands was discussed previously.[117] It was on the same type of factors that the Court and its Chamber built their decisions with regard to the Kerkennahs[118] and Seal Island,[119] respectively.

With regard to the geographical relationship of one coast to another, the Decisions usually classify this relationship as one of oppositeness or adjacency. The importance of this classification is self-evident, for, whatever the method selected for delimiting the boundary line, this line has to reflect the geographical correlation between the relevant coasts. Thus, in both the *Tunisia–Libya* and the *Gulf of Maine* Decisions, the rationale for the use of different methods was based on the change in the geographical relationship between the coasts of the parties from one of adjacency in the vicinity of the land boundary terminus to one of 'oppositeness' or semi-oppositeness in the area further seaward.[120]

The Decisions also emphasize the importance of this distinction between opposite and adjacent coasts when the strict equidistance line is used. It is generally admitted that problems such as the cutting-off effect of concave coasts or the distorting effects of off-lying features tend to be aggravated where the coasts are adjacent, rather than opposite; for whereas with opposite coasts the effect of two basepoints is more limited since control over the line is quickly passed to other points on the coasts, with adjacent coasts a distorting feature, such as a promontory or island, can continue to influence the equidistance line over large distances out to sea.[121]

The final element pertinent for reflecting geography is the determination of the 'relevant area' of delimitation. This is necessary for circumscribing the area where the geographical relationship between the coasts can be identified, and also for applying the factor of proportionality, as will be seen later. However, in the determination of the 'relevant area' the Decisions seem generally to distinguish between two geographical situations. In situations where the coasts of two adjacent States

[116] See *infra*, Ch. 9 s.2(*e*).
[117] *Supra*, Ch. 3, s.3(*b*).
[118] *Tunisia–Libya* case, paras. 128, 133.
[119] *Gulf of Maine* case, para. 222.
[120] *Tunisia–Libya* case, paras. 126–7; *Gulf of Maine* case, paras. 206–7.
[121] This point proved to be decisive in the Court's rejection in the 1969 Judgment of the equidistance method. See *supra*, Ch. 3 s.3(*a*); see also the Court's analysis in the *Tunisia–Libya* case, para. 126; the *Gulf of Maine* case, para. 210.

are open to the high seas no precise definition of the seaward extension of the relevant area is needed. Thus in the *Guinea–Guinea Bissau* Arbitration the Tribunal adverted to the position of the parties' coastlines in the open Atlantic Ocean and, consequently, held that the non-determination of the outer limit 'can have no effect on the reasoning of the Tribunal'.[122]

In another geographical context, however, where a small maritime region is surrounded by several bordering States in an opposite/adjacent relation to each other, any determination of the segments of the 'area relevant to the delimitation' between the two States would partly overlap with the pretensions of other States. Faced with this problem, the majority of the Decisions seem to uphold a policy which, although not precluding them from identifying, in rough terms, the 'area of concern', does on the other hand preserve the rights of third States. This will be further considered in connection with third States' claims.

2. *Economic factors*

The argument regarding economic factors may take three forms: (1) the overall, relative wealth of the parties involved, which was discussed earlier; (2) the existence of natural resources across overlapping claims; and (3) established or historic economic usage. The latter two are considered successively.

(*a*) *Natural resources across overlapping claims* In the early debates in the International Law Commission, attention was drawn to the possibility of a mineral deposit overlapping a delimitation line, and the question was raised as to whether this could be a circumstance requiring modification of the boundary line.[123] Mouton proposed that in the case of hydrocarbon deposits the primary concern should be the preservation of the unity of the deposit.[124] Arguably, this may also apply to a common sedentary or mobile fishery ground. However, the Commission did not take up the point, and although it was raised in the Geneva Conference[125] it played no role in the debates.

In the *North Sea* cases the Court saw the problem as one for the parties to deal with, despite the fact that it was central to the dispute between them.[126] The Court merely directed the parties in their negotiations to

[122] Para. 96. See also para. 130(3)(C).

[123] *Yearbook of the ILC* (1950), i. 233.

[124] M. Mouton, 'The Continental Shelf', 58 *Recueil des cours* (1954), i. 347, at 422.

[125] UNCLOS I, *Official Records*, vi. 93, 95 (UK and Uruguay).

[126] In his Sep. Op. Judge Jessup pointed out that the problem of exploitation of the oil and gas resources in the North Sea 'was in the front of the minds of the parties but . . . none of them was prepared to base its case squarely on consideration of this factor' (*North Sea* cases, 72). Later on (p. 76) he added, '[a]lthough the arguments in the pleadings were defected by the parties away from outright reliance on the location of hydrocarbons under

take account of any 'common deposits' which might be found to straddle the boundary, and more generally, in the *dispositif*, required that account be taken of the natural resources of the area 'so far as known or readily ascertainable'.[127] No further guidelines were given as to how to employ this factor, especially as to how the parties were to seek to draw a reasonable accommodation between the three competing policy goals involved, that is, the achievement of maximum productivity of resources, avoidance of harmful and uneconomical exploitation, and equitable distribution of resources. Judge Ammoun emphasized that 'if the preservation of the unity of deposits is a matter of concern to the Parties, they must provide for this by a voluntary agreement'.[128] That would cast doubt on the existence in international law of the rule suggested by Mouton, a view expressed by some writers.[129]

However, this general yet positive statement of the Court on the relevance of hydrocarbon deposits[130] suffered a degree of uncertainty in the 1982 *Tunisia–Libya* Judgment. The Court enigmatically stated, 'As to the presence of oil wells in the area to be delimited, *it may, depending on the facts*, be an element to be taken into account in the process of weighing all relevant factors to achieve an equitable result.'[131] No precise criteria were laid down to assess the degree of weight to be given to this factor independently, or in relation to other factors. Nor is it clear whether the Court's statement was based on its conviction of the relevance of common deposits *per se*, or was rather associated with its emphasis in that case on the conduct of the parties involved regarding their respective oil concessions.[132]

In the *Gulf of Maine* case, the Chamber, reflecting on the submissions of the parties on Georges Bank, admitted that it had 'borne constantly in mind the problem of determining the final segment . . . when . . . establishing the previous segments'[133] and said '[t]his Bank is the real subject of the dispute . . . from the viewpoint of the potential sources of the

the North Sea, their bilateral and trilateral negotiations were specifically related to such resources and indicated that more was known about their location than the pleadings indicate'.

[127] Para. 101.

[128] His Sep. Op., 149.

[129] See e.g. D. Padwa, 'Submarine Boundaries', 9 *ICLQ* (1960), 628, at 645; E. Brown, *The Legal Status of Hydrospace* (Stevens & Sons, London, 1971), 67; W. Onorato, 'Apportionment of an International Common Petroleum Deposit', 26 *ICLQ* (1977), 324, at 325. See however, R. Lagoni, 'Oil and Gas Deposits across National Frontiers', 73 *AJIL* (1979), 215.

[130] For the differing interpretations of the Court's dictum on this issue, see e.g. Brown, *The Legal Status of Hydrospace*, 63; Lagoni, 'Oil and Gas Deposits', 241.

[131] Para. 107 (emphasis added).

[132] See *infra*, s.3(*b*)(iii)(4).

[133] Para. 226.

subsoil and . . . of fisheries'.[134] Nevertheless, it concluded such factors were 'ineligible for consideration as criteria to be applied in the delimitation process itself',[135] reflecting its finding that criteria drawn from geographical factors were to be applied.[136]

Whatever the correct interpretation might be, the conclusion from the Decisions is that, unlike, for instance, geographical factors, the relevance of common deposits might be considered only in the general assessment of the equity of a predetermined boundary line.

It should, however, be observed that although the language of the Decisions avoids any express reference to 'a just and equitable share' of resources,[137] the Decisions in practice do draw a line dividing an existing or potential resource in a fair manner. In *Tunisia–Libya* the Court knew that, on the basis of its proposed line, all the Tunisian producing wells would be on the Tunisian side, and so, too, for the Libyan wells. And, in the *Gulf of Maine* case the Chamber drew a line across Georges Bank which left to each party its major locations of sedentary catches, at least for scallops and lobsters.[138] The conclusion of the Chamber concerning the division of the main areas in which the subsoil was being explored for mineral resources was virtually the same.[139] In *Libya–Malta*, the Court was aware of the parties' competing interests in the Medina Bank, the area of promising hydrocarbons. The Court's resultant line did divide the area more or less equally between the parties.[140]

(*b*) *Resources associated with traditional or 'historical' dependency* The consideration of this factor invokes two interrelated policy goals. The stability of historical or long-standing patterns of economic activity producing dependency ought not to be disturbed in the interest of stable boundaries. On the other hand, if such economic considerations which might change over a period of time were allowed to play an overriding role in the delimitation process, the outcome might lead to refashioning of geography. Therefore, a reasonable balance must be achieved. The Decisions indicate that the factor of historic use and economic dependency could be used as a supportive element operating in conjunction with the main criteria based on geographic factors.

[134] Ibid.
[135] Para. 232.
[136] Para. 59.
[137] In fact the Court in the *North Sea* cases rejected explicitly the German argument of a 'just and equitable share' of the seabed on the grounds that this doctrine 'is quite foreign to, and inconsistent with, the basic concept of continental shelf entitlement'. See paras. 16–20.
[138] Para. 238.
[139] Para. 239.
[140] The Court in principle acknowledged that natural resources of the continental shelf, 'so far as known or readily ascertainable, might well constitute a relevant circumstance'. Para. 50.

The *Grisbadarna* case provides an example of a line being shifted to avoid the division of the important Grisbadarna bank which the Court determined to allocate to Sweden.[141] The Court based its solution on equitable considerations supported by 'all of several circumstances of fact'. The principal circumstance was that 'lobster fishing in the grounds of Grisbadarna has been carried on for a much longer time, to a much longer extent and by [a] much larger number of fishers subjects of Sweden than . . . subjects of Norway' and that Sweden had performed various executive acts 'owing to her conviction that these regions were Swedish', whereas Norway has shown 'much less solicitude' in this regard.[142]

The other example is the decision of the ICJ in the *Anglo-Norwegian Fisheries* case, whereby the Court upheld the straight baselines system of Norway, partly on the basis of that State's constant and sufficiently long practice, linked with economic dependency.[143]

Once again in the *Fisheries Jurisdiction* case, the Court held that, under customary law existing in the early 1970s, a coastal State may claim preferential rights when it is in a situation of dependency on coastal fisheries.[144] In this case, Iceland was deemed to have acquired preferential rights because of its clear dependence on its coastal fisheries and because these fisheries were subject to intensified exploitation.[145]

None of the above cases was of course concerned with the delimitation of extended areas of shelf or exclusive zone. But even in such extensive areas the relevance of established patterns of economic dependency is difficult to deny if a decision is ever to be reached with some degree of predictability. Its relevance is *a fortiori* in EEZ delimitation, where the regime covers not only the seabed but also the superjacent waters.

In *Tunisia–Libya*, the Court examined the Tunisian argument that 'fishing resources derived from its claimed "historic rights" and "historic waters" area must necessarily be taken into account as supplementing its national economy in eking out its survival as a country'.[146] The Court, however, observed that this claim is 'more nearly related to the concept of the exclusive economic zone'.[147] In other words, it is possible that in delimiting overlapping claims of EEZ or FZ, traditional fishery rights may well be considered in the final fixing of a boundary line initially drawn on geographic grounds.

[141] The *Grisbadarna* case, English text in 4 AJIL (1910), 226.

[142] Ibid. 233.

[143] See *infra*, Ch. 7 s.3(*a*).

[144] *Fisheries Jurisdiction* (*UK* v. *Iceland*), Merits, 23.

[145] Ibid. at 26–7. But the Court also noted that particular regard should be had to the position of the fishermen of the UK who had been engaged in the same fishery over a considerable period of time, establishing economic dependency on the same fishing ground (28).

[146] Para. 106. [147] Para. 100.

In the *Gulf of Maine* case the United States stressed the need for consideration of the continuous human presence which took the form especially of fishing, and of the conservation and management of fisheries, but also included other activities such as navigational assistance, rescue, research, and defence. The United States argued that all these activities, which exceeded in duration and scale those of Canada, had to be regarded as a major relevant circumstance. The United States' argument tended therefore towards the invocation of historic rights, although, as the Chamber noted, it did not use such an expression.[148] The Chamber, however, felt that socio-economic circumstances were ineligible for consideration as criteria to be applied in the delimitation process itself.[149] Nevertheless, it was prepared to recognize their relevance in assessing 'the equitable character of a delimitation first established on the basis of criteria borrowed from physical and political geography'.[150]

3. *The factor of security*

There is little doubt that exclusive jurisdiction is partly justified by the necessity to protect the coastal State against unregulated and unsupervised activities of foreign installations off its coast, constituting potential threats to its security.[151] This factor, however, seems to have a limited role in the decided cases.

In the *Anglo-French* Arbitration, France had argued that the application of the equidistance method in the Channel Islands region would be inequitable, *inter alia*, because it would involve severing the French coast into two zones separated by a United Kingdom zone. This, in turn, would be prejudicial to the vital interests of France in the defence of its territory. France referred in particular to the fact that it would be deprived of any right to control activities in the whole Hurd Deep Zone and that this would involve serious inconveniences and risks for French submarines stationed at Cherbourg, as well as affecting France's military supervision of the approaches to its territory.[152] The United Kingdom argued that France's security, defence, and navigational interests were not entitled to be given weight.[153] The Court expressed the view that such considerations

[148] Para. 233.
[149] Para. 232.
[150] Ibid.
[151] See e.g. the Truman Proclamation on the Continental Shelf (1945), para. 4; repr. in 40 *AJIL* (1945) (Suppl.), 45; Reports of the Special Rapporteur François on the Regime of the High Sea and on the Continental Shelf, *Yearbook of the ILC* (1950), ii. 51, and (1953), ii. 6.
[152] Paras. 161–3.
[153] Paras. 175–6.

might be relevant in principle, although it attributed to them very little weight in the Channel Islands case.[154]

In the other decisions, however, there seems to have been a tendency to relate security to the principle of non-encroachment, which indicates that a delimitation line should not result in allowing one State to have control over the maritime territory that lies immediately off the territory of another State. This principle has been recognized as relevant by international adjudication ever since the 1969 *North Sea* cases.[155] In the 1982 *Tunisia–Libya* case Judge Aréchaga linked the security needs of the coastal State to the principle of non-encroachment. He referred to the

> immediate and almost instinctive rejection by all coastal States of the possibility that foreign States, or foreign companies or individuals, might appear in front of their coasts, outside their territorial sea but at a short distance from their ports and coastal defences, in order to exploit the sea-bed and erect fixed installations for that purpose.[156]

And in *Libya–Malta* the Court confirmed this approach. The distance between the coasts of the two States

> is an obviously important consideration when deciding whether, and by how much, a median line boundary can be shifted without ceasing to have an approximately median location, or approaching so near to one coast as to bring into play other factors such as security.[157]

Security, therefore, according to these Decisions, becomes relevant in association with the decision-maker's power to shift a provisional line in order to reflect equity, taking into consideration the distance between the two States. This is of some significance, particularly in restricted sea areas.

4. *Conduct of the parties*

It is generally accepted that a State's conduct may produce legal consequences in its relations with other States.[158] In maritime boundaries the

[154] The Tribunal stated that 'the weight of such considerations in this region is, in any event, somewhat diminished by the very particular character of the English Channel as a major route of international maritime navigation serving ports outside the territories of either of the Parties. Consequently, they cannot be regarded by the Court as exercising a decisive influence on the delimitation of the boundary in the present case. They may support and strengthen, but they cannot negative any conclusions that are already indicated by the geographical, political and legal circumstances of the region which the Court had identified.' Para. 188.

[155] Para. 101. See also the *Gulf of Maine* case, para. 157; the *Guinea–Guinea Bissau* Arbitration, para. 110.

[156] His Sep. Op. in the *Tunisia–Libya* case, para. 72.

[157] Para. 73. See also the *Guinea–Guinea Bissau* Arbitration, para. 124.

[158] *Temple of Preah Vihear* case, ICJ Reports (1962), 32 ff.

application of this principle may take the extreme form of acquiescence or estoppel, where the conduct of the parties over a given period of their relationship constituted acquiescence by one of them in the application of a specific line advocated by the other. Recognizing this possibility in principle, the Chamber in the *Gulf of Maine* case observed that the concepts of acquiescence or estoppel 'both follow from the fundamental principles of good faith and equity'.[159]

At a lower, and perhaps more frequent, level the conduct of the parties might have resulted in a *de facto* or *modus vivendi* line corresponding to a common acceptance in practice. The application of this particular form of conduct as well as its supporting argument is well established in the *Tunisia–Libya* case. Emphasizing that nothing in the case was equivalent to 'tacit agreement' or to a claim on the basis of 'estoppel', the Court ruled that it was, nevertheless, its duty

> to take into account whatever indicia are available of the line or lines which the Parties themselves may have considered equitable or acted upon as such—if only as an interim solution affecting part only of the area to be delimited.[160]

Thus, in the actual circumstances of the case, the Court determined the first segment of the boundary line off Ras Ajdir, partly on the basis of a *modus vivendi* line for fisheries enforcement, which had been tacitly accepted by the colonial powers acting in Libya and Tunisia. This line was also found generally to correspond with another *de facto* line, resulting from the manner in which both parties had granted offshore oil and gas concessions.[161]

Similarly, in *Guinea–Guinea Bissau* the Tribunal believed itself justified in considering the 'southern limit' as a factor 'it should take into account with a view to a delimitation tending to achieve an equitable result'.[162] The consideration of the said limit (proposed by France and Portugal for determining the land boundary domain) was entirely based on factors relating to the conduct of the parties (and their predecessors) in treating the limit as a *modus vivendi* boundary for the area adjacent to the coastline since 1886.[163]

5. *Claims of third States*

The factor of the claims of third States, actual or prospective, was stressed in the 1969 Judgment and it figured prominently in all the subsequent

[159] Para. 130. The Chamber dedicated a significant part of its reasoning to the discussion of the arguments of both parties, Canada and the USA, regarding the existence of acquiescence or *modus vivendi* in the delimitation of Georges Bank. See paras. 126–253.
[160] Para. 118.
[161] For details, see *infra*, Ch. 8 s.4(*b*)(ii)(1), (2).
[162] Para. 106.
[163] Paras. 105–6. See *infra*, Ch. 5 s.3(*c*).

UNIV DUNELM

cases. This factor has an a priori relevance in EEZ delimitation. Because the distance criterion is an essential element of the EEZ institution and its application is the basis for legal title, the relevance of other delimitations (or claims) in the same region is bound to be more pronounced. This is particularly true in the restricted areas of enclosed seas, such as the Mediterranean, where the proximity of the coastal States and the application of the distance principle imply a higher incidence of overlapping claims.[164]

In the *Guinea–Guinea Bissau* Arbitration the Tribunal agreed that '[a] delimitation designed to obtain an equitable result cannot ignore the other delimitations already or still to be made in the region'.[165]

The impact of this factor on the question of 'the relevant area' has already been pointed out. It is pertinent to refer here to its relevance in two other cases. First, a third State claim may serve as a 'special circumstance' or 'relevant factor' for the assessment of the equity of a given method of delimitation. In the *North Sea* cases the Court stated:

> In the case of the North Sea in particular, where there is no outer limit boundary to the continental shelf, it happens that the claims of several States converge, meet and intercross in localities where . . . the bed of the sea still unquestionably consists of continental shelf. A study of these convergences . . . shows how inequitable would be the apparent simplification brought about by a delimitation which, ignoring such geographical circumstances, was based solely on the equidistance method.[166]

Thus the Court ruled in the *dispositif* that in applying the factor of proportionality account should be taken 'of the effect, actual or prospective, of any other continental shelf delimitation between adjacent States in the same area'.[167]

In contrast to this situation, the application of the equidistance method in the relatively open area of the 'Atlantic Region' was not considered by the 1977 Decision as inequitable even when the possibility of a prospective boundary with a third State was not excluded. Invoking the 1969 Judgment, the United Kingdom argued that in assessing what would be an equitable delimitation as between itself and France, account should be taken of the prospective effects of the delimitation of the continental shelf with the Republic of Ireland.[168] In its assessment of this argument the Court adverted to the fact that it was entrusted with the task of deciding the course of the boundary between the United Kingdom and France in the area as far as the 1,000-metre isobath.[169] The fact that there might be area still open to delimitation beyond that limit marked a distinction between the circumstances of this case and that of the *North Sea*.

[164] See e.g. analysis *infra*, Ch. 9 s.2.
[165] Para. 93.
[166] Para. 89.
[167] Para. 101(D)(3).
[168] Para. 24.
[169] Para. 27.

That problem [the existence of a third-party claim] would normally find its appropriate solution by negotiations directly between the States concerned, negotiations which may indeed be called for by the prolongation of their maritime zones beyond the 1,000-metre isobath to 200 nautical miles.[170]

Further on, the Court spoke of the possibility of redressing inequities by the use of available space. In other words, if something was taken away from one country in a particular area that State might be able to redress what it had lost in that area by extending its claim in another.[171]

The dictum of the 1969 Judgment was endorsed in the *Guinea–Guinea Bissau* Arbitration. The Tribunal stated that

a valid method [of delimitation] consists of looking at the whole of West Africa and of seeking a solution which would take overall account of the shape of its coastline. This would mean no longer restricting considerations to a *short coastline* but to *a long coastline*.[172]

Second, besides recognizing the normal situation where an existing or prospective boundary may limit the 'area relevant to the delimitation', the Court in the last of its cases went so far that it felt justified in refraining from adjudicating on an area in dispute between the parties on the ground that it was also subject to a third-State claim. The Court ruled that

in replying to the question put to it in the Special Agreement as to the principles and rules of international law applicable to the delimitation of the areas of the continental shelf appertaining to each of the Parties, [the Court] will confine itself to areas where no claims by a third State exist.[173]

The reasoning for, as well as the full impact of, this latest development will be considered later.[174] Suffice it to state here that, according to this Judgment, a third-State claim operates beyond its usual function of testing, among several other factors, the equitableness of the result achieved by any particular line. It has become an 'imposed rule' of jurisdiction-limitation. This finding obviously contrasts with the Decisions' previous stance in similar circumstances. In *Tunisia–Libya*, the interests of third States did not prevent the Court from 'hypothesizing' the 'area of concern' subject

[170] Para. 28.

[171] Para. 200. However, it is of interest to note that the Court did not seem to have fully dismissed the argument of the UK. In the hearings the Court asked the parties to indicate on a map how the prospective boundary with Ireland would encroach upon the area attributed to the UK in application of a median line (information obtained from an interview with Professor Bowett, who appeared in the case as Counsel for the UK).

[172] Para. 108.

[173] Para. 22.

[174] See *infra*, Ch. 9 s.2(*g*).

to the dispute between the parties. The Court's actual words are cited below.[175]

Similarly, in the *Guinea–Guinea Bissau* Arbitration, the fact that the lateral limits of the parties' maritime jurisdiction were respectively overlapping with third States' claims did not deter the Tribunal from forming 'an approximate idea of the zone to be considered, based on an approximate evaluation'.[176]

6. *Proportionality*

The origin of the factor of proportionality lies in the *North Sea* Judgment and has been applied in almost all subsequent cases. The Decisions reveal that proportionality is not only justified as an aspect of equity but is considered a necessary reflection of State coastal sovereignty. As the Court in the *Libya–Malta* case put it, 'Proportionality is certainly intimately related both to the governing principle of equity, and the importance of coasts in the generation of continental shelf rights.'[177]

The relevance of proportionality according to the Decisions may take one of two forms. First, it ensures that a reasonable degree of proportionality is maintained between the extent of the continental shelf/exclusive zone area appertaining to the coastal States concerned and the lengths of their respective coasts.[178] Therefore, to quote *Tunisia–Libya*, the Court took note of the relationship between the lengths of the relevant coasts of the parties, and compared that relationship with the ratio between the areas of the continental shelf attributed to each party. On the basis of the figures for distances (measured by the general direction of the coastlines) and ratios, the Court concluded that the result of the delimitation contemplated would 'meet the requirements of the test of proportionality as an aspect of equity'.[179]

Proportionality in this sense can in many cases be effected with reasonable certainty and objectivity. Although the Decisions formally assert that the course of the boundary does not depend on any 'nice calculations of proportionality',[180] they do in fact apply mathematical ratios for verifying the equity of the result of the delimitation contemplated. Indeed one cannot see why courts should be reluctant to accept a precise calculation when the circumstances of the case so allow. The more precise the ratio, the better it reflects the geography of the area, being the predominant aspect of equity.

[175] See *infra*, text at n. 183.
[176] Para. 94.
[177] Para. 55.
[178] See the *North Sea* cases, para. 98.
[179] Para. 131.
[180] See e.g. the *Anglo-French* Arbitration, paras. 26–7.

However, even within the context of this relatively simple function, proportionality is a difficult test in two particular cases adverted to by the Decisions. First, proportionality can only be used very relatively when the area 'relevant' to delimitation between the parties concerned overlaps with the claims of a third State.[181] This problem is particularly persistent in areas of narrow seas where, as noted earlier, the existence of a third-State interest may adversely affect the calculation of the length of the coastlines and the expanse of the area between the two States.[182]

It is sufficient to state here that the Court seems to accept this problem as an inevitable consequence of geography on the one hand and States' exclusive jurisdiction on the other. In the *Tunisia–Libya* case, after referring to the fact that the delimitation of the seaward limit of the area to be taken into account was only hypothetical and could not be precise (because of potential Maltese claims), the Court maintained:

> if it were not possible to base calculations of proportionality upon hypotheses of this kind, it is difficult to see how any two States could agree on a bilateral delimitation as being equitable until all the other delimitations in the area had been effected.[183]

The second difficulty relating to the first form of proportionality is where it is not possible to determine a well-identified 'coastal front' due to the existence of numerous islands in the area of delimitation. The Australian–Papua New Guinean islands in the Torres Strait[184] and the Greek islands in the Aegean Sea[185] are excellent examples. International adjudication has not yet dealt with this problem in its extreme form. Statements such as 'proportionality does not imply "refashioning of nature"'[186] and 'proportionality is particularly relevant when nature yields coasts of comparable equality'[187] may be taken to imply that proportionality in such areas as mentioned above is an impractical exercise.[188] However, the jurisprudence of the Decisions also affirms that proportionality does not necessarily mean absolute conformity with the ratio of the coastal fronts. To use the language of the 1982 *Tunisia–Libya* Judgment, 'the essential aspect of the criterion of proportionality is simply that one must compare like with like'.[189] Considering proportionality in

[181] Indeed in the *Libya–Malta* case the Court found it 'inappropriate' to apply proportionality, because of the fact, *inter alia*, that 'the area to which the Judgment will in fact apply is limited by reason of the existence of claims of third States'. Para. 74.

[182] See *supra*, s.3(*b*)(iii)(5).

[183] Para. 130.

[184] See *infra*, Ch. 5, text at n. 216.

[185] See *infra*, Ch. 10.

[186] See *supra*, text at nn. 71–2.

[187] See e.g. the *Anglo-French* Arbitration, para. 101; the *Tunisia–Libya* case, para. 56.

[188] See e.g. *infra*, Ch. 10 s.4(*d*)(v).

[189] Para. 130. See also para. 104.

the very broadest terms, it is possible in such circumstances to envisage a mathematical criterion that takes account of the total length of a State's 'coastal front', including all or some (depending on the balance of the other relevant factors) of the islands. One way of effecting this criterion is explained in detail later.[190] It should be emphasized here that the allocation of shelf or exclusive zone areas in such situations is not meant to be solely a reflection of the resultant ratio. Within the bounds set forth by the Decisions for the application of proportionality, it appears that such a ratio does not afford an autonomous criterion or method of delimitation. It can only be considered as one factor which enters into the equation leading to an equitable delimitation.

The second form of proportionality is when it is intended as a means of identifying and then correcting the kind of distortion—disproportion—that could arise from using a method inappropriate for taking adequate account of some kinds of coastal configuration or other features. In the words of the 1977 *Anglo-French* Arbitration,

> The factor of proportionality may appear in the form of the ratio between the areas of continental shelf to the lengths of the respective coastlines . . . But it may also appear, and more usually does, as a factor for determining the reasonable or unreasonable—the equitable or inequitable—effects of particular geographical features or configuration upon the course of an equidistance-line boundary.[191]

Proportionality in this second sense suffers a great deal of subjectivity and uncertainty. The Decisions reveal very wide discretion in the way they think disproportionality would be adjusted. For example, the logic of giving half-effect to the Scilly Isles in order to rectify the disproportionality resulting from the Isles' location (21 miles) west of the Cornish peninsula, and yet giving full effect to Ushant Island, lying 10 miles west of the tip of Brittany,[192] is quite elusive. Equally subjective is the Chamber's decision in the *Gulf of Maine* case to count as part of the Canadian coast the coasts of the Bay of Fundy up to the point where the waters are no more than 12 miles from the coasts off both sides of the Bay. One has little doubt that the Chamber was concerned here to obtain an 'acceptable ratio' between the length of the parties' respective coasts to support its line, rather than rectifying the line itself by logically applying the proportionality test.[193]

190 See *infra*, Ch. 10 s.4(*d*)(iv).

191 Para. 100.

192 Para. 251.

193 The reasons given by the Chamber are indeed unconvincing. In para. 221 the Chamber merely contended that 'the fact that the two coasts opposite each other on the Bay of Fundy are both Canadian is not a reason to disregard the fact that the Bay is part of the Gulf of Maine, nor a reason to take only one of these coasts into account for the purpose of calculating the length of the Canadian coasts in the delimitation area'. The Chamber,

(*c*) *The choice of a method: the utility of 'equidistance–special circumstances'*

The above discussion suggests that the Decisions establish 'equitable principles' by identifying the relevant circumstances of each case. Equity is then supposedly effected by choosing a method or methods best reflecting those relevant circumstances. This final phase of the decision-making process certainly requires discretion, for a court may find it appropriate to use different methods in different sectors of a continuous boundary line, usually in response to a change in the geography of the area of delimitation.[194]

It is worrying, however, that the Decisions' practice in this context seems to indicate an enormous use of discretionary power, initiated by emphasizing the elusive principle of equity. Given that the Decisions have also exercised very wide discretion in selecting the relevant circumstances and in weighing one factor against another, the result is undeniably difficult to predict with any real accuracy. It is frequently observed that unpredictability in the Decisions of the ICJ has often been a source of worry to those wishing to seek a third-party settlement.[195] The noticeable phenomenon of resorting to *ad hoc* arbitration or conciliation commissions, or even to a chamber rather than the Court in full,[196] to resolve maritime boundary disputes reinforces this observation.

None the less, there are at least two *settled* factors regarding the choice of an appropriate method or methods. First, the Decisions agree that, in delimitation between adjacent States, the status and direction of the land boundary terminus is very important in guiding the adjudicator to choose the method of delimitation, at least in the inner-shore sector of the

however, never mentioned why 12 miles, not 6, 8, or 24, was the appropriate criterion for including the relevant coasts of the Bay of Fundy. Judge Schwebel took this point in his Dissent to the majority judgment, 354 ff.

194 See e.g. *supra*, text at n. 120.

195 The paucity of cases before the ICJ, especially in the 1960s and 1970s, has been interpreted as 'a crisis of confidence' in the Court and its ability to produce predictable judgments. As Jenks pointed out: 'the essence of the question is confidence; confidence in the integrity and predictability of the Courts and Tribunals administering the law.' Quoted by R. Anand, 'Role of International Adjudication', in L. Gross (ed.), *The Future of the International Court of Justice* (Oceana Publications, New York, 1976), i. 1, at 14.

196 Although, under Article 27 of the ICJ Statute, a Judgment given by the Chamber 'shall be considered as rendered by the Court', it was made clear in the Special Agreement between the United States and Canada to submit their boundary dispute to a chamber of the Court that the dispute 'will not go before a chamber of the Court which includes in its composition a judge considered by either of them to be "unacceptable"' (see D. McRae, 'Adjudication of Maritime Boundary in the Gulf of Maine', 15 *CYIL* (1977), 292, at 295). Therefore, it did not come as a surprise that the Chamber which dealt with the case was composed of Western judges only (see the *Gulf of Maine* case, 247); for the view seemed to be that, with such a composition, a chamber's decision would be more predictable than that of the full Court.

delimitation area. To cite the *Guinea–Guinea Bissau* Arbitration, the Tribunal stressed that this factor

> is in itself relevant because, as the International Court of Justice emphasised in the *Libya/Tunisia* case: '. . . there is also the position of the land frontier, or more precisely the position of its intersection with the coastline, to be taken into account.'[197]

Second, the Decisions also agree in acknowledging that, among the relevant circumstances, it is especially the politico-physical geography that will determine the appropriate method in the sense of leading to an equitable result. In the *Anglo-French* Arbitration the Court of Arbitration considered that 'the appropriateness of the equidistance method or any other method for the purpose of effecting an equitable delimitation is a function or reflection of the geographical and other relevant circumstances of each particular case'.[198]

Within this framework, it is proposed to enquire how the Decisions' *practice* reflects the status of the 'equidistance–special circumstances' rule, a rule enhanced by long-standing State practice[199] and given more precise formulation in the 1977 *Anglo-French* Arbitration.[200] This enquiry is a priori in view of the new 'trend' in the matter of title to continental shelf/exclusive zone. As mentioned earlier, the Court emphasized the establishment by the 1982 Convention of the principle of State entitlement, without further condition and irrespective of the formation of the seabed, to a continental shelf/exclusive zone, up to a distance of 200 miles. If the distance principle has become the sole criterion of title within this limit, this surely enhances the a priori application of the 'equidistance–special circumstances' rule.

In an article written in 1979, Blecher argued, primarily on the basis of the 1977 Award's findings, that equidistance emerges as 'a substantive principle' of equity.[201] It is interesting to pursue this argument by considering the subsequent practice of the Decisions regarding the enhancement or otherwise of this argument.

The Court states that it is not required in law even as 'a first step' to examine the effects of a delimitation by application of the equidistance *method*.[202] In practice, however, the 'equidistance–special circumstances'

[197] Para. 106.
[198] Para. 97. See also the *Tunisia–Libya* case, para. 110.
[199] See *infra*, s.4; Ch. 5 s.4.
[200] *Supra*, Ch. 3 s.3(*b*).
[201] M. Blecher, 'Equitable Delimitation of Continental Shelf', 73 *AJIL* (1979), 60, at 68–73.
[202] See e.g. the Court's statement in *Libya–Malta*, reiterating the 1982 *Tunisia–Libya* decision, para. 43.

rule has been the most applicable rule, at least for a greater part of the relevant area.

The 1977 *Anglo-French* Award's equation between the equidistance–special circumstances rule and customary law and the Court's actual application of this rule was described earlier. In the *Tunisia–Libya* case the Court noted the equity of equidistance, at least in the area affected by the change in the direction of the Tunisian coastline. The Court found:

> The major change in direction undergone by the coast of Tunisia seems to the Court to go some way, though not the whole way, towards transforming the relationship of Libya and Tunisia from that of adjacent States to that of opposite States, and thus to produce a situation in which the position of an equidistance line *becomes a factor to be given more weight* in the balancing of equitable considerations than would otherwise be the case.[203]

In the *Gulf of Maine* case, deciding that the equitable criteria applicable should be 'more specifically derived from geography', the Chamber found that within this framework

> it is *inevitable* that the Chamber's *basic choice* should favour a criterion *long held to be as equitable* as it is simple, namely that in practice, while *having regard to the special circumstances of the case*, one should aim at *an equal division* of areas where the maritime projections of the coasts of the States between which the delimitation is to be effected converge and overlap.[204]

Considering the difference in length between the respective coastlines of the States concerned, the Chamber then reaffirmed 'the necessity of applying to the median line as initially drawn a correction which, though limited, will pay due heed to the actual situation'.[205]

In *Libya–Malta*, the equidistance–special circumstances rule wholly determined the result. The *dispositif* reads:

> an equitable result may be arrived at by drawing, as a first stage in the process, a median line every point of which is equidistant from the low-water mark of the relevant [coasts of Libya and Malta], that initial line being then subject to adjustment in the light of the above-mentioned circumstances and factors.[206]

As mentioned earlier, the Decisions in principle construed 'equidistance' as a geometrical method not implying any privileged legal status in relation to other methods; yet this way of viewing the situation almost affords the equidistance–special circumstances rule the force of an equitable rule in practice. Considering the fact that the above quotations were associated with 'opposite' coasts, it might be argued that the Decisions observe that, in contrast with adjacent States, equidistance between States with opposite

[203] Para. 126 (emphasis added).
[204] Para. 79 (emphasis added).
[205] Para. 218.
[206] Para. 79(C).

coasts has often been commended as an equitable method. It is submitted, however, that this argument loses sight of two factors.

First, although Article 6 of the Shelf Convention adopted this classification, it aimed to point out a factual but not legal distinction which might assume some juridical significance.[207] The distinction is very much a question of degree and is not, therefore, especially relevant for considering the equity inherent in that rule, as long as 'equidistance' is considered, be the parties classified as 'opposite' or 'adjacent', in the light of the circumstances prevailing. The elimination of this distinction from Article 12 of the Territorial Sea Convention[208] altogether substantiates this conclusion.

Second, even in the cases where the Decisions seemed to apply different criteria and methods, the basic rationale behind those criteria and methods did not differ substantially from a qualified equidistance method. For instance, in both *Libya–Tunisia* and the *Gulf of Maine* the resultant lines were simply approximations to a geometrical method, with obvious preference for bisected angles, based on what is effectively a distance principle.[209] Such an approach is obviously no more than a different way of effecting 'equidistance–special circumstances'. The use of either approach might of course lead to slightly different boundary lines. The differences, however (taking the aforesaid cases as an example), are submitted to have resulted from the decision-maker's own perception of 'equity', but not from a qualitative difference between the two approaches in principle. When the attempt to draw a compromise boundary results implicitly from 'apportionment',[210] then the use of geometry—basically by bisecting lines of latitude and longitude—may serve that purpose. When a more reasonable approach is attempted, the use of geometry implicit in the 'equidistance–special circumstances' rule may yield a different but more predictable result. As the lines suggested respectively by Judges Oda and Gros seem to show, the difference is attributable to the latters' precise use of basepoints (thus reflecting the geography of the coasts more precisely) as well as to some predictability regarding the use of particular geographical features and the factor of proportionality.[211]

Why then did the aforementioned Judgments fail to specify 'equidistance–special circumstances' as a positive, general rule for the delimitation of maritime boundaries? The reason, in this writer's view, is twofold.

207 See *supra*, Ch. 3 s.3(*b*)(iii); see also Judge Gros, his Dissent to the *Gulf of Maine* Judgment, 387.

208 See *supra*, Ch. 2 s.3(*a*).

209 See, respectively, analysis in *infra*, Ch. 5 s.3(*b*) and Ch. 9 s.2(*f*).

210 This is obvious, for instance, in the Court's inexplicable decision to give the Kerkennah Islands half-effect and ignore Jerba; and the Chamber's decision to take account of the Bay of Fundy. See, respectively, *supra*, s.3(*b*)(iii)(1); text at n. 193.

211 See *Tunisia–Libya*, Judge Oda, Dissent, Ch. 8; *Gulf of Maine*, Judge Gros, Dissent, paras. 42–8.

First, the Court seems to be 'stuck' with its Decision of 1969, whereby Article 6 of the Shelf Convention was incorrectly viewed as denoting a hierarchical classification, with many uncertainties as to the proper meaning of its clauses.[212] However, by 1969 the Court had yet to establish any positive rule on delimitation. State practice was only sporadic; certain aspects of the continental shelf, especially its outer limits, were unclear, and, most importantly, the emergence of a homogeneous zone of 200 miles from the coast was far beyond legal speculation. However, as has been pertinently noted, 'it would, no doubt, have been unrealistic to expect the [ICJ] to depart from its earlier 1969 Judgment and recognize that the [equidistance–special circumstances] rule embodied in Article 6 of the 1958 Convention still offered a well-tested formula for the attainment of an equitable solution'.[213]

The second reason is related to the Court's treatment of the basic source from which the equidistance–special circumstances rule has emerged[214] and evolved,[215] namely State practice. This will be elaborated after discussing patterns of State practice.

4. STATE PRACTICE

(*a*) *Introduction*

Perhaps the most concise way of describing contemporary State practice is through the 1977 *Anglo-French* Award's statement 'that up to date a large proportion of the delimitations of the continental shelf have been effected by the application of the equidistance method, or, not infrequently, of some variant of that rule'.[216] The validity of this statement has been reinforced to a remarkable degree, considering the number of negotiated boundary agreements concluded since 1977. As will be seen below, this is equally true with regard to boundary agreements establishing EEZ or single maritime boundaries.

In the following analysis of State practice, particular attention is paid to the factors which States take account of in effecting 'equitable boundaries'. This enquiry may lead to an identification of some State *opinio aequitatis* with regard to the equidistance–special circumstances rule, which the Decisions tend, nevertheless, to disregard. The reasons for the Decisions'

[212] See *supra*, Ch. 3 s.3(*a*).
[213] E. Brown, 'The Tunisia–Libya Continental Shelf Case: A Missed Opportunity', 7 *MP* (1983), 142, at 161.
[214] See *supra*, Ch. 2 s.2.
[215] See *infra*, s.4(*b*).
[216] Para. 85. See also the Report of the Jan Mayen Conciliation Commission, 20 *ILM* (1981), 797, at 824.

failure to consider the implications asserted by State practice will be discussed afterwards.

(*b*) *Patterns of State practice*

(i) Continental shelf delimitation

So far, over eighty-four continental shelf boundary agreements are reported to have been concluded.[217] The following analysis will focus on three important regions, namely the Gulf area, the North Sea, and the South China Sea.

State practice in the Gulf area is an important source of enquiry for several reasons. The Gulf is a semi-enclosed sea which opens out into the Gulf of Oman and the Arabian Sea through the Strait of Hormuz. At its widest point it is 150 miles across; at the Strait of Hormuz it is 28 miles wide. There are a number of islands and a few are located near the middle. Disputes concerning claims of sovereignty over some islands have existed and still exist. The Gulf has hydrocarbon resources, apart from living resources. Between 1958 and 1974, seven agreements on the continental shelf boundary were concluded.[218] None of the States in question is party to the 1958 Shelf Convention. Therefore, with this wide range of circumstances, their practice may indicate evidence of customary law on the point.

Regarding the applicable principles, equidistance was generally followed for delimiting the boundary line, although not all agreements refer to this explicitly.[219] The two special features of these agreements are the treatment of islands and of discovered or prospective hydrocarbon deposits.

1. In one agreement, all the islands on either side of the median line were ignored.[220] In another agreement, islands were ignored in some sections, but those located in the middle were used to establish points for the boundary line.[221] In the Iran–Saudi Arabia Agreement, the dispute relating to sovereignty over the islands of Farsi and Al-Arabiya (both located near the median line) was resolved by giving Farsi to Iran and Al-Arabiya to Saudi Arabia. The median line was adjusted in order to give each island an arc of 12 miles of territorial sea. The median line was again

[217] S. Jagota, *Maritime Boundary* (Martinus Nijhoff, Dordrecht, 1985), 71.

[218] For full analysis see A. El-Hakim, *The Middle Eastern States and the Law of the Sea* (Manchester University Press, Manchester, 1979), 86–107.

[219] In five agreements (Iran–Saudi Arabia, 1968; Iran–Qatar, 1969; Iran–Bahrain, 1971; Iran–Oman, Iran–UAE, 1974) reference was made to the determination of the line in 'a just and accurate' or 'just, equitable and precise manner'. In Abu Dhabi–Qatar, 1969, no reference was made to the principle followed.

[220] Iran–Qatar, 1969. El-Hakim, *The Middle Eastern States*, 100; *LIS* 25.

[221] Bahrain–Saudi Arabia, 1958; El-Hakim, *The Middle Eastern States*, 86; *LIS* 12.

adjusted by giving the Iranian island Kharg, located about 17 miles from the coast, half-effect.[222] In one agreement, an island located near the median line was given a territorial sea arc of 3 miles and the boundary line was adjusted accordingly.[223]

2. In respect of known deposits the solutions varied between sharing the revenues of the common deposits (thus having no effect on the boundary line) or adjusting the median line so as to allocate to one party all or part of the common deposit. An example of the first solution is found in the Abu Dhabi–Qatar Agreement. Point B on the boundary was located on a known petroleum field, Al Bunduq, which was to be under Abu Dhabi's jurisdiction but revenues from which were to be shared equally between the parties.[224] The second solution is found in the Agreement between Bahrain and Saudi Arabia. Although the Agreement provides that the line is delimited 'on the basis of the median line', Points 12–14 of the boundary are determined by reference solely to geographical co-ordinantes, apparently to include the Fasht Bu Safah Hexagon (an area of proven petroleum resources) under the jurisdiction of Saudi Arabia.[225] A deflection from the median line is also reflected in the 1968 Agreement between Iran and Saudi Arabia. In 1965 the two States reached an Agreement based on the median line but it was never ratified, because the Iranians, after new oil resources were discovered north of the 1965 proposed line, felt that the Agreement did not provide equal sharing of hydrocarbons. According to the present Agreements, Points 8–14 of the 1965 Agreement are modified, 'giving only a slight net gain in seabed area to Iran, but presumably a substantial increase in Iran's share of estimated oil resources'.[226]

In respect of prospective common deposits, four of the seven agreements include a common clause on the manner in which parties should exploit such deposits. Following earlier State practice in the North Sea, the clause provides that both parties should endeavour to reach agreement as to the manner in which the operation on both sides of the line could be co-ordinanted or utilized. In five agreements, however, an original clause prohibits directional drilling by either side within certain distances from the boundary line.[227]

[222] El-Hakim, *The Middle Eastern States*, 91–2; *LIS* 24.

[223] Abu Dhabi–Qatar, 1969. El-Hakim, *The Middle Eastern States*, 97; *LIS* 18.

[224] El-Hakim, *The Middle Eastern States*, 97; *LIS* 18.

[225] *Supra*, n. 221. However, the Agreement provides for equal sharing of income derived from the exploitation of this area.

[226] El-Hakim, *The Middle Eastern States*, 92; R. Young, 'Equitable Solutions for Offshore Boundaries: The 1968 Saudi Arabia–Iran Agreement, 64 *AJIL* (1970), 152, at 155.

[227] In Iran–Saudi Arabia the distance is 500 metres. In the Agreements respectively between Iran and Qatar, Bahrain, Oman, and UAE drilling is restricted within 125 metres from the boundary line.

The political/economic/geographical characteristics of the North Sea are similar to those of the Gulf.[228] The contribution of the agreements relating to the shelf boundary in this area is therefore of no less importance to the development of international law and State practice in maritime delimitation. Between 1964 and 1978, the five States bordering the North Sea concluded a series of continental shelf boundary agreements.

The Agreement between Norway and the United Kingdom, 1965, is 'based, with certain minor divergencies for administrative convenience, on [an equidistant line]'.[229] The Norwegian Trough (a geomorphological feature limiting the natural prolongation of the Norwegian Shelf) was ignored.[230] The British Shetland Islands were given full effect. Similarly, the 1965 Agreement between Norway and Denmark describes the boundary as median.[231] The Norwegian Trough was also disregarded.

The partial Boundary Agreements between the Federal Republic and The Netherlands (1964) and Denmark (1965), respectively, established a continental shelf boundary up to 26 miles, on the basis of equidistance.[232] But the parties were unable to agree on the principle to be used for delimiting their remaining boundaries. Subsequently, they agreed to refer the dispute to the ICJ, which delivered its Judgment in 1969. Following this Judgment, further negotiations between the parties led in 1971 to the completion of their respective shelf boundaries.

The latter boundaries are located on semi-equidistant turning-points, measured from specific basepoints.[233] The underlying factors in those two Agreements are naturally the general guidelines given by the Court in its 1969 Judgment.[234] Another factor apparently influencing the resulting lines is the consent by Germany to leave Denmark and The Netherlands in possession of some areas for which exploration licences had already been granted by the two States before the 1969 Judgment.[235] Following these two Agreements, Germany was in a position to negotiate a boundary with the United Kingdom. The resulting line is an 8-mile negotiated line extending slightly west of a true equidistance line between the two States, thus falling closer to the United Kingdom by 20 miles. Its southern and northern termination points coincide respectively with the point intersecting the lines dividing the continental shelves of the United Kingdom,

[228] For a concise description of the politico-geographical characteristics of the North Sea, see the *North Sea* cases, Merits, para. 3.

[229] Art. 1, *LIS* 10 (Revised), 2.

[230] The Trough is between 20 and 81 n.m. wide, has a depth of up to 371 fathoms, and lies close to the Norwegian coast, both towards the west and in the south.

[231] Art. 1, *LIS* 10: 6.

[232] See ibid. 18, 21–2.

[233] See *infra*, n. 234.

[234] *Supra*, Ch. 3 s.3(*a*).

[235] Jagota, *Maritime Boundary*, 111; Symonides, 'Delimitation of Maritime Areas', 44.

The Netherlands,[236] and Germany, and that intersecting the lines dividing the continental shelves of the United Kingdom, Denmark,[237] and Germany.

Finally in 1978 Norway and the United Kingdom signed a Protocol extending the shelf boundary established under the Agreement of 1965 in a northerly and north-westerly direction. The principle of equidistance is mentioned in the Protocol with reference to the 1965 Agreement.[238]

All in all, therefore, eleven shelf boundary agreements have been negotiated in the North Sea, revealing a useful and constructive implementation of the equidistance–special circumstances rule. Six agreements mention the 'equidistant' or 'median' line expressly. Four establish the boundary on the same principle. The variety of circumstances involved includes giving full effect to islands, taking account of administrative convenience, the unity of known hydrocarbon structures, ignoring geomorphological features, and the special circumstance arising from the concavity of the German coast. The fact that 'equidistance' is not completely rejected in the 1971 Agreements between Germany, Denmark, and The Netherlands[239] illustrates the flexibility inherent in 'equidistance' when substantially modified to secure equity. Finally, in the Agreements detailed provisions have been made for the exploitation of potential single deposits extending across the boundary line.[240]

Off the 'Southern Flank' of the North Sea the territorial sea/continental shelf boundary between France and Spain in the Bay of Biscay was delimited in 1974. The boundary provides a good example of modified equidistance, owing to differences in the projection and lengths of the coastlines. The parties agreed without difficulty to use the equidistance line up to a certain point inside the area forming approximately a right angle with relatively equal lengths of coastline from the terminus of the land boundary. This first segment was agreed to terminate at the point where the equidistance line bisected the diagonal of an arc drawn between a point where the French coast changed its direction towards the north-

[236] The 1965 Shelf Boundary Agreement between The Netherlands and the UK was amended by a Protocol signed on 25 Nov. 1971. Although no principle of delimitation is mentioned in either instrument, the line is delimited according to the equidistance principle. The line extends for 25 n.m. See *LIS* 10: 14–15.

[237] A Protocol amending the Agreement of 1966 between Denmark and the UK was also signed on 25 Nov. 1971. The boundary extends for a distance of 11 n.m., equidistant from the two States' respective territories. Ibid. 10–11.

[238] Jagota, *Maritime Boundary*, 113.

[239] In the Denmark–Germany Shelf Agreement, Point S1 is equidistant. The average variation in distance between the rest of the negotiated points and the respective basepoints is less than 9 n.m. In respect of the Netherlands–Germany Agreement, Point E4 is equidistant. The rest of the points have an average variation in distance of 12 n.m. In both Agreements the points are drawn from selected basepoints on coastal islands. The rest of the coastline is ignored. See *LIS* 10, tables v and vi.

[240] See J. Horigan, 'Utilization of Petroleum Reservoirs Extending across Sub-sea Boundary Lines of Bordering States in the North Sea', 7 *NRL* (1974), 67.

east and a corresponding point on the Spanish coast. The terminus coincided with the two States' 12-mile territorial sea.[241] On the remaining boundary the parties could not agree on a specific principle. Eventually, they compromised by 'splitting the difference'. First, they extended the equidistance line to a point further seaward, and then apportioned the remaining area, approximately 35,670 sq.m., using a ratio 9 to 5, in accordance with the length of each party's coastline.[242] Within this area, the Shelf Agreement established a common zone, the resources of which were to be divided equally between the parties.

In the South China Sea six agreements establish the continental shelf boundary in the Strait of Malacca and the Andaman Sea. Although the applicable principle is not mentioned in any agreement, the median line generally applies between the baselines, except where modified by negotiation. In the 1969 Agreement between Indonesia and Malaysia, defining their shelf boundary in the Strait of Malacca and South China Sea, the median line northwards from Tanjong Datu Point is adjusted westwards in favour of Malaysia, apparently to reflect the difference between the small, scattered islands of Indonesia and the long opposite coast of Malaysia.[243] In 1971 the two States agreed to extend their shelf boundary in the Northern Malacca Strait.[244] The line is equidistant from the respective baselines and extends up to the 'common point' of the Indonesia–Malaysia–Thailand shelf boundaries. The Agreements on this tripoint, together with the Thailand–Indonesia and the Thailand–Malaysia Shelf Boundary Agreements, were concluded in the same year.[245] The tripoint is a modified equidistance point ignoring the Indonesian archipelagic baselines drawn in this area.[246]

The Thailand–Indonesia boundary has been determined on the basis of equidistance, except in a small segment where the line is closer to Thailand than to the Indonesian baselines.[247] The Thailand–Malaysia line has also been established on the basis of equidistance with a slight modification to give effect to the Malaysian island P. Langkawi, which is much larger than the opposite Thai island group of Butang.[248] Finally, this line was extended in 1973 to delimit the two States' respective continental shelf in

[241] In fact, Spain was claiming at the time only 6 miles of territorial sea plus 12 miles of contiguous zone. For details see *LIS* 83.

[242] For the text of the Agreements and the annexe on the common zone, see 80 *RGDIP* (1976), 366–72.

[243] Indonesia–Malaysia (1969), *LIS* 1.

[244] Indonesia–Malaysia (1971), *LIS* 81.

[245] Ibid. 1.

[246] Ibid. 2.

[247] Ibid. 4.

[248] Ibid. 6.

the Andaman Sea. The line is equidistant, except that the northern terminal point is about 27.2 miles closer to Indonesia.[249]

All the above agreements include a common clause on single hydrocarbon structures, ensuring that the parties 'shall seek to reach agreement as to the manner in which the structure shall be most effectively exploited'.

Finally, another two important examples in the Asian agreements should be mentioned: the Japan–Republic of Korea and the Australia–Indonesia Agreements. The first establishes the shelf boundary between Japan and the Republic of Korea in the Korea Strait. The boundary follows a median line between the main islands as well as the midway islands of Tsushima and Cheju,[250] except at Point 3, where the boundary is closer to the Republic of Korea. The boundary starts well south of the island of Takeshima, title to which is disputed between the parties. A second agreement between the parties establishes (without prejudice to the question of sovereignty) a Joint Development Zone[251] in an area of 24,092 sq.m., most of which appears to be on the Japanese side of the median line between Japan and the Republic of Korea and Japan and China. China does not recognize this arrangement.[252]

The delimitation of the continental shelf between Australia and Indonesia started in the 1971 Agreement, defining the boundary in part of the Arafura Sea, and was completed in the 1972 Agreement, extending the boundary in the Arafura and Timor Seas.[253] The first segment of the line is drawn on the basis of equidistance. The other two segments do not follow the median line. They follow a negotiated line between the true median line and the deepest trough near the Indonesian coast, known as the Timor Trough. Another factor relevant to this line is the petroleum concessions already awarded by Australia.[254] A third agreement was concluded between Australia (for Papua New Guinea) and Indonesia which defined the shelf boundary between the adjacent coasts of Indonesia and Papua New Guinea. The entire boundary has been drawn on the basis of equidistance.

[249] Malaysia–Thailand (1973), *LIS* 93: 5.

[250] See *infra*, Ch. 10 s.4(*b*)(i)(1).

[251] Agreement of 30 Jan. 1974, *LIS*, 93.

[252] For details see Ying-jeou Ma, *Legal Problems of Seabed Boundary Delimitation in the East China Sea* (University of Maryland, School of Law, Baltimore, 1984), 31.

[253] See J. Prescott, *Australia's Maritime Boundaries* (Dept. of International Relations, Canberra, 1985), 103–18.

[254] The Agreement contains a provision enabling the existing holder of a permit or licence granted by Australia in the area of the seabed which lies to the north of the boundary line to apply within nine months to Indonesia for an appropriate production-sharing contract in accordance with Indonesian law. Prescott, *Asstralia's Maritimẽ Boundaries*, 105; Jagota, *Maritime Boundary*, 90.

(ii) EEZ delimitation

If the terms employed by agreements are taken as a criterion of classification, there appear to be only two agreements which mention an EEZ boundary explicitly. These are the 1980 Agreement between France and Tonga[255] and the 1980 Agreement between France and Mauritius.[256] Both Agreements provide that the boundary is the median line. Although most agreements delimiting both the seabed and water-column jurisdictions use the term 'maritime' rather than EEZ boundary, within the 200-mile limit, they are all technically EEZ boundary agreements; and are, consequently, relevant to the present analysis. For practical purposes, however, the pattern of State practice envisaged from such agreements will be discussed in the next chapter.[257]

However, legislation promulgating EEZ claims is deemed to be a good source indicating State practice, since all these claims are recent and were mostly promulgated after the well-documented debate on the delimitation question at UNCLOS III. The most recent collection of legislation—published in 1986[258]—reflects the trend in legislation promulgating EEZ claims since 1973. Of the eighty-seven States and territories reported as having passed legislation, sixty-one appear to have made specific reference to the issue of delimitation. Of these, fifteen refer specifically to the 'equidistance' or 'median line'; fifteen to the same principle in the absence of agreement; three refer in general terms to international law; three mention 'equitable principles'; five refer to 'agreement on the basis of international law'; and twenty refer to 'agreement' or 'negotiation'. However, of these eighty-seven claimants the present writer has been able to identify thirty-three States and territories which have used 'equidistance' in their negotiated shelf or EEZ or single boundary agreements with neighbouring States.

These agreements and unilateral claims seem to confirm, therefore, that the legal basis underlying EEZ entitlement, namely the distance principle,[259] points out the a priori use of the equidistance–special circumstances rule.

(*c*) *Judicial treatment of State practice*

Despite this considerable reliance by States on the equidistance–special circumstances rule as a primary means of boundary delimitation, the

[255] Agreement of 1980, establishing the EEZ boundary between Tonga and the French colony Wallis and Futuna Islands. See Jagota, *Maritime Boundary*, 93.

[256] Agreement of 1980, defining the EEZ boundary between the island of Réunion and Mauritius. See *LIS* 95.

[257] See *infra*, Ch. 5 s.4(*b*).

[258] R. Smith, *Exclusive Economic Zone: An Analysis and Primary Documents* (Martinus Nijhoff, Dordrecht, 1986).

[259] See *infra*, Ch. 4 s.2(*e*); s.3(*a*).

evidential value of this practice for the clarification of the rules of law on delimitation does not seem to be fully appreciated. The reason seems to be largely contingent on the ICJ's treatment of State practice. In the *Libya–Malta* case the Court admitted that it 'has no doubt about the importance of State practice in this matter'.[260] Reiterating its position in the 1969 and 1982 Judgments,[261] however, the Court quickly observed, 'Yet that practice . . . falls short of providing the existence of a rule, prescribing the use of equidistance, or indeed of any method, as obligatory.'[262]

The reason behind the Court's relative indifference towards State practice seems to stem from several interrelated factors, the most important of which are submitted to be the following:

1. *A negotiated agreement is not necessarily a reflection of legal obligation.* In directly negotiated agreements a boundary frequently results at least in part from an examination of valid politico-economic considerations not strictly limited to the application of rules of law. For instance, whilst observing that the United States had concluded three delimitation agreements with its neighbours based on equidistance,[263] a State Department official concluded that '[f]undamentally, however, these three agreements are negotiated agreements, rather than based on any particular legal theory'.[264] The Court's suspicion of the legal value of the rule of 'equidistance–special circumstances' appears to be partly related to this factor. As stated in the *North Sea* cases, 'There is no lack of other reasons for using the equidistance, so that acting, or agreeing to act in a certain way, does not of itself demonstrate anything of a judicial nature.'[265]

However, to contend that because of this factor inter-State practice cannot be described as a valid source of law (or at least evidence of the act of creating that law) is to draw an unrealistic, clear-cut distinction between the term 'source' and that of social origin or other 'causes' of law.[266] The proof of legal conviction or *opinio juris* is of course essential for the crystallization of political, i.e. State, conduct into a binding rule of law. But it does not follow that, because the application of 'equidistance–special circumstances', as it were, has been motivated by factors other than purely legal consent there cannot be any evidence of a rule of law. In inter-State practice in particular, there are always facts through which

260 Para. 44.

261 See the *North Sea* cases, para. 79; the *Tunisia–Libya* case, para. 107.

262 Para. 44.

263 See *infra*, Ch. 5 s.4(*b*).

264 Testimony of M. Feldman, Deputy Legal Adviser for the US Dept. of State, before the Senate, S. Exec. Rep. No. 96., 49, 96th Cong., 2nd Sess. (1980), at 22.

265 Para. 76. See also *Libya–Malta*, para. 44.

266 This failure to distinguish between the two terms is responsible, according to Oppenheim–Lauterpacht, for the existence of differences of opinion among writers as to the kinds and numbers of the sources of international law (*International Law* (Longman, London, 1954), i. 24). For criticism, see O'Connell, *International Law*, 7–8.

such consent can become possible. Far from being alien impositions, these facts alone make meaningful the notion that States opt for submitting themselves, through the process of usage, to certain rules of international law.[267] In this respect, one has to bear in mind that judicial procedure 'is only a strategy among strategies' used by States in conducting their international relations.[268] Yet one should not suggest that a judicial decision cannot be relied upon to give a consistent, logical explanation of the law, simply because of this political element inherent therein.

2. *The rules and methods applicable are not always proclaimed.* A further reason which hinders the evidential value of State practice is that, in many delimitation agreements, the principle or method applicable is not explicitly stated in the text of the agreement; or, as is often the case, the reason behind the adjustment or partial modification of the equidistance line is not explained.[269] This is considered a logical consequence of reason (1) given above. States are free to reach a solution which they regard as 'fair', without any obligation to declare clearly the principles or reasons for reaching that solution. It is submitted, however, that this does not justify precluding altogether the relevance of State practice. As attempted in this study, it is possible by analysing boundary agreements to discern the applicable rules, methods, and factors that States deem relevant in the establishment of their negotiated boundaries.[270]

3. *The restricting effect of State practice.* Finally, the actual approach of the Court towards the question of delimitation as a whole cannot be ignored. As seen earlier, the Court has adamantly refused to acknowledge any 'primordial' rule of delimitation other than the abstract principle of equity.[271] In view of the obscurity inherent in this principle, it may well be that the Court relies on it more to preserve a virtually unfettered discretion in identifying the relevant factors and selecting the criteria and methods that would satisfy its sense of justice.[272] In conformity with this

[267] Whilst admitting that 'State practice' and 'custom' are not synonyms, O'Connell pertinently advised: 'the one is rather the obverse of the other, "practice" being suggestive of the formative process, "custom" of its completion' (*International Law*, 9). Admitting the existence of this link between the two concepts, one finds it difficult to argue that direct, inter-State boundary agreements are not valid sources of law, simply because they include politico-economic expediencies towards the final establishment of, in our example, a maritime boundary.

[268] Anand, 'Role of International Adjudication', 1.

[269] See e.g. Agreements discussed *supra*, nn. 219, 243–9.

[270] See e.g. *supra*, s.4(*b*)(i), and (ii); *infra*, Ch. 5 s.4(*b*). The analysis provided by the US Office of the Geographer Series, *Limits in the Sea*, is illustrative of the possibility of effectively identifying the 'general trend' in State practice.

[271] *Supra*, s.2(*d*).

[272] Most of the dissenting opinions of the cases express this concern. e.g. Judge Gros in the *Gulf of Maine* has protested that this 'inordinate use of equity would lead to government

approach the Court would not wish to recognize any *substantive* rule of delimitation, even if it was deriving its legal force from State practice, or even, for that matter, from considerations of 'reasonableness' and clarity. Most certainly, 'equidistance–special circumstances' could have been an effective 'precedent' which would have reduced the Court's judicial freedom of appraisal considerably.

5. APPRAISAL

The records of the Group of 7 suggest that the controversy between the 'Equidistance Principle Group' and the 'Equitable Principles Group' has underlain the discussion of the applicable rules on maritime delimitation. This controversy, however, emanated largely from States' self-interest and had very little to do with any reasonable, let alone legal, analysis of the principles under discussion. Obviously, there were some misconceptions on both sides as to the proper meaning of these principles. Due to the influence of the 1969 *North Sea* Decisions, for instance, 'equidistance' was perceived as a strict, geometrical method. In this atmosphere the assumption of 'oppositeness' between these two principles continued; and finally resulted in an empty formula which simply maintains the status quo established by the previous law on delimitation; or at least allows for its development through judicial and State practice.[273]

There is little doubt that 'equidistance–special circumstances' has been considered by States the predominant rule not only for the delimitation of the limited areas of territorial waters[274] but also for the more extended areas of maritime jurisdiction.[275] The recent development, embodied in the recognition of 'distance' as the basis of entitlement to both continental shelf and EEZ within 200 miles, has certainly enhanced the role of 'equidistance' as an equitable approach for delimiting overlapping jurisdictions.[276]

Adjudication, however, contends that 'equity' is the only positive rule of law applicable to boundary delimitation. The Decisions might have contributed a great deal in pointing out those elements (or circumstances) which contribute to current equity doctrine, but the problem rests in the

of judges, which no State would easily accept' (p. 385); in *Libya–Malta* Judge Oda has expressed his concern that the Court 'is in danger, in its application of equity, of taking the principle of equity for what it subjectively feels to be equitable in a particular case' (para. 1).

[273] In the *Gulf of Maine* case, the Chamber interpreted the absence of a substantive formula of delimitation as serving 'to open the door to continuation of the development effected in this field by international case law'. Para. 95.

[274] *Supra*, Ch. 2 s.2.

[275] See *supra*, s.4(*b*); *infra*, Ch. 5 s.4(*b*).

[276] See *supra*, s.3(*b*)(ii)(1); *infra*, Ch. 5 s.3(*a*).

Decisions' translation of equity simply as a subjective appreciation of these circumstances. The *Tunisia–Libya* and *Libya–Malta* Decisions relating to islands or to a third-party claim, as well as the Chamber's calculation of coastal lengths, illustrate elements of inconsistency and unpredictability which are similarly found in the other Decisions; and they have a debilitating effect on the concepts of justice and equity. The pronouncement that a decision accords with justice and equity will not alter the subjectivity inherent in this approach if it is not accompanied by fairly precise reasons for the conclusion reached. The use of some substantive, yet fairly flexible, rules such as 'equidistance–special circumstances' at least as auxiliary means for testing this 'assumed equity' will, it is submitted, redress this subjectivity.

The assumption in favour of 'equidistance–special circumstances' is based primarily on State practice. It indicates a wide range of applications of 'equidistance' in a more flexible form. Equidistance is modified because considerations of fairness and equity so require. The special circumstances necessitating such modification relate primarily to island location, coastal length and projection, or simply to simplification of the line for administrative convenience. In other words, 'equidistance', as borne out by State practice, has been found equitable and efficient only through its formulation in a more comprehensive two-element rule, namely the 'equidistance–special circumstances' rule. It is in this complex form that the rule embodied in the Geneva Conventions is submitted to have attained the same status attributed by adjudication to the notion of equity. The Court not only fails to understand how 'equidistance–special circumstances'—employed since the emergence of the maritime law on delimitation—has always been interpreted in connection with equity, but also appears, as Judge Oda puts it, 'to offer equity as the approved antithesis' thereto.[277]

This is not to suggest that adjudication should not be allowed an independent and creative role in the formation of international law. This is a necessary requirement for maintaining the role of the judiciary in enhancing peaceful settlements of disputes among States.[278] In that sense, equity might be introduced as a governing, or supervisory, principle to monitor the administration of justice; and hence the actual criteria and methods for implementing this justice do not lose their independent existence. Their legality should, therefore, be intrinsically clarified by something more than the suggestion that the result is equitable.

[277] His Dissent in the *Libya–Malta* case, 127.

[278] O'Connell pertinently observed, '[i]f international law is seen as an organic growth reflecting more the life of the international community than the conscious operation of the government will, then the creative role of the judiciary is important, and its pronouncements upon the emanations from basic principles [are] significant'. *International Law*, 11.

There might be various reasons which justified the Court's refusal to equate—as the 1977 Arbitral Decision did—'equidistance–special circumstances' with customary law. They are partly related to the misleading construction of the rule in the Geneva Conventions[279] and partly to the relative evidential value of State practice in reflecting the emergence and development of this rule.[280] But it is feared that the Court's major concern was to avoid the recognition of any substantive rule restricting its wide margin of appreciation normally associated with the application of such an elusive norm as equity.[281] This might prove good for the convenience of the decision-maker. But it affords considerable room for differences of opinion in the application of equity to problems of maritime delimitation; and thus eliminates the element of predictability essential for the survival and credibility of the Court's ability to administer the law.

[279] See *supra*, Ch. 1 s.3(*b*)(i); Ch. 2 s.3(*a*),(*b*); Ch. 3 s.2.

[280] See *supra*, s.4(*c*).

[281] This was strongly observed by Judge Gros: 'the course taken since [the 1982 *Tunisia–Libya* Judgment] has been to indulge in an equity beyond the law, detached from any established rules, based solely on whatever each group of judges seised of a case declares itself able and free to appreciate in accordance with its political or economic views of the moment.' His Dissent in the *Gulf of Maine* case, para. 47.

5

Delimitation of Maritime Zones by a Single Boundary

1. INTRODUCTION

Until recently, the idea of formulating legal principles and rules for multi-purpose maritime boundaries was a matter of legal speculation. This is because there was no need to delimit a single boundary[1] when the only extended claim to coastal State jurisdiction recognized as valid was the continental shelf. Ironically, the 1982 Convention perpetuates this situation; for although it establishes the right of a coastal State to claim an exclusive economic zone up to 200 miles, together with an identical criterion for delimiting the continental shelf,[2] it seems to ignore the possible impact of this development on the question of delimitation. Instead, it contains, as detailed earlier, separate Articles for the delimitation of the territorial sea, the shelf, and the exclusive economic zone boundaries.

Relevant questions thus arise: should the shelf and the exclusive economic (or fishery)[3] zone boundaries necessarily coincide? And, if not, do any legal or practical considerations favouring such a solution exist? Finally, what are the legal principles and rules that may apply to a delimitation by a single boundary?

The answer to the first question seems obvious. Functionally, the EEZ has been designed essentially to reserve for the coastal State the right to exercise jurisdiction over the exploitation of fishery resources. Certainly, many States attach importance to the new powers of pollution control within the EEZ,[4] but, that apart, the other jurisdictions related to the

[1] The term 'single maritime boundary' appears to relate exclusively to the delimitation by one line of at least *two different legal regimes*, such as continental shelf and exclusive economic or fishery zone. The territorial sea (or contiguous zone) may also be involved as a third legal regime (e.g. the *Guinea–Guinea Bissau* Arbitration, *infra*, s.3(*c*)).

[2] See *infra*, s.2(*e*).

[3] Whilst an EFZ jurisdiction is not a fully fledged EEZ claim its validity in customary law does not seem to be questionable. Its inclusion for the purpose of studying the concept of a single boundary is *a fortiori*, since it reflects a more independent existence *vis-à-vis* the shelf than that established by the EEZ regime.

[4] See pt. XII, s.5 of the 1982 Convention, especially Art. 211 (pollution from vessels), 220 (enforcement by coastal states).

zone are either provided for by the shelf regime[5] or will not be technically or economically feasible in the near future, at least for the majority of the world community.[6] In contrast, the jurisdiction in the continental shelf is related almost exclusively to hydrocarbons. Theoretically at least, with respect to delimitation, the relevant circumstances determining the equity of an EEZ delimitation may vary in some respects from those related to an equitable shelf delimitation. For instance, while the considerations which call for abating the effects of an incidental feature would normally apply to an EEZ delimitation as much as to the continental shelf,[7] there may be differences: a feature may have an ecological (environmental) effect on allocation of certain fishery species but not on that of the mineral resources in the subsoil. Therefore, separate EEZ (or EFZ) and shelf boundaries, though they may prove impractical,[8] are not incompatible with the fundamental principle underlying the law on delimitation, namely that the delimitation is to be equitable in the light of relevant circumstances.

The following sections will attempt to consider the rest of the questions posed above: section 2 examines the relevant provisions of the 1982 Convention which may help to clarify the basis for applying the concept of a single maritime boundary. This can be done by enquiring into the legal relationship between the EEZ and shelf regimes as embodied in that Convention. Sections 3 and 4 respectively examine the contribution of judicial decisions and State practice to the clarification of the rules applicable to delimitation by a common boundary.

2. THE PROBLEM UNDER THE 1982 CONVENTION

(*a*) *The relationship between the shelf and the EEZ*

The preoccupation of jurisprudence with the relationship between the shelf and EEZ regimes is perhaps best illustrated in Judge Oda's statement that the question of 'whether the sea-bed, at least within 200 miles

[5] Art. 56(3) provides that the 'rights set out in this article with respect to the sea-bed and subsoil shall be exercised in accordance with Part VI [continental shelf]'. The explicit reference in Art. 56(1)(b)(ii) to the coastal State rights in the EEZ with regard to marine scientific research may be said to have no equivalent in the part on the continental shelf. Its application to the continental shelf, however, is implicit in Art. 246, which provides that coastal States 'have the exclusive right to regulate, authorize and conduct marine scientific research in their exclusive economic zone and on their continental shelf'.

[6] e.g. the production of energy from water currents and winds (Art. 56(1)(a)). See generally L. Wick and W. Schmitt (eds.), *Harvesting Ocean Energy* (The UNESCO Press, Paris, 1981).

[7] *Supra*, Ch. 4 s.3(*a*).

[8] *Infra*, s.2(*g*).

of the coast, has been incorporated in the regime of the Exclusive Economic Zone or whether it should still come under the separate regime of the Continental Shelf in parallel with the Exclusive Economic Zone was far more essential than generally thought'.[9] An answer, if any,[10] would of course largely determine the law applicable to a single maritime boundary. In his Dissent in the *Gulf of Maine* case, Judge Gros has attempted to find an answer in the light of two complementary sources of enquiry, the one historical and the other derived from the language and structure of the 1982 Convention.[11]

In his enquiry into the historical source, Judge Gros takes as his point of departure the 1945 Truman Proclamation on the continental shelf. A comparison between this Proclamation and the contemporaneous Proclamation on coastal fisheries in the high seas[12] is of interest. Whilst the former provides explicitly for the 'division' of the resources of the seabed with other nations, the latter contains no provision at all for boundary delimitation. Rather, it opts for protecting and conserving fisheries in areas in which more than one nation participates.

Judge Gros also briefly reviews the history of the emergence of the continental shelf in the 1958 Conference and the 200-mile zone in the 1970s, which led to its adoption in UNCLOS III. With respect to the resulting 1982 Convention, Judge Gros compares the main features of the regime of the continental shelf with those related to the EEZ and concludes that the intention of the drafters was to keep the two regimes independent of each other. The necessity of harmonizing States' interests, he observes, requires the autonomy of both regimes: 'Comparison of Articles 56–62 and 73–74 (Zone) and 76, 77, 78, 81 and 83 (Continental Shelf) seems to leave only this alternative: either two legal regimes, or chaos.'[13]

Assuming that this conclusion is correct,[14] it can be said that the 1982

[9] Judge Oda, his Dissent to the *Tunisia–Libya* Judgment, ICJ Reports (1982), 223.

[10] In the *Gulf of Maine* case, the Chamber observed that the parties had 'simply taken it for granted that it would be possible, both legally and materially, to draw a single boundary for *two different jurisdictions*' (ICS Reports (1984), para. 27 (emphasis added)). It is pertinent, however, to point out that the issue before the Chamber was the delimitation of continental shelves and EFZs, but not EEZs. In other words, the possible impact on delimitation of the common denominator, i.e. the EEZ seabed jurisdiction, was not an issue before the Chamber. See however, *infra*, n. 85.

[11] Judge Gros, his Dissent to the *Gulf of Maine* Judgment, 372 ff.

[12] Respectively, Presidential Proclamations Nos. 2667 and 2668 (1945) repr. in 40 *AJIL* (1946) (Supplement), 45–7. See E. Collins and M. Rogoff, 'The International Law of Maritime Boundary Delimitation', 34 *MLR* (1982), 1, at 20–1.

[13] Judge Gros, *Gulf of Maine* case, 374.

[14] Indeed, there are many issues in the Convention organized on the basis that the two institutions, even within 200 miles, are legally distinguishable. The provisions regarding control of pollution in the EEZ were alluded to earlier. The duties specified in Art. 62 with regard to the optimum utilization of the living resources of the EEZ and the allocation of

Convention presumes the possibility of two boundaries, one for the continental shelf and another for the exclusive economic zone.[15] The independence of the two institutions, however, does not prevent *per se* the possibility of combining their delimitation by a single boundary. It is reasonable, therefore, to look into other aspects of comparison between the two zones in order to ascertain which solution under the Convention as a whole seems to be most reasonable.

(*b*) *The legal nature of the 'water column' in both zones*

Under the 1958 Shelf Convention, the continental shelf is an area of high seas, subject to sovereign rights by the coastal State.[16] Article 3 of the Convention declares that the rights of the coastal State over the shelf 'do not affect the legal status of the superjacent waters as high seas, or that of the air space above those waters'.

The juridical nature of the exclusive economic zone has been described at UNCLOS III as 'one of the most controversial issues facing the Conference'.[17] Different views were represented, each reflecting the importance of the issue at stake.[18] However, the prevailing view, which seems to accord with the resulting provisions of the 1982 Convention,[19]

any surplus do not exist in the continental shelf regime. This is partly explained by the difference between renewable (zone) and non-renewable (shelf) resources. Thus, under Art. 77(1), the Convention refers to sovereign rights only for the purpose of exploring the shelf and exploiting its resources. The reference to conservation and management found in Art. 56(1) (zone) has been left out. Part of the explanation may also lie in the relevance of certain 'socio-economic factors' related to the exploitation of the living resources which deserve special treatment. These provisions, however, do not apply to sedentary fisheries, as they fall—quite artificially, it is submitted—under the shelf regime. There are other notable differences between the two institutions to be found in the Convention. Art. 56(2) states the duties of the coastal State in the EEZ. There is no equivalent in the part on the shelf regime. Art. 77 refers only to the rights of the coastal State over the shelf. Under the shelf regime, all States are entitled to lay pipelines on the shelf, but, in accordance with Art. 79(3), their delineation is subject to the consent of the coastal State. Under the EEZ regime no such consent is explicitly mentioned.

[15] There are indeed very few writers who doubt that a separate delimitation of shelf and EEZ can be logically and legally feasible. See e.g. P. Peters and G. Tanja, 'Lateral Delimitation of Continental Shelf and Exclusive Economic Zone', 86 *Diritto marittimo* (1984), 463, at 475.

[16] C. Colombos, *The International Law of the Sea* (Longman, London, 1967), 77; R. Churchill and A. Lowe, *The Law of the Sea* (Manchester University Press, Manchester, 1988), 121–3.

[17] Memorandum by the President of the Conference, Amerasinghe, UNCLOS III, *Official Records*, vol. viii, at 68; J. Quéneudec, 'La Zone économique', 79 *RGDIP* (1975), 321, at 342 ff.

[18] See D. Attard, *The Exclusive Economic Zone in International Law* (Clarendon Press, Oxford, 1987), 61–7.

[19] The 1982 Convention establishes an intricate mechanism for balancing the coastal States' EEZ rights with those of the international community. The pivot of this mechanism is found in Arts. 56 and 58. Together they provide an elaborate framework for accommo-

holds that the EEZ is *sui generis* a zone of national jurisdiction, different from that governing the territorial sea and the continental shelf or the high seas.[20] Thus, under the 1982 Convention, the term 'exclusive economic zone' is excluded from the areas to which the regime of high seas applies. Article 1 of the 1958 Convention on the High Seas states that the term 'high seas' 'means all parts of the sea that are not included in the territorial sea or in the internal waters of a State'. To this is added in the 1982 Convention 'the exclusive economic zone and the archipelagic waters'.[21]

In view of this development, one wonders whether the legal status of the shelf in customary law is still that determined in the 1958 Conventions. In retrospect, it is possible to state that, except for the Truman Proclamations,[22] the distinction between the overlying waters of the shelf and the seabed and subsoil was not absolute. The so-called Latin American approach is often cited as a landmark in the evolution of the concept of the continental shelf.[23] Their concept of an 'epicontinental' shelf was based on the indissoluble economic, physical, and biological unity of the continental shelf and its superjacent waters.[24] In doctrine, the first introduction of the shelf concept in international law discussions was based on the assumption that the waters above the shelf were of the utmost importance to fisheries.[25] This relationship was formulated as early as 1916 by Captain Storni, who had emphasized the shelf's importance to fisheries in the 'Argentinian Sea' and the need for adequate controls.[26]

In this respect, it is of interest to note that the United States' earlier

dating rights of interested parties. Under this framework, all States enjoy within the EEZ: (1) Freedom of navigation and overflight; (2) Freedom to lay submarine cables and pipelines; and (3) 'other internationally lawful uses of the sea related to these freedoms' (Art. 58(1)). Nevertheless, the Convention never refers to these freedoms as consonants of the legal status of the EEZ as high seas.

[20] e.g. Aguilar, Chairman of the Seabed Committee, UNCLOS III, *Official Records*, v (1976), 153; Oda, his Dissent to the *Tunisia–Libya* Judgment, 228; Quéneudec, 'La Zone économique', 342.

[21] Art. 86(1).

[22] *Supra*, text at n. 12.

[23] Commenting on a paper presented by J. Suarez to the Brazilian Bar Association (in which he stressed the unity between the shelf and the superjacent waters), Professor O'Connell pointed out that 'the modern concept of the continental shelf dates from this paper'. 'Adumbrations of the Continental Shelf Doctrine', in *Mélanges offerts à Charles Rousseau* as quoted in D. J. Attard, *The Exclusive Economic Zone in International Law* (Oxford University Press, Oxford, 1987), 130.

[24] e.g. the Mexican Presidential Declaration of 1945; the Argentine Decree No. 14,708 of 1946; Panama Decree No. 449 of 1946; the Chilean Presidential Declaration of 1947; the Presidential Decree of Peru No. 781; Costa Rican Decree No. 116 of 1948; El Salvadoran Constitution of 1950; the 1952 Declaration on Maritime Zone (Chile–Ecuador–Peru).

[25] M. Mouton, *The Continental Shelf* (Martinus Nijhoff, The Hague, 1952), 46.

[26] S. Storni, *Intereses argentinos en el mar* as quoted in Attard, *Exclusive Economic Zone*, 3.

efforts, which resulted in the 1945 Proclamations, reflected this integrated view of jurisdiction over the adjacent seabed and water column. It is reported, for instance, that President Roosevelt was in full agreement with a letter received from the Secretary of the Interior relating to the desirable policy of the United States with regard to the continental shelf. It stated, *inter alia*, that

> The continental shelf extending some 100 or 150 miles from our shores forms a fine breeding place for fish of all kinds; it is an excellent hiding place for submarines, and since it is a continuation of our continent, it probably contains oil and other resources similar to those found in our States.[27]

The early deliberations of the ILC on the continental shelf did not distinguish between the superjacent waters and seabed of the continental shelf.[28] This may have stemmed somewhat from the fact that the integrated view of jurisdiction had long since applied to the territorial sea. Whilst the resulting 1958 Shelf Convention distinguished between the legal status of the 'two components' of the shelf,[29] it was inevitable from the nature of the activities involved that exploration and exploitation activities would have some effect on the legal status of the superjacent waters as high seas. Article 5(1) of the said Convention expressed this fact by providing that:

> The exploration of the continental shelf and the exploitation of its natural resources must not result in any *unjustifiable interference* with navigation, fishing or the conservation of the living resources of the sea. (emphasis added)

The word 'unjustifiable' is certainly significant in the attempt to qualify reference to 'high seas' and ultimately to accommodate potentially conflicting uses.

However, the gap between the argument for integrating the shelf and its superjacent waters and the 'duality' approach of the 1958 Convention seems to be bridged by the 1982 Convention, under which explicit refer-

[27] M. Whiteman, *Digest of International Law* (Dept. of State Publication, Washington, DC, 1963–70), iv. 946. For Roosevelt's comment, see 947. See also previous bills which were aimed at extending the jurisdiction of the USA to—quoting from the 1938 Copeland Bill—'all waters and submerged land adjacent to the coast of Alaska . . . and lying within the limits of the continental shelf'. Drafted also along similar lines was the Dimond Bill of 15 Nov. 1937. See P. Jessup, 'The Pacific Coast Fisheries', 33 *AJIL* (1939), 129, at 129–30, 135.

[28] See comments by members of the Commission, *Yearbook of the ILC* (1953), i. 126–8. e.g. Sandstrom mentioned, among the special circumstances that might affect the delimitation of the continental shelf, 'navigation and fishery rights', ibid. i. 126. Later on before the Fourth Committee many States argued that the distinction made by the Commission between the shelf and the resources of the water column was illogical and unwarranted. See e.g. statements by the delegates of Peru, Ecuador, and Uruguay, UNCLOS I, *Official Records*, vol. v, respectively, pp. 5, 17, 18.

[29] Art. 3.

ence to the juridical status of the shelf's superjacent waters as high seas[30] is omitted. Article 78(1) of the 1982 Convention adopts the clause provided for in Article 3 of the 1958 Convention but it excludes the words 'as high seas'. Obviously, this omission was made necessary by the introduction of the EEZ regime, which is expressly excluded from areas to which the regime of the high seas applies.[31] The impact of the EEZ regime on the water column of the continental shelf within the 200-mile distance[32] cannot therefore be denied.

(*c*) *The basis of title in both zones*

Another difference which the Convention seems to establish lies in the sources of rights attributed to either zone. According to Article 77(3) of the 1982 Convention—which is based on Article 2 of the 1958 Shelf Convention—the rights of a coastal State over the shelf do not depend either on occupation or on any express proclamation. As paraphrased by the ICJ, those rights exist '*ipso facto* and *ab initio* without there being any question of having to make good a claim to areas concerned'.[33] As the section on the EEZ contains no similar provision, it has been advanced that, *a contrario*, the rights of a coastal State over its 200-mile offshore jurisdiction are inoperative until it expressly proclaims such a zone. Furthermore, Article 57 of the Convention states that the EEZ 'shall not extend beyond 200 miles'. This language, it might be argued, seems to imply that the coastal State may decide on a breadth less than 200 miles.[34] The practical implication of this interpretation is obvious. Unless the coastal State takes some positive legal action to establish an EEZ or EFZ, the living resources of the waters overlying its shelf may be exploited by the nationals of other States without permission.

Yet one should realize that the *ipso facto* and *ab initio* finding seems to be an *ex post facto* rationale for an overwhelming State practice establishing continental shelf claims all over the world, a development which

[30] Art. 3 of the 1958 Shelf Convention.

[31] Art. 86(1). For the same reason, the language of Art. 24 of the 1958 Territorial Sea Convention was also modified. Art. 24 provides for certain coastal State jurisdictions 'in a zone of high seas contiguous to its territorial sea'. Under the 1982 Convention, the explicit indication of the juridical nature of the zone as part of the high seas is omitted (Art. 33(2)).

[32] In cases where the continental shelf is extended beyond that distance, maybe up to 350 miles (Art. 76, paras. 5, 6), the superjacent waters retain their high-seas status. Art. 86 provides that '[t]he provisions of this Part [High Seas] apply to all parts of the sea that are not included in the exclusive economic zone, in the territorial sea or in the internal waters . . . or in the archipelagic waters of an archipelagic State'.

[33] The *North Sea* cases, ICJ Reports (1969), para. 39.

[34] There are a few States which claim EFZs of less than 200 miles, e.g. Finland (12 n.m.); Malta (25 n.m.). For a convenient tabulation of recent claims, see R. Smith, *Exclusive Economic Zone: An Analysis and Primary Documents* (Martinus Nijhoff, Dordrecht, 1986), 12–23.

was legally and politically irreversible and which was also reflected in the 1958 Shelf Convention.

In this respect, however, it seems reasonable to state that the same irreversible movement is taking place with regard to the EEZ. The rapid evolution of State practice in this regard is striking. UNCLOS III is recognized as having accelerated the integration of the EEZ into customary international law. A proposal establishing the zone was presented at the Second Session in 1975 and reappeared in the subsequent drafts leading to the final act of the 1982 Convention.[35] As a consequence, State practice in this area has changed rapidly as the 200-mile concept has become widely accepted. A recent official publication reveals that fifty-one States had established a full 200-mile EEZ between 1974 and the date of adoption of the Convention.[36] The same publication shows that, as of November 1986, sixty-nine States had created a 200-mile EEZ, while twenty States had decreed an EFZ of the same breadth, making a total of eighty-nine States. To this number may be added the thirteen States still claiming a 200-mile territorial sea.[37]

These indications reveal that assertions of 200-mile jurisdiction have contributed significantly not only to the general acceptance of the Convention's EEZ provisions but also to the gradual emergence of that zone as a concept of customary international law.[38] Thus, while the ICJ held in 1982 that the EEZ may be regarded as part of international law,[39] by 1985 the same Court had no hesitation in recognizing that the EEZ had been implanted in customary law.[40] It would not therefore seem to be too long before States recognize a coastal State's EEZ jurisdiction as an *ipso facto* and *ab initio* right.

(*d*) *The significance of identical delimitation provisions*

As explained in the previous chapter, the 1982 Convention contains two separate provisions for delimiting the shelf and EEZ respectively, although the wording of these two provisions is identical. The possibility that the peculiarity of each regime may require a delimitation provision different from the other was almost never brought up at UNCLOS III.[41]

[35] For details see e.g. W. Extavour, *The Exclusive Economic Zone* (Institut Universitaire de Hautes Études Internationales, Geneva, 1979), 135–62; Oxman, articles cited *supra*, Ch. 4 n. 1.

[36] *LSB* 2 (Mar. 1985).

[37] Ibid. 8 (Nov. 1986), 28.

[38] Writing in 1979, Professor Brownlie considered only the 200-mile EFZ to be part of customary law, *Principles of Public International Law* (Clarendon Press, Oxford, 1979), 221.

[39] *Tunisia–Libya* case, paras. 82 and 100.

[40] *Libya–Malta* case, ICJ Reports (1985), para. 31.

[41] As far as could be ascertained, this possibility was mentioned only once in passing by the Chairman of Group 7 in his report on the work of the Group in its 7th Session. He

In the *Gulf of Maine* case the Chamber referred to the significance in the symmetry of the said provisions, without indicating its particular influence on the delimitation issue.[42]

Could this symmetry be interpreted as implying the drafters' wish to favour a single boundary solution, at least as a first step in the process of delimitation?[43] It is submitted that the answer should be in the negative for at least two obvious reasons. First, the similarity between the provisions is limited to abstract principles. The provisions contain no substantive rules of delimitation.[44] The application of both provisions does not, therefore, necessitate any a priori similarity in the delimitation principles. As noted earlier, the autonomy of both zones may require that certain considerations which are relevant to one zone cannot automatically be regarded as such for the delimitation of the other.[45] This was expressly stated by the Chamber[46] and could not be challenged on the basis of the Convention's identical, but 'empty', formulae. Second, it is clear that the said provisions of the 1982 Convention, although practically identical, deal with separate zones. Neither of them can be said to be legally applicable to the delimitation of a third and quite independent line delimiting both zones. The mere identity of the two provisions bears no significance with regard to the 'equitable criteria' relevant to the delimitation of the two zones either separately or by a single line. As one commentator pertinently observes, if it was the Conference's intention to have a single boundary, 'it would have been a simple drafting matter to say so'.[47]

(*e*) *The significance of the 200-mile limit*

A major rule in the 1982 Convention is that the legal basis for coastal State jurisdiction over the EEZ and the shelf within 200 miles is to rest not on any geophysical concept of prolongation but on geographical adjacency measured by distance. What is the legal significance of this development in the delimitation of a single boundary? This pertinent question was put by Judge Oda to the parties in the *Tunisia–Libya* case.[48]

stated that reference might be made to the fact that 'we have been considering the same set of criteria to be applicable both to the [shelf and EEZ]. One could, perhaps, also examine whether some kind of distinction in this respect, as related to the applicable criteria of delimitation, offered elements conductive to our search for a comparison' (NG7/11).

[42] See *infra*, text at n. 98.

[43] For instance, this was the view put forward by Canada in the *Gulf of Maine* case, *Mémoire du Canada* (1982), 128–9.

[44] *Supra*, Ch. 4 s.2.

[45] *Supra*, s.1.

[46] See *infra*, s.3(*b*).

[47] D. Bowett, *The Legal Regime of Islands in International Law* (Oceana Publications, New York, 1979), 189.

[48] Judge Oda's Dissent to the *Tunisia–Libya* Judgment, 232.

Tunisia held that it was difficult to conceive of different boundaries inside the 200 miles.[49] Libya, on the other hand, considered that the two boundaries need not necessarily coincide.[50] The reason behind this Libyan answer is of interest.[51] It brings to the fore the preliminary question of the impact of the 200-mile distance criterion on the delimitation of the continental shelf. The real issue is how far this distance criterion is able to *detach* the 'legal continental shelf' from its geological/geomorphological origin.

The answer provided by adjudication is considered below. But as far as the 1982 Convention is concerned, the relevant provisions do not seem to assist in giving a precise answer.

Article 76(1) of the 1982 Convention reads:

> The continental shelf of a coastal State comprises the sea-bed and subsoil of the submarine areas that extend beyond its territorial sea throughout the natural prolongation of its land territory to the outer edge of the continental margin, or to a distance of 200 nautical miles . . . where the outer edge of the continental margin does not extend up to that distance.

The history of this paragraph is well known and does not need to be restated.[52] It was included to harmonize a State's title to the continental shelf with its entitlement to a 200-mile exclusive economic zone. To complement this harmonization—which was a prime aim of the Conference's consensus policy—one might argue that the Convention indicates strongly the relevance of the distance criterion to shelf delimitation.[53] One might argue, on the other hand, that Article 76 relates to the definition of the continental shelf, and not to delimitation, which is dealt with in a separate Article.[54] To clarify any doubt as to the independence of this provision from the delimitation issue, paragraph 10 of Article 76 expressly asserts that: 'The provisions of this article are without prejudice

[49] Ibid.

[50] Ibid.

[51] Libya argued that 'the delimitation . . . ought not, in the majority of cases, to be different. Nevertheless, there may be factors relevant to fishing, such as established fishing practices, which have no relevance to shelf resources; and, conversely, there may be factors relevant to shelf resources—such as geological features controlling the extent of a natural prolongation—of no relevance to fishing. It therefore follows that the two boundaries need not necessarily coincide.' Ibid.

[52] e.g. P. Gardiner, 'Reasons and Methods for Fixing the Outer Limit of the Legal Continental Shelf beyond 200 Nautical Miles', 11–12 *RIRI* (1978), 145; J. Lupinacci, 'The Legal Status of the Exclusive Economic Zone in the 1982 Convention on the Law of the Sea', in F. Orrego (ed.), *The Exclusive Economic Zone: A Latin American Perspective* (Westview Press, Boulder, Colo., 1984), 75, at 84 ff.

[53] e.g. *Gulf of Maine* case, Mémoire du Canada, 123.

[54] e.g. Judge Mbaye, his Sep. Op. to the *Libya–Malta* Judgment, 95; Professor Bowett, *The Legal Regime of Islands*.

to the question of delimitation of the continental shelf between States with opposite or adjacent coasts.'

Indeed this argument is reinforced by the fact that the old concept of natural prolongation is not completely abandoned in the Convention. Paragraph 1 still retains 'the natural prolongation' expressly; and this is further supplemented by the paragraph's direct reference to the situation of States possessing a continental margin with an outer edge extending to less than 200 miles.[55]

However, the development of the law on boundary delimitation, especially since the 1977 *Anglo-French* Arbitration, has tended to detach the concept of coastal State jurisdiction from a purely physical meaning and attach to it a politico-juridical identity.[56] The alignment of the shelf regime with that of the EEZ within 200 miles can be regarded as a characteristic of this development. This alignment seems at the same time to have enhanced the legal 'synthesization' of the shelf regime, which is henceforth to be primarily defined by a certain distance rather than by the geophysical characteristics of the seabed and subsoil.

Assuming this conclusion to be plausible, it remains to be seen how essential the legal basis of a title is for clarifying the rules on delimitation. A comparison between Article 6 of the 1958 Shelf Convention and Article 83 of the 1982 Convention might prove appropriate. Article 6 provides for a substantive clause on delimitation, namely the 'equidistance-special circumstances' rule. Article 83, unlike Article 6, provides no substantive, self-contained formula. In cases of overlapping or mutually interfering continental shelf rights, it might be necessary to consider also Article 76. In such a situation, it is submitted that the Court's 1969 statement that 'Articles 1 and 2 of the [Shelf Convention] do not appear to have any direct connection with inter-State delimitation'[57] no longer provides an appropriate analogy. In view of the vacuum in Article 83 (found equally in Article 74), the structuring of a close relation between the legal basis of a title and the delimitation rule appears to be essential. As Judge Mbaye pertinently noted,

> It is through a comparison of the titles which underlie the conflicting rights, and of the methods of evaluating them, that the solution to this conflict may be found. The course indicated by the new provisions is such that this comparison will almost always lead to a 'solution' attributing rights.[58]

[55] Judge Sette-Camara, his Sep. Op., *Libya–Malta* case, 68–9.
[56] *Supra*, Ch. 4 s.3(*b*)(ii)(1).
[57] The *North Sea* cases, para. 40.
[58] Judge Mbaye, Sep. Op., *Libya–Malta* case, 96.

The Court itself ignored in the 1969 Judgment its statement quoted above and opted for regarding 'natural prolongation'—a factor of entitlement—as the prime principle for delimitation.[59]

Thus, on the basis of the 1982 Convention, it seems appropriate to say that, when an assessment of conflicting shelf rights is made by reference to the distance principle, the delimitation of these conflicting rights will be determined primarily by geography (and not by geology or geomorphology), and the method will tend to be some form of modified equidistance. Also relevant is the alignment of the EEZ seabed rights with shelf rights within 200 miles. Both factors seem *a fortiori* to reinforce the relevance of the distance criterion for delimiting a single maritime boundary.[60]

(*f*) *The relevance of existing shelf boundaries to claimants of EEZs*

A major problem associated with a single maritime boundary is the situation where a continental shelf boundary already exists. In many cases of course States will agree to utilize the old shelf boundary as the new single maritime boundary.[61] But what if either State is not content with such a result? This may well happen when the State wants to negotiate a new boundary that would, for instance, reflect a favourable special circumstance such as the location of fish stocks or traditional fishing activity. Such factors are considered to be 'more nearly related to the concept of the exclusive economic zone'.[62] The present controversy between France and Spain regarding a single boundary in the Bay of Biscay is a recorded example.[63]

It is obvious that, when two governments want to carry out a delimitation by treaty, the consideration of these and other associated difficulties would be part of the whole 'package deal' that the negotiators may come up with. In principle, co-operative schemes can be devised to share jurisdiction, or to grant access for one State to an area subject to the jurisdiction of another. The Australia–Indonesia Agreements regarding their seabed boundary and fishery jurisdiction line in the Timor and Arafura Seas are a good example. The seabed boundary between the two States was completed in 1972, partly by taking into account the natural boundary created by the Timor Trough.[64] In 1981 the two States reached

[59] See *supra*, Ch. 3 s.3(*a*).
[60] See dissenting opinions of Judges Aréchaga and Evensen in the *Tunisia–Libya* case, 115, 319, respectively.
[61] See e.g. *infra*, s.4(*b*).
[62] *Tunisia–Libya* case, para. 100.
[63] See *infra*, Ch. 9, text at nn. 245–6.
[64] See *supra*, Ch. 4, text at n. 254.

a Memorandum provisionally establishing a modified equidistance fishery line, thus partly deviating from the Timor Trough natural seabed boundary. It was agreed, however, that the provisional line was not to affect traditional fishing by Indonesian fishermen in the vicinity of specific Australian reefs and islets straddling or near the median line.[65]

The situation, however, is different if the problem is referred to a tribunal which is not empowered to reach an *ex aequo et bono* decision. Could the tribunal, on the basis of the 1982 Convention, take the existence of the old continental shelf boundary into account as a relevant factor? Would the parties' request for a new boundary be interpreted, by operation of law, as abrogating the existing shelf boundary? The Convention's delimitation rules are notoriously unhelpful. The matter seems to rest partly on finding an answer to the open question referred to above, namely, whether the regime of the continental shelf within 200 miles has been, for practical as well as legal purposes, superseded by the regime of the EEZ. As pointed out earlier, the 1982 Convention seems to preserve the autonomy of the two regimes; and the Court's decisions seem to confirm this approach.[66] But the *incorporation* of the shelf rights into the 200-mile EEZ appears largely to assimilate the relevant factors upon which the question of delimitation should be approached, at least as far as the seabed and subsoil rights are concerned. This might be an important 'indicium' towards the equity of the old boundary for the new situation.

In fact, certain provisions of the 1982 Convention seem to suggest that, during the debate in UNCLOS III, it was assumed that the old boundary should as a matter of principle continue to govern the EEZ boundary. The argument that this is implicit in the Convention's identical formulae for delimitation was mentioned earlier. More convincing, it seems, is the clause embodied in Article 74(4) of the 1982 Convention, which might be taken as implying the Conference's wish to preserve the validity of existing continental shelf arrangements *vis-à-vis* EEZ delimitations. The paragraph reads, 'Where there is an agreement in force between the States concerned, questions relating to the delimitation of the exclusive economic zone shall be determined in accordance with the provisions of that agreement.'

As is well known, the difficulty in interpreting this clause lies in determining the proper meaning of the word 'agreement in force between the parties concerned',[67] especially in view of the fact that Article 83(4)

[65] P. Bassett, *Australia's Maritime Boundary* (Australian Foreign Affairs Record, Dept. of Foreign Affairs, 1984), 186.

[66] *Tunisia–Libya* case, para. 100. In *Libya–Malta* the Court considered that, although the two regimes are linked, this 'does not mean that the concept of the continental shelf has been absorbed by that of the exclusive economic zone'. Para. 33.

[67] See e.g. R. D. Hodgson and R. W. Smith, 'Boundaries of the Economic Zone', in E.

(shelf delimitation) repeats *mutatis mutandis* the same provision. The interpretation just referred to, however, seems to find support in the origin of the provision in a working paper 'containing certain basic principles on an economic zone and on delimitation'[68] submitted by the delegations of Australia and Norway to the Seabed Committee.[69] Under this proposal Australia and Norway were concerned with the problem of delimiting a new 'economic zone [or] Patrimonial seas and . . . seabed areas'[70] in view of the already existing continental shelf boundary; and, accordingly, their proposal was meant to provide for concordance between the existing and potential boundaries between neighbouring States.[71]

Further support for this interpretation may be found in Article 311 of the 1982 Convention. Whilst its first paragraph asserts the prevalence of the 1982 Convention, 'as between States Parties, over the Geneva Convention', paragraph 2, on the other hand, states that 'This Convention shall not alter the rights and obligations of States Parties which arise from other agreements compatible with this Convention.' In other words, shelf delimitations effected, for instance, on the basis of Article 6 of the 1958 Shelf Convention cannot be abrogated on the basis of the 1982 Convention. The importance of this paragraph is highlighted by the attempts made at the Conference to draft paragraph 1 so as to make the 1982 Convention 'supersede' or 'replace' the 1958 Conventions. In the face of opposition from many States, the compromise arrived at was to apply the word 'prevail' in Article 311(1).[72]

It thus seems plausible to state that, while the Convention does not eliminate the possibility of having two separate boundaries, it does establish a legal presumption in favour of utilizing the old shelf boundary for the new exclusive economic or fishery zone boundary as well.

Miles and S. Gamble (eds.), *Law of the Sea: Conference Outcomes and Problems of Implementation*, Proceedings of the 10th Law of the Sea Institute, Kingston (Ballinger, Cambridge, Mass., 1977), 183, at 197–203; B. Vukas, 'The LOS Convention and the Sea Boundary Delimitation', in Vukas (ed.), *Essays on the Law of the Sea* (Zagreb, 1985), 147, at 178–81.

[68] Doc. A/AC.138/SC. II/L.36, UN General Assembly, *Official Records*, iii (1973), Supplement No. 21 (A/9021), 77–8.

[69] The proposal was later reproduced as part of the 'Main Trends' represented by the Chairman of the Second Committee in 1975. It was only then that the provision included in the Article of the ISNT dealing with the delimitation of the economic zone was also included in the Article on the delimitation of the continental shelf. A/CONF. 62/WP. 8/pt. II; see Hodgson and Smith, 'Boundaries of the Economic Zone', 193–200.

[70] Para. 1(b) of Doc. A/AC. 138/SC. II/L.36.

[71] Vukas, 'The LOS Convention', 180.

[72] J. Evensen, 'The Delimitation of EEZs and Continental Shelves', in C. Rozakis and C. Stephanou (eds.), *The New Law of the Sea* (Elsevier Science, Amsterdam, 1983), 146–7.

(g) *Problems of implementing two separate boundaries*

Perhaps the strongest argument in favour of a single maritime boundary lies in its pragmatism. For obvious reasons no State would wish to go through the long and tiring process of direct or third-party settlement twice—once for the continental shelf and once for the exclusive economic or fishery zone. States may be prepared in principle to reach an agreement which would in one way or another mitigate some of the problems ensuing from drawing two different boundaries, but this would certainly produce intricate issues between the parties to which the 1982 Convention provides no direct solution.[73] Two types of problem may illustrate the preference for the single line solution.

1. *Problems of negotiating two different boundaries*. The existence of political factors will make the negotiation of separate boundaries for the shelf and the EEZ extremely difficult. A few possible difficulties will illustrate this point. Are the parties able to agree upon a scheme to manage overlapping areas? Do their established hydrocarbon concessions or other established interests in the area overlap? Is either party prepared to risk the possibility that the one will erect oil drilling installations in the other's rich fishing ground—a real possibility when there are two boundaries, one (zone) based on equidistance and the other (shelf), for instance, on natural prolongation?[74] Such problems do of course occur even in single shelf or EEZ boundaries; but they are certainly aggravated if two, rather than one, lines are delimited. These are some of the problems which show that negotiation of a single maritime boundary is more feasible.

2. *Problems of resolving jurisdictional acts*. Assuming two separate boundaries based on the above different criteria, the situation would lead to a 'clash' of jurisdiction within the same maritime area. For instance, one State may wish to establish an artificial island on its EEZ (and accordingly exercise its rights relating to protection of its custom, fiscal, health, safety, and immigration regulations),[75] but within areas which belong to the continental shelf of the other State. The latter is also recognized as possessing the same rights within its continental shelf.[76] Must the first State seek the prior permission of the second? And if

[73] The provisions for the peaceful settlement of disputes provided in pt. xv of the Convention are of course relevant.

[74] e.g. the Australia–Indonesia shelf boundary and (provisional) fishery zone line in the Timor and Arafura Seas. *Supra*, text at nn. 64–5; the Australia–Solomon Islands Agreement, *infra*, n. 189.

[75] Art. 60(2).

[76] Art. 80 of the 1982 Convention provides that Art. 60 (artificial islands, installations and structures in the exclusive economic zone) applies *mutatis mutandis* to the regime of the continental shelf.

so, does the latter have the right to withhold such an authorization, on the grounds, for instance, that its sedentary fishery regulations would be interfered with? The same problem applies equally to jurisdictional acts such as the establishment of specific conservation and management measures. The Convention is silent on these and other similar jurisdictional problems.[77] These difficulties would of course be avoided by the application of a single boundary for both the shelf and the superjacent waters.

(*h*) *Conclusion*

The 1982 Convention indicates that the regimes of the continental shelf and the exclusive economic zone reflect major differences and that accordingly some measure of juridical autonomy and distinctiveness should be retained within the general framework of State jurisdiction. On the other hand, the Convention tends to assume that a degree of integration between the two regimes is inevitable. This integration seems to lessen the differences between the factors regarded as relevant to the delimitation of the two zones, which in turn tends to exert a decisive influence on the concept of the single maritime boundary as a legally permissible, though not necessarily mandatory, approach to the problem of dual State jurisdiction. Indeed, in practical terms, the single maritime solution seems to be the logical outcome of the extension of coastal State jurisdiction over the resources of the EEZ and the alignment of that jurisdiction with pre-existing rights over the continental shelf.

3. THE APPLICATION OF A SINGLE MARITIME BOUNDARY BY ADJUDICATION

(*a*) *The relevance of the 'distance principle' in the shelf cases*

Conceptually, the EEZ, as defined in the 1982 Convention, incorporated the continental shelf up to a maximum distance of 200 miles from the shore. The continental shelf, as also redefined in that Convention, extends to the same distance, but as the minimum limit that a coastal State can enjoy. The development of this 'common denominator' is the most obvious element in the argument for a single boundary.[78]

As pointed out earlier,[79] the relative demise of the natural prolongation

[77] Extavour, *The Exclusive Economic Zone*, 223–8; Attard, *The Exclusive Economic Zone in International Law*, 216–17.
[78] See *supra*, s.2(*e*).
[79] *Supra*, n. 56.

concept began with the 1977 *Anglo-French* Arbitration. But the 'trend towards the distance principle' as the basis of coastal State jurisdiction over the shelf and EEZ was explicitly recognized by the Court in its Judgment in the *Tunisia–Libya* case. In particular, the Court noted that, in so far as the 1982 Convention provided in certain circumstances that 'distance from the baseline, measured on the surface of the sea, is the basis for the title of the coastal State, it departs from the principle that natural prolongation is the sole basis for the title'.[80]

This development in the basis of title over the shelf might be argued to have rendered the alignment of the criteria relating to the delimitation of the shelf and the EEZ a logical as well as legal necessity. This has been aptly emphasized by the Court in its Judgment in *Libya–Malta*. In an unprecedented statement on the applicable criterion of delimitation the Court stated that, as the 1982 Convention demonstrated,

> the two institutions—continental shelf and exclusive economic zone—are linked together in modern law. Since the rights enjoyed by a State over its continental shelf would also be possessed by it over the seabed and subsoil of any exclusive economic zone which it might proclaim, one of the relevant circumstances to be taken into account for the delimitation of the continental shelf of a State is the legally permissible extent of the exclusive economic zone appertaining to that same State.[81]

And, after asserting the autonomous character of both zones, the Court delivered its most relevant dictum to the issue under consideration:

> It follows that, for juridical and practical reasons, the distance criterion must now apply to the continental shelf as well as to the exclusive economic zone; and this quite apart from the provision as to distance in paragraph 1 of Article 76.[82]

This unequivocal elevation of the distance principle into the operative criterion and the discarding of the natural prolongation concept represents a major detachment from the thinking of the previous cases, which, as shown earlier, viewed the matter on a case-by-case basis.[83]

Specifically, the Court's line of reasoning might be seen as irreconcilable with the rejection of the distance principle by the Chamber in the *Gulf of Maine* case only twelve months before the *Libya–Malta* Judgment.[84] One is tempted, however, to submit that this rejection by the Chamber of the 'distance principle' is only at face value. For, as demonstrated below, the Chamber's preference for 'neutral criteria' for effecting a single boundary can only lead to the primacy of the factor of geography, which

[80] *Tunisia–Libya* case, para. 48. See *infra*, Ch. 9 s.2(*e*).
[81] *Libya–Malta* case, para. 33.
[82] Ibid. para. 34.
[83] *Supra*, Ch. 4 s.3(*b*)(ii)(1).
[84] See e.g. Bowett, 'Exploitation of Mineral Resources and the Continental Shelf' 1, at 4.

is best reflected by geometrical methods. Indeed, this latter solution can only enhance the primacy of the distance principle. The Chamber's 'equitable criteria' and 'methods' seem to be basically in accord with this line of reasoning.

(*b*) *The* Gulf of Maine *case*

The fundamental question facing the Chamber in attempting to comply with the parties' request for a single boundary[85] was how considerations of the equities relating to the continental shelf were to be compared with and balanced against those relating to the fishing zone. This was of course Judge Ago's question put to the parties to the case: if the method for the delimitation of the continental shelf differs from the method for the delimitation of the fishing zone, on what basis is the one to be preferred over the other?[86]

Judge Gros appeared to consider that the answer to this question was a precondition to any delimitation of a single boundary, and consequently criticized the Chamber for not adequately covering this prerequisite.[87] However, the majority judgment took a pragmatic stance, which seems already to have influenced the approach of the Tribunal in the *Guinea–Guinea Bissau* Arbitration.[88] The Judgment took the view that, firstly, there was no rule prohibiting the Chamber from complying with the parties' request.[89] Secondly, it held that, because of the dual function of the single line to be delimited, it would be equitable to give primacy to those 'neutral criteria' and methods and put aside considerations that were relevant to one element alone (continental shelf or fishing zone). It was this stance that the Chamber firmly took throughout its Judgment.

In reviewing the rules of international law on maritime delimitation, the Chamber stated that delimitation had to be effected by applying equitable criteria and by using practical methods capable of ensuring,

[85] According to Art. 2(1) of the *compromis*, the request before the Chamber was the delimitation of the parties' respective continental shelf and fishery zone by a single boundary. However, on 10 Mar. 1983—i.e. after the proceedings had begun—the USA proclaimed an exclusive economic zone (Proc. No. 5030). Canada, for its part, pointed out that the rights and jurisdiction it exercised within its 200-mile fishing zone 'are substantially similar to the rights and jurisdiction associated with the concept of the exclusive economic zone'; and 'it has already put in place most of the elements of that zone'. (See the *Gulf of Maine* case, *Counter-Memorial* submitted by Canada (1983), para. 464.) The Chamber, however, went on deciding the case on the basis of the parties' original request, namely 'a single boundary for two different jurisdictions', without taking notice of this development. See L. Legault and B. Hankey, 'From Sea to Seabed: The Single Maritime Boundary in the Gulf of Maine Case', 79 *AJIL* (1985), 961, at 976.

[86] *Gulf of Maine* case, para. 161.

[87] Judge Gros, his Dissent in ibid. 362 ff.

[88] *Infra*, s.3(*c*).

[89] Para. 27.

with regard to the geographic configuration of the area and other relevant circumstances, an equitable result.[90]

In its effort to identify the equitable criteria or practical methods to be adopted,[91] the Chamber considered the applicability of the delimitation provisions of the 1958 and 1982 Conventions. It observed that Article 6 of the 1958 Shelf Convention, which was in force between the parties, was not applicable to the case under consideration as it was not confined to the delimitation of the shelf, but extended to the determination of a single maritime boundary encompassing the shelf and the superjacent waters.[92] Envisaging 'equidistance' as a method subject to exceptions, the Chamber did not find any evidence to suggest that the said method had become part of customary law and therefore applicable to every maritime delimitation.[93] In this respect, the Chamber remained faithful to the jurisprudence of the Court in the previous cases.[94]

Turning to the 1982 Convention, the Chamber observed that 'certain provisions of the Convention, concerning the continental shelf and the exclusive economic zone . . . may . . . be relevant to the present case'.[95] In particular, the adoption in the Convention of the 200-mile regime was of legal significance. In its opinion, the regime, 'even in some aspect [bearing] the mark of the compromise surrounding [its] adoption, may nevertheless be regarded as consonant at present with general international law on the question'.[96] In respect of the delimitation provisions, the Chamber took the view that the absence of any substantive rule in paragraph 1 of Articles 74 and 83 did not negate the importance of the provisions for the case under consideration: '[a]lthough the text is singularly concise it serves to open the door to continuation of the development effected in this field by international case law.'[97] Most interesting, in its view, were the parallels between the two texts

> [i]n a case like the present one, where a single boundary line is to be drawn both for the sea-bed and for the superjacent fishery zone, which is included in the exclusive economic zone concept. The identity of the language which is employed, even though limited of course to the determination of the relevant principles and rules of international law, is particularly significant.[98]

The probative character of these observations seems to suggest that the Chamber considered that the 1982 Convention points out indicia in favour of the concept of a single boundary.

However, in respect of the Convention's 200-mile criterion for the definition of the shelf and the EEZ, the Chamber opted to qualify its relevance to delimitation, for, in the Chamber's opinion, the boundary

[90] Para. 112. [91] Paras. 109–11, 114. [92] Paras. 115–21.
[93] Paras. 122–5.
[94] *Supra*, Ch. 4 s.3(*c*). [95] Para. 94.
[96] Ibid. [97] Para. 95. [98] Para. 96.

'results from a rule of law, and not from any intrinsic merit in the purely physical fact'. It is correct 'to say that international law confers on the coastal State a legal title to an *adjacent* continental shelf or to a maritime zone *adjacent* to its coast; it would not be correct to say that international law recognizes the title *conferred on the State by the adjacency* of that shelf or that zone, as if the mere natural fact of adjacency produced legal consequences'.[99] The Chamber therefore considered itself free to select appropriate equitable criteria and practical methods which would lead to an equitable delimitation.

The Chamber observed that a criterion or 'relevant factor' found 'equitable' or even 'reasonable' for the delimitation of the shelf did not automatically mean it possessed 'the same properties in relation to the simultaneous delimitation' of the shelf and the superjacent fishery zone.[100] The dual aspect of the delimitation, the Chamber emphasized, '[rules] out the application of any criterion found to be typically and exclusively bound up with the particular characteristics of one alone of the two natural realities that have to be delimited in conjunction'.[101]

Putting this principle into effect, the Chamber dismissed as irrelevant any geophysical characteristics that can be observed in the area's seabed. For whilst such characteristics might have special effect in determining the delimitation of the continental shelf, there would be no reason to extend their effect to the delimitation of the superjacent waters.[102] Correspondingly, the Chamber remarked that a 'natural boundary' that might be derived from the ecological characteristics of the water column might be relevant to the division of the superjacent waters but was not appropriate to shelf delimitation.[103]

Consequently, the Chamber concluded that a delimitation by a single line could only be carried out by the application of a 'neutral' criterion, or combination of criteria, which did not 'give preferential treatment to one of these two objects to the detriment of the other, and at the same time is such as to be equally suitable to the division of either of them'.[104]

The kind of criteria chosen to reflect such 'neutrality' and 'equity' were not difficult to predict. In its view, criteria derived from the geography of

[99] Para. 103.

[100] Para. 192.

[101] Para. 193.

[102] Ibid.

[103] Ibid. The USA argued that Georges Bank constituted an integrated oceanic/ecosystem and that the living resources of Georges Bank were common pool resources in economic terms. Any delimitation dividing the Bank, the USA argued, would thus result in wasteful competition and harm to the resources that international law was designed to protect. See the *Gulf of Maine* case, *Memorial* submitted by the USA (27 Sept. 1982), 142–4; pt. III, ch. I s.2.

[104] Para. 194.

the coasts possessed a 'neutral character' and were therefore appropriate to the case.[105] Within this framework, the Chamber advanced the criterion

that in principle, while having regard to the special circumstances of the case, one should aim at an equal division of areas when the maritime projections of the coasts of the States between which delimitation is to be effected converge and overlap.[106]

However, in order to attain equity, the Chamber observed that the application of this criterion must not be rigid, but had to be adjusted in the light of the geographical circumstances of the case. The adjustment had to be made by applying criteria of an 'auxiliary nature':

a fair measure of weight should be given to a by no means negligible difference within the delimitation area between the lengths of the respective coastlines . . . It also . . . is held equitable partially to correct any effect of applying the basic criterion that would result in cutting off one coastline, or part of it, from its appropriate projection across the maritime expanses . . . or . . . the necessity of granting some effect, however limited, to the presence of a geographical feature such as an island or group of small islands lying off a coast, when strict application of the basic criterion might entail giving them full effect or, alternatively, no effect.[107]

Given this correlation between one 'basic criterion' and several 'auxiliary' considerations, it is difficult for any lawyer to conceive a significant difference between the Chamber's line of reasoning and that achieved by the application of the 'equidistance–special circumstances' rule. The Chamber's resultant boundary line reinforces this contention clearly, as seen earlier.[108]

However, with respect to the practical methods, the Chamber noted that they had to reflect the chosen criteria;[109] in particular those methods considered appropriate for reflecting coastal geography and suitable for the delimitation of a single boundary: 'In the outcome, therefore, only geometrical methods will serve.'[110] The Chamber emphasized, however, the necessity of not going too far in drawing a boundary based on 'strict compliance with the canons of geometry'. As a single boundary delimiting, *inter alia*, areas whose exploitable resources were found 'in a volume of liquid in movement, forming the habitat of mobile fauna', a degree of simplification is 'an elementary requisite to the drawing of any delimitation line in such an environment'.[111]

Putting its findings into effect, the Chamber noted the change in the geographical relationship of the coasts within the Gulf of Maine. In the north-eastern sector of the Gulf the prevailing relationship was that of

[105] Para. 195. [106] Ibid. [107] Para. 196. [108] *Supra*, Ch 4 s.3(*c*).
[109] Para. 199. [110] Para. 201. [111] Para. 203.

lateral adjacency. In the sector nearest to the closing line of the Gulf the prevailing relationship was one of oppositeness.[112] This difference in relationship necessitated, the Chamber noted, 'that the boundary within the Gulf must be drawn in two segments'.[113]

Consequently, the Chamber used two different methods to draw the boundary. In the first segment, it drew perpendiculars from the starting-point given by the parties[114] to the two basic coastal lines and bisected the angle so formed.[115] For the second segment, the Chamber drew a median line, which was moved laterally to take into account the difference between the respective lengths of the parties' coastlines and the presence of the Seal Islands.[116]

With regard to the final segment, lying outside the Gulf proper, the Chamber held that the determination of its direction must depend on the first two segments within the Gulf; however, there was 'from the geographical point of view . . . no point of reference, outside the actual shores of the Gulf that can serve as a basis for carrying out the final operation required'.[117] Under these circumstances, the Chamber felt justified in resorting again to a geometrical method. It drew a perpendicular to the closing line of the Gulf, the starting-point coinciding with the point where the modified median line intersected the closing line of the Gulf.[118] The decisive criterion described by the Chamber for the terminus of the perpendicular was that the delimitation must equitably divide the area in which the maritime projections of the coasts of the parties overlapped.[119] This was where the last point of the perpendicular reached within the overlapping 200-mile zones of the respective parties.[120]

The last remaining task of the Chamber was to ascertain whether the resulting lines could be considered intrinsically equitable in the light of all the relevant circumstances.[121] With regard to the first two segments, such verification was not necessary since their guiding parameters were provided by the physical and political geography of the area.[122] With respect to the third segment, factors other than those of geography were strongly present

[112] Para. 206.
[113] Para. 207.
[114] According to Art. 2 of the *compromis*, the two parties requested the Chamber to start the boundary from a specific 'starting point', referred to by the Chamber as Point A, nearly 40 n.m. south-west of the international terminus in Grand Mannan Channel. See para. 20 and map 1, p. 269.
[115] Paras. 209–14.
[116] Paras. 215–23.
[117] Para. 224.
[118] Ibid.
[119] Para. 228.
[120] Ibid.
[121] Para. 320.
[122] Para. 231.

in the parties' arguments with regard to Georges Bank, the principal stake in the dispute. In this respect, the United States emphasized its alleged long history of fishing and other marine activities in the area.[123] Canada laid great emphasis on what it considered to be the socio-economic aspects, in particular the maintenance of existing fishing patterns vital to coastal communities.[124]

The Chamber said that it could not adopt these positions of the parties.[125] Until very recently, the fisheries in question were high seas and as such freely open to fishing by nationals of all countries. Consequently, in the Chamber's view, it was 'evident that the respective scale of activities concerned with fishing—or navigation, defence or, for that matter, petroleum exploration and exploitation—cannot be taken into account as a relevant circumstance or . . . an equitable criterion to be applied in determining the delimitation line'.[126] However, as indicated earlier,[127] by drawing a line which granted to Canada approximately one-sixth of the Bank, including the resource-rich 'North-East Peak' and most of the 'Northern Edge', and the United States the remaining area, the Chamber wanted undoubtedly to make sure that its division—in view of the circumstances alleged above—would seem reasonable.

(*c*) *The* Guinea–Guinea Bissau *Arbitration*

In the *Guinea–Guinea Bissau* Arbitration, the two parties requested the Tribunal to determine whether the 'maritime boundary' between them had been fixed by a Convention of 1886, and if not, to identify the course of the line 'delimiting the maritime territories' of the parties. Although the term 'single' was not used in the *compromis*[128] the boundary was to divide the territorial sea, the continental shelf, and the EEZ, and the Tribunal characterized the request as one for a 'single' line.[129] The Tribunal did not appear to be troubled by the theoretical problem posed by this kind of delimitation and proceeded to delimit the boundary without reference to the separate nature of the regimes involved or any fusion of them. Like the Chamber in the *Gulf of Maine* case, the Tribunal gave overriding significance to geography as the 'neutral' factor, and discounted the relevance of other factors. Thus, it also avoided any need

[123] Para. 233. See *supra*, Ch. 4 s.3(*b*)(iii)(2).

[124] Para. 234.

[125] Para. 235.

[126] Para. 237.

[127] See *supra*, Ch. 4, text at n. 138.

[128] Art. 2 of the *compromis* between the parties simply refers to 'the boundary between the maritime territories'.

[129] 25 ILM (1986), para. 42.

to weigh or balance the equities of the factors peculiar to one of the regimes concerned against those peculiar to the other.

The Tribunal, having found that the 1886 Convention did not fix the maritime boundary between the parties,[130] proceeded to determine the course of a single maritime boundary. At the outset it felt it necessary to repeat after the Chamber that in cases concerning a multi-purpose delimitation customary law provides 'only a few basic legal principles, which lay down guidelines to be followed with a view to achieving an essential objective'.[131]

In respect of the case under consideration,

> the essential objective consists of finding an equitable solution with reference to the provisions of Article 74, paragraph 1, and Article 83, paragraph 1 of [the 1982 Convention]. This is a rule of international law which is recognized by the Parties and which compels recognition by the Tribunal. However, in each particular case, its application requires recourse to factors and the application of methods which the Tribunal is empowered to select.[132]

Consequently, in order to place a delimitation on an equitable and objective basis, the Tribunal felt that equitable criteria must be primarily derived from the geography of the coastlines involved. In this respect, it found that an important factor was the configuration and direction of the parties' coastlines, including the coastal islands of the area and the Bijagos archipelago.[133]

The Tribunal noted that, taken together, the coasts of the parties, including islands, were concave; and this characteristic was accentuated if the presence of Sierra Leone was considered.[134] Envisaging equidistance in its strict form, the Tribunal found that the application of this method in this situation would cut off Guinea's maritime area in front of its coasts and would tend to enclave it between the maritime areas appertaining to Guinea Bissau and Sierra Leone.[135]

Under these circumstances, the Tribunal found it equitable not to limit itself to the coasts of the parties proper, but to consider the 'long coast' by focusing upon the entire West African region.[136] In its view, such an approach would lead to the conclusion that an equitable delimitation had to be carried out by following a direction which took overall account of the concave shape of the West African coastline, and would be adaptable to present or future delimitations in the region.[137] In considering the method which best took into account the general direction of the West African coastline, the Tribunal found that a system 'using the maritime façade and, for this reason, selecting a straight line joining two coastal

[130] Paras. 1, 84. [131] Para. 88. [132] Ibid. [133] Para. 98.
[134] Para. 103. [135] Para. 104. [136] Para. 108. [137] Ibid.

points on the continent' would give more weight to the general direction of the coastline, 'at the risk of starting from a line crossing through islands and even encroaching on the continent'.[138] The most appropriate straight line was found to be that joining Almadies Point in Senegal and Cape Shilling in Sierra Leone, which reflected the overall configuration of the said coastline more faithfully.[139]

Accordingly, the Tribunal found that an equitable solution would be a straight line with a bearing of 230°, *grosso modo* perpendicular to the Almadies–Shilling line.[140] It considered 'that such a line would reduce the risk of enclavement to a minimum'.[141] The starting-point of the perpendicular was determined to coincide with the point where the 'southern limit' met the aforesaid Almadies–Shilling line. This 'southern limit', established by Article 1 of the 1886 Convention, was held by the Tribunal to be equitable for the delimitation of the first segment,[142] starting from the extremity of the land boundary as far as the elevation of the Guinean island of Alcatraz, and ending 12 miles west of that island.[143]

The Tribunal concluded its Award by examining the equitable character of its boundary in the light of the relevant circumstances. The first circumstance related to the geophysical character of the area's seabed. Referring to the distance principle embodied in Article 76(1) of the 1982 Convention, the Tribunal observed that the definition of the continental shelf by reference to distance, 'without derogating from the rule of natural prolongation, reduces its scope by substituting it in certain circumstances specified in the above-mentioned paragraph of Article 76'.[144] In any event, the Tribunal found that there was no possibility of invoking this factor, since the shelves of the parties comprised a single whole without sufficiently marked division.[145]

Another factor which the Tribunal held to be relevant was proportionality between the extent of maritime areas to be allocated and the length of the coastlines. It is of interest to note that in respect of this factor the Tribunal took into account the islands which had been included in the parties' coastal lengths and for this reason it concluded that, in the particular case, proportionality was already achieved as the inclusion of the islands resulted in the coasts of the two parties having the same length.[146]

The Tribunal's consideration of the factor of the conduct of the parties, demonstrated in the 'southern limit', was mentioned above.[147] Finally, the Tribunal considered aspects related to security. It held that the implications of this factor were already resolved in that it had attempted

[138] Para. 110.
[139] Ibid. [140] Para. 111. [141] Ibid.
[142] Para. 106. [143] Para. 111. [144] Para. 115. [145] Para. 117.
[146] Para. 120. [147] See *supra*, text at nn. 142–3.

to ensure that each State controlled the maritime areas in front of its coasts and in their vicinity.[148]

(*d*) *Appraisal*

The *Gulf of Maine* and *Guinea–Guinea Bissau* Decisions singled out coastal geography as the common factor whose importance, according to these Decisions, was enhanced by the special aspects of the cases under consideration, namely, that they divided both seabed and superjacent waters by a single line.

The emphasis on geography as an aspect of the fact-intensive approach is understandable. This was already implicit in the earlier cases.[149] But it is submitted that the two Decisions' unequivocal preference for 'neutral criteria' (in effect derived from geography) to the exclusion of other factors related either to the seabed or the water column showed an over-preoccupation with 'the novelty' of the common boundary requested by the parties. It might be logical to assume that the concept of a single boundary should have increased the number of variables involved in the boundary settlement. It inevitably brings into play a much broader range of interests, activities, and functions, relating to both the seabed and overlying waters.[150] For instance, the criteria derived from the unity of ecosystems or ecological regimes or from the necessity of reducing hazards of pollution are areas of major concern to present-day coastal State jurisdiction and could be very cogent in the argument on delimitation.[151]

This risk of ignoring such factors and other similar ones in the name of neutrality tends to discourage States from requesting delimitation by a single maritime boundary.[152] This is particularly true in view of the earlier Decisions' assurance that the legal debate concerning maritime boundaries was open to any factor that might help achieve, or identify, an equitable solution.[153]

The paradox in the Chamber's line of reasoning lies in its recognition of the independent existence of each of the zones involved and its exclusion at the same time of all factors that underlie this independence. As Judge Gros pertinently stated, '[o]ne cannot simultaneously recognize the existence of two different realities and decide to ignore the difference in

[148] Para. 124.

[149] *Supra*, Ch. 4 s.3(*b*)(i).

[150] This view was taken by Canada in its pleadings: Canadian *Counter-Memorial*, para. 469; Legault and Hankey, 'From Sea to Seabed', 968.

[151] See D. Colson, 'Environmental Factors: Are They Relevant to Delimitation?', paper given at the 19th Annual Conference of the Law of the Sea Institute (Cardiff, 1985), 1, at 5.

[152] Attard, *The Exclusive Economic Zone in International Law*, 220.

[153] See *supra*, Ch. 4 s.3.

determining the boundary except on the supposition that words can be used to suppress a problem rather than deal with it'.[154]

4. STATE PRACTICE

Before analysing patterns of State practice in negotiated boundary agreements, a relevant point in the contribution of State practice to clarifying the law on single maritime boundaries must be considered, namely the 'general trend' in respect of the impact of the new EEZ claims on existing continental shelf rights.

(*a*) *The impact of new 200-mile EEZ claims*

In respect of unilateral practice, it seems certain that the large majority of EEZ claimants have not fused the two institutions. Indeed, States explicitly declare that their EEZ claims do not affect the shelf regime. For instance, the Republic of China made a declaration of its economic zone in 1979. It specifically declared that the sovereign rights enjoyed by the Republic over its continental shelf 'shall not be prejudiced in any manner by the proclamation of the present exclusive economic zone or the establishment of such zones by any other State'.[155] Similarly, the EEZ Proclamation of the United States declares that the Proclamation 'does not change existing United States policies concerning the continental shelf'.[156] Other States have retained the autonomy of the two regimes in their legislation. These include Bangladesh,[157] Burma,[158] Cambodia,[159] Colombia,[160] the Dominican Republic,[161] Fiji,[162] France,[163] Iceland,[164] Indonesia,[165] Pakistan,[166] Portugal,[167] Sri Lanka,[168] Vanuatu,[169] and Venezuela.[170]

[154] Judge Gros, his Dissent to the *Gulf of Maine* Judgment, 368.
[155] Repr. in 1 *CYILA* (1981), 152.
[156] Presidential Proclamation No. 5030, The White House, Office of the Press Secretary, 10 Mar. 1983.
[157] Act No. XXVI (1974), ss. 5, 7.
[158] Act No. 3 (1977), Arts. 12, 17.
[159] Decree of 1982, Arts. 5, 6.
[160] Law No. 10 (1978), Arts. 7, 10.
[161] Law No. 573 (1977), Arts. 4, 7.
[162] Law No. 18 (1977), ss. 6, 7.
[163] Law No. 76-655 (1976), Arts. 1, 2.
[164] Law No. 41 (1979), Arts. 3, 5.
[165] Declaration (1980), Arts. 1, 3.
[166] Territorial Waters Act of 1976, ss. 5, 6.
[167] Act No. 33/77 (1977), Arts. 2, 9.
[168] Law No. 22 (1976), s.5, subss. 3, 6.
[169] Act No. 23 (1981), pt. 5 (8, 9, 10).
[170] Act No. 2, 291 (1978), Art. 3.

However, with regard to bilateral State practice it is interesting to note that two independent results seem to emerge that might persist in parallel indefinitely, and may thus characterize the overall 'policy' of State practice on common boundaries.

In the first place, it seems evident in this practice that the rules and methods which have developed in the context of territorial sea and shelf boundaries have provided the legal framework for the negotiations. Indeed, the 'novelty' of the concept of a single boundary seems to have little influence in this regard. For instance, in his testimony before the Senate concerning three single maritime boundary agreements concluded by the United States and its neighbours, an administration official stated that, in negotiating these agreements, the United States followed 'principles relating to maritime boundaries which flow from the Geneva Convention on the Continental Shelf, and from the World Court decision in 1969 . . . and from the opinion of the Court of Arbitration in 1977 in the Anglo-French case'.[171] As will be seen shortly, this also seems valid for the majority of negotiated single boundary agreements.

In the second place, whilst this practice indicates that States are prepared in principle to *negotiate* a single maritime boundary, it is not equally clear that the same impetus will exist in respect of delimitation via adjudication. The *Gulf of Maine* and *Guinea–Guinea Bissau* Decisions might be argued to be the 'pioneering' examples. One is not sure, however, that the 'principle' of 'neutral criteria' initiated in these decisions will encourage other States to follow suit. States may prefer to risk the settlement of their shelf and EEZ boundaries separately. This, for example, appears in the drafting history of the Libya–Malta *compromis* which led to the submission of their boundary dispute to the ICJ.[172] In 1976 Libya submitted a draft Special Agreement which referred specifically to the delimitation of the shelf and the EEZ.[173] Malta opted for restricting the Court's judgment to the delimitation of the shelf and rejected the inclusion of the EEZ.[174] The resulting *compromis* referred only to the delimitation of the shelf. The negotiations between the parties and the subsequent omission of any reference to the EEZ (despite Malta's obvious interests in the *Kannizzati* fishery)[175] tends to indicate that States may not wish to risk an unpredictable decision affecting all their water-column and seabed jurisdictions.

[171] Testimony of M. Feldman, Deputy Legal Adviser, before the US Senate, S. Exec. Rep. No. 96, 49, 96th Cong., 2nd Sess. (1980).

[172] *Infra*, Ch. 9 s.2(*f*).

[173] *Memorial* submitted by the Libyan Arab Jamahiriya, vol. i (1983), 66.

[174] *Memorial* submitted by the Republic of Malta, vol. i (1983), 26.

[175] See ibid. 18–19 and map 4 (showing *Kannizzati* lines as drawn by Maltese fishing authorities for allocation to Maltese fishermen) and map 5 (showing area covered by certain Maltese fishing activities (longline fishing and trawling), ibid. vol. iii, maps.

(*b*) *Patterns of State practice*

The overall conclusion which one draws from analysis of the relevant State practice is that in a large majority of cases States have been satisfied that the equidistance–special circumstances rule leads to an equitable result. This can be considered a logical consequence of States accepting the considerations which they regard as relevant in territorial sea and shelf (or EEZ) boundaries, to govern agreements on single maritime boundaries. What is worth emphasizing in respect of these agreements is that the impact of geomorphology as a possible circumstance leading to the adoption of a modified equidistance or negotiated line is nil. Instead, considerations relating to socio-economic and/or conservation factors appear to be most prominent in determining the equity of the negotiated boundary line. The following analysis of a sample of negotiated boundaries,[176] reflecting different regions and various circumstances, illustrates the above remarks.

Whilst both the Truman Proclamation on the Continental Shelf and United States EEZ Proclamation declare that the boundaries of the United States shall be in 'accordance with equitable principles',[177] its boundary agreements with Cuba, Mexico, and Venezuela were negotiated on the basis of equidistance and the use of islands as basepoints.[178]

The 1977 Agreement with Cuba establishes a seabed–water-column equidistance boundary, giving full effect to the Florida Keys and Dry Tortugas. Although the boundary line was agreed to be provisional until both parties resolve their differences on the geographical co-ordinates, the validity of the principles involved is not expected to alter.[179] The 1978 Agreement with Venezuela applies a simplified equidistance line and gives full effect to islands, including the off-lying Venezuelan island of Aves.[180] The 1978 Agreement with Mexico establishes simplified equidistance boundaries in the Gulf of Mexico and off the Pacific coast up to the outer limit of their respective 200-mile jurisdiction in order to straighten them as much as possible. However, the United States–Mexico Agreement has yet to be ratified by the Senate, which has been reluctant to do so because American geologists have vigorously protested against giving effect to islands whose inclusion has given Mexico an important section in

[176] For a more comprehensive survey see S. Jagota, *Maritime Boundary* (Martinus Nijhoff, Dordrecht, 1985); J. Prescott, *The Maritime Political Boundaries of the World* (Methuen, London, 1985).

[177] *Supra*, nn. 12, 156 respectively.

[178] M. Feldman and D. Colson, 'The Maritime Boundaries of the United States', 75 *AJIL* (1981), 729, at 742 ff.

[179] The Treaty set provisional boundaries for a period of two years from 1 Jan. 1978, later extended for another two years. See K. Schmitt, 'The Problem of Maritime Boundaries in U.S.–Mexican Relations', 22 *NRJ* (1982), 139, at 147.

[180] See *LIS* 91.

the central Gulf of Mexico with great potential for petroleum deposits.[181] The argument of the State Department, however, seems to suggest that the main concern of the United States negotiators was to reach agreement that could be advantageous to United States fisheries interests in the Gulf and the Pacific.[182]

The equidistance principle has also been the rule applicable in the delimitation of the maritime boundaries between India and its neighbours.[183] The India–Sri Lanka Agreement of 1976 provides for a single boundary with each party having 'sovereign and exclusive jurisdiction over the continental shelf and the exclusive economic zone as well as over their resources, whether living or non-living, falling on its own side of the boundary'.[184] The parties 'apparently have agreed to a modified equidistance line and/or to one created by the selective choice of relevant basepoints'[185] to diminish the number of turning-points that would have resulted from the strict application of the equidistance method. The same principle appears to have been followed in the boundaries with the Maldives in 1976, Indonesia in 1974 (extended in 1977), and Thailand in 1978. The last two Agreements refer only to 'continental shelf' boundaries. It was, however, agreed with both Indonesia and Thailand that when they established their own EEZs the agreed boundaries would also apply respectively to the exclusive economic zones.[186]

Article 1 of the 1980 territorial sea/continental shelf Boundary Agreement[187] between Burma and Thailand states that the boundary is an equidistance line formed by a series of straight baselines connecting nine points. The Article however adds that 'in the event that Thailand establishes her exclusive economic zone, this same line shall also constitute the boundary between the exclusive economic zone of Burma and the exclusive economic zone of Thailand'.[188]

The 1978 Agreement between Australia and the Solomon Islands provides that both the shelf and the Australian FZ and the Solomon Islands' EEZ boundaries should follow the same median line between the outermost islands on each side.[189] Similarly, the 1982 Agreement between

[181] S. Exec. Rep. No. 96 (1980); Schmitt, 'The Problem of Maritime Boundaries', 147.

[182] See argument of M. Feldman before the Senate, S. Exec. Rep. No. 96 (1980); Szekely, 'A Commentary With the Mexican View on the Problem of Maritime Boundaries in U.S.–Mexico Relations', 22 *NRJ* (1982), 154, at 158.

[183] See generally P. Rao, *The New Law of Maritime Zones with Special Reference to India's Maritime Zones* (Millind Publications, New Delhi, 1983), 150–4.

[184] Art. V(2); See *LIS* 77.

[185] Analysis of the US Office of the Geographer, *LIS* 77: 7.

[186] Jagota, *Maritime Boundary*, 79–83.

[187] See *LIS* 102.

[188] Art. 1(3). Thailand established its exclusive economic zone on 23 Feb. 1981. Burma had established its zone on 9 Apr. 1977. *LIS* 36, respectively 167, 31.

[189] Bassett, *Australia's Maritime Boundary*, 190. However, the two boundaries depart

Australia and France is established on the basis of equidistance and extends up to 200 miles.[190] The continental shelf boundary appears to be the same, as the Agreement provides that for the purposes of shelf rights (supposedly beyond 200 miles) the boundary may later be extended.

The practice of States in the Caribbean is an important source of enquiry, not only because of the number of maritime boundaries established in this semi-enclosed basin, but because of the variety of the factors involved.[191] Nine negotiated agreements have been concluded delimiting marine and submarine boundaries, either among island States or involving an island State. In all the agreements the principle of delimitation is mentioned expressly. Thus the Agreements between Cuba and Mexico (1977),[192] Colombia and Haiti (1978),[193] Colombia and the Dominican Republic (1981),[194] France and St Lucia (1982),[195] and Costa Rica and Panama (1982)[196] refer to the equidistance principle directly as the basis for the boundary line. In the Agreement between Venezuela and the Dominican Republic (1979)[197] the preamble refers to delimiting the boundary 'justly, accurately, and on the basis of equitable principles'. The boundary line, however, appears to be determined on the basis of equidistance.[198] In the 1977 Agreement between Cuba and Haiti the delimitation is to be made 'on the basis of the principle of equidistance or equity, as the case requires'.[199] The boundary is a simplified median line, giving no effect to the straight baseline system of Haiti which closes, *inter alia*, the Gulf of Gonava.[200]

The 1978 Agreement between The Netherlands and Venezuela, establishing the maritime boundary between the Dutch Antilles and the Venezuelan mainland and islands, refers to the boundary as 'just, precise and based on equitable principles'.[201] The boundary consists of four sectors.

from one another near the northern end where the seabed boundary extends beyond 200 miles.

[190] The Agreement provides for the seabed–water-column boundary between the Australian Coral Sea Islands and Lord Howe and Norfolk Islands and French New Caledonia on the one hand, and Heard and McDonald Islands and the French Kerguelen Islands on the other hand, ibid. 190.

[191] K. Nweihed, 'EZ (Uneasy) Delimitation in the Semi-enclosed Caribbean Sea: Recent Agreements between Venezuela and Her Neighbours', 8 *ODIL* (1980), 1, at 4.

[192] Art. 1. Small offshore islands and cays were used in determining the equidistance line, *LIS* as 104.

[193] Art. 1, *New Directions*, viii. 76.

[194] Art. 1, ibid. viii. 78.

[195] Preamble, Jagota, *Maritime Boundary*, 105.

[196] Art. 1, *LIS* 97.

[197] *New Directions*, viii. 80.

[198] Jagota, *Maritime Boundary*, 103.

[199] Art. 1, *New Directions*, viii. 69–70.

[200] See *supra*, Ch. 1 n. 106.

[201] Nweihed, 'EZ (Uneasy) Delimitation', 23–9.

In sectors A, B, and C the boundary is a negotiated line. In sector D it is the median line. The Agreement includes a common deposit clause, whereby drilling by either party up to 1 n.m. from the boundary line is prohibited. Article 1 of the maritime Boundary Agreement between Colombia and Panama describes the boundary as the median line, 'with the exception of some minor deviations that have been approved to simplify the route'.[202] However, the boundary commences equidistant from the adjacent mainland territories, but when dividing the opposite territories of the mainland of Panama and the various cays belonging to Colombia, it follows no specific principle. It consists of a series of straight lines in a stepped formation. Several Colombian cays appear to have been ignored.

Of interest are the Soviet maritime Boundary Agreements with Finland and Poland. To date they represent the only examples whereby States agree formally to convert their former shelf boundaries to single maritime boundaries.[203] The original Agreements provided for modified equidistance boundaries.[204]

The following maritime boundary agreements represent almost the only negotiated lines which do not apply equidistance in their entirety.[205] These are the Agreements between Ecuador and Peru,[206] Peru and Chile,[207] and Colombia and Ecuador.[208] The boundary lines are drawn on parallels of latitude from a specific point near the land boundary terminus. The coastlines of the States involved exercise no influence on these lines. No boundary delimitation principles have been specified in these treaties. The boundaries appear to have been negotiated on the basis of acceptable solutions agreed upon by the parties.

As was seen earlier, a number of States have agreed on a joint zone solution either across the boundary line or without the delimitation of the entirety of their respective jurisdiction.[209] What impact these joint zones (exclusively related to hydrocarbons) will have on prospective exclusive zone or single boundary delimitation between the same States is an interesting question. If one were to speculate, a solution would seem to be largely contingent on the relative importance of the fishery resources contained in the exclusive zone for either or both of the States involved. Perhaps it will become important for the States to develop a mechanism

[202] Agreement of 1977, *LIS* 79: 6–7.

[203] USSR–Poland, 13 Mar. 1986; USSR–Finland, 24 Nov. 1986, *LIS* 108: 26.

[204] USSR–Poland, 19 Oct. 1973, *LIS* 55; USSR–Finland, 19 Oct. 1973, *LIS* 56.

[205] See also the 1977 Boundary Agreement between Colombia and Costa Rica, delimiting their maritime zones in the Caribbean, *LIS* 84, analysis, p. 5.

[206] Declaration of 18 Aug. 1952 between Chile, Ecuador, and Peru in *LIS* 88.

[207] Ibid. see analysis in *LIS* 86.

[208] Maritime Boundary Agreement of 1975, *LIS* 69.

[209] *Supra*, Ch. 4 s.4(*b*).

whereby their respective fishery and hydrocarbon interests may coexist. A collective solution, such as that taken by the European Commission within the framework of the EEC for, *inter alia*, the North Sea,[210] is a complicated but relevant example.

By analogy to several negotiated single boundary agreements, States may feel impelled to treat the question of fishing as one of resource conservation and control rather than of delimitation. Thus, in order to protect the existing fishing rights of the local fishermen, the 1976 'Maritime Boundary'[211] Agreement between Kenya and Tanzania establishes a common fishing zone within 12 miles on either side of the boundary line.[212] The line is defined as equidistant from specified baselines in the Pemba Channel. To the north of the Pemba Channel, the boundary follows a parallel of latitude to an unspecified seaward limit. Both States can extend their jurisdiction to a full 200-mile limit.

The 1973 Argentina–Uruguay Boundary Agreement establishes 'an equidistance line, determined by the adjacent coasts method'[213] to delimit the continental shelf. However, rather than delimit their exclusive fishery zone, the Treaty establishes a common fishing zone seaward of 12 miles and further provides that the 'volumes of catch, by species, shall be determined equitably, in proportion to the ichthyological resources of each of the Parties, evaluated on the basis of scientific and economic criteria'. The Treaty provides also that the parties 'shall respectively exercise appropriate functions of control and supervision on both sides of the [boundary line]'.[214] Article III of the 1978 Agreement between Colombia and the Dominican Republic establishes a joint scientific research and fishing zone, which straddles the boundary, extending within 20 n.m. on each side thereof.[215]

The 1978 Australia–Papua New Guinea Agreement provides a comprehensive scheme for boundary and other related issues between the two States. It deals with questions of sovereignty over islands, maritime boundaries concerning the respective territorial sea, continental shelf, and the fishery zone, and the establishment of a 'Protected Zone' for protecting the traditional rights of the islands and coastal inhabitants.[216]

210 See *infra*, Ch. 9 nn. 4, 220.

211 Although described as dividing the 'Territorial Sea' boundary, the Agreement provides, in effect, for a multi-purpose boundary. The boundary extends beyond 12 miles to cover the Tanzanian claimed territorial sea of 50 miles. Furthermore, the Agreement does not specify the boundary terminus, apparently for future shelf/EEZ boundary delimitations. Kenya claims a 200-mile EEZ. See *LIS* 92.

212 Jagota, *Maritime Boundary*, 97.

213 Art. 70, *LIS* 70.

214 Art. 76.

215 Agreement of 13 Jan. 1978, *LIS* 105.

216 The Agreement is repr. in 18 *ILM* (1979), 291 ff. For full analysis see H. Burmester, 'The Torres Strait Treaty: Ocean Boundary Delimitation by Agreement', 76 *AJIL* (1982), 321; K. Ryan and M. White, 'The Torres Strait Treaty', 7 *AYIL* (1976–7), 87.

The Treaty establishes the continental shelf boundary extending to approximately 1,200 miles. The shelf boundary is a modified median line. The fifteen Australian islands which lie to the north of the shelf boundary have been given a 3-mile territorial sea around them. In the area of overlap close to the mainland of Papua New Guinea the territorial sea boundary follows the median line. The fishery jurisdiction zone boundary follows the shelf boundary, except in the centre of the Torres Strait where it diverges so as to tie the three Australian island groups off the coast of Papua New Guinea with the fisheries jurisdiction zone of Australia. This was done in recognition of the importance of fishing to the inhabitants of these Australian islands close to Papua New Guinea. The Treaty establishes a Protection Zone cutting across the fishery and the shelf boundaries for protecting the socio-economic life of the aborigines inhabiting islands in the Torres Strait. Neither Australia nor Papua New Guinea had proclaimed a full exclusive economic zone.

In the Agreement between Iceland and Norway (Jan Mayen)[217] (putting into effect the recommendations of the 1981 Conciliation Commission[218]) the parties agreed to extend the existing EEZ boundary to the continental shelf, thus effectively converting it to a single boundary. The 1980 EEZ Boundary Agreement provided that Iceland's 200-mile EEZ had full effect and therefore formed the boundary where the distance between Iceland and Jan Mayen was less than 400 miles. However, the 1982 Shelf Agreement established a joint fishery commission to manage joint fish stocks in the Jan Mayen area. It also laid down co-operative arrangements for the exploration and exploitation of hydrocarbons in a rectangular area which cut across the boundary line.[219]

(*c*) *Conclusion*

Whilst States tend to maintain the intrinsic autonomy of continental shelf and EEZ regimes, such an autonomy does not seem to exert any special influence on the rules relating to their delimitation by a common boundary. Indeed, their practice in this regard reflects a continuation of their 'experience' in territorial sea and shelf delimitations. The equidistance–special circumstances rule has proved adequate in implementing what States agree upon as equitable. In comparison, however, the socio-economic factor has played a more prominent role, naturally because of the relevance of fishery resources in this kind of boundary.

[217] Agreement on the Continental Shelf between Iceland and Jan Mayen, 22 Oct. 1981. In force 2 Jan. 1982. English text in 21 *ILM* (1982), 1222.

[218] See *supra*, Ch. 3 n. 72.

[219] For details see R. Churchill, 'Maritime Delimitation in the Jan Mayen Area', 9 *MP* (1985), 16, at 22–3.

The last five agreements considered above are of interest. They demonstrate (1) that States do not feel a single boundary is essential or obligatory, and (2) that it is possible to devise mechanisms which alleviate the problems arising from islands on the 'wrong side' (Australia–Papua New Guinea) or overlapping jurisdictions (Iceland–Jan Mayen). Furthermore, the 'common fishery zone' solution in the Australia–Papua New Guinea, Argentina–Uruguay, and Kenya–Tanzania Agreements illustrates that the question of fisheries might be dealt with as one of resource allocation and management, and not necessarily as one of delimitation. The credibility of such solutions depends on the readiness of the States involved to allocate their political, economic, and scientific resources to deal with ensuing problems of supervision, resource allocation, quotas, registration, etc. These solutions cannot, however, be regarded as providing sufficient evidence that factors relating to fishing such as socio-economic or historic dependency are of reduced importance in the location of the delimitation line.[220] The consideration of such factors by negotiating solutions such as those provided in the aforesaid agreements is but part of the whole arrangement which finally enables the States involved to agree on a maritime boundary.

[220] See e.g. Collins and Rogoff, 'The International Law of Maritime Boundary Delimitation', 1.

PART II

Practice of States in the Mediterranean Sea

6

Delineation of Baselines in the Mediterranean

1. INTRODUCTION

The Mediterranean coastline totals 13,890 miles, but its distribution is irregular, with almost 8,077 miles along the northern sector contrasting with only about 3,107 miles along the southern sector and the remainder along the Levantine coast. This variability in length between the two sectors is due to differences in configuration. The southern sector is relatively uniform, running generally east–west with only one major indentation—from Cape Bon on the Tunisian coast along the Libyan coast to approximately Benghazi in eastern Libya—whereas the northern sector is highly irregular, running in a wide variety of directions, but predominantly in either an east or west direction. The northern sector is also marked by the presence of islands, varying in their size, proximity to the mainland coast, and political status. However, the geography of the 'local' coasts of the States abutting the northern Mediterranean varies considerably. Some, such as the coasts of Greece and Yugoslavia, are highly indented and dotted with islands; others, like those belonging to Italy on the Adriatic Sea and Spain on the Alboran Sea, are generally regular, with a limited number of islands lying in their vicinity.

Because of this geographic diversity it is not surprising to find that the nineteen Mediterranean States and territories have adopted various practices in relation to the choice of their respective baselines. Twelve States use the low-water line in combination with straight and closing baselines to enclose coastal configurations, islands, juridical bays, and historic waters; four adopt only the low-water line; and three do not adopt a specific baseline officially, although the geography of their coasts gives them little choice to depart from the standard rule of the lower-water line. In all only seven have officially determined 'the low-water line' by reference to a specific low-tide datum.[1]

Having said that, a critical study of the application of these baselines on the various Mediterranean coasts is a matter to which we must now turn.

[1] Egypt and France, lowest low water; Lebanon, Spain, Syria, and Turkey, lowest low tide; UK (Gibraltar and the Sovereign Base Area in Cyprus), mean low-water springs.

For reasons of convenience, however, enclosures on the basis of a 'historic' title in the Mediterranean will be described while examining the claim of Libya over the Gulf of Sert as a 'historic bay' in the next chapter.

2. THE LOW-WATER LINE SYSTEM

Cyprus, Lebanon, Monaco, Israel, and the United Kingdom (the Sovereign Base Area in Cyprus) all adopt the low-water line. Although the prerequisites for using the straight baseline method probably exist for segments of the coast of Cyprus, Article 3 of the Law of 1964[2] states that the territorial waters of the Republic are measured from the low-water line. The coasts of Monaco and Israel are good examples of the application of the low-water line, for they are not deeply indented, nor is there a fringe of islands in their immediate vicinity. According to the *Ordonnance souveraine* of 1973, the territorial waters of Monaco are measured from 'la ligne de base formée par la laissée de basse mer longeant la côte'.[3] The baseline for the measurement of the territorial sea of Israel is determined by the low-water line.[4]

The practice of Lebanon[5] posed, until very recently, a special problem, for there was no specific legislation regarding its territorial sea limits.[6] In 1983, Lebanon fixed its territorial waters at 12 n.m. measured from the low-water line.[7]

The two areas of the coast of Cyprus belonging to the United Kingdom (the Sovereign Base Area) mainly lie within two smoothly curved segments formed by Larnaca Bay (the eastern area), and the Akrotiri and Episkopi Bays (the western area). According to section 3(1)(2) of the 1960 Treaty between Cyprus and the United Kingdom[8] the delimitation

[2] Law No. 45 of 6 Aug. 1964. Text in *WEDLS* (1983) (Binder I) (Cyprus), 5. There are two major indentations in the north-western coast of Cyprus: the Bays of Norphon and Khysochou, with widths of 18 and 14 miles respectively. Although they do not meet the semicircle test, the Bays are fringed by several rocks and islets such as Mazall and Kakoskallon. See *Mediterranean Pilot*, 5 (1976), 101–14.

[3] *Ordonnance souveraine* No. 5,094 of 14 Feb. 1973. Text in *WEDLS* (1983) (Binder II) (Monaco), 5.

[4] Art. 2 of Territorial Waters Law No. 5717 of 23 Oct. 1956, text in *UNLS* (ST/LEG/SER.B/18) (1976), 26. See Art. 1 of the Interpretation Ordinance No. 5741, 1981, *Laws of the State of Israel* (1981), 370.

[5] The coast of Lebanon is generally quite regular. There are, however, a few islet and shoal patches near the coast of Tripoli in the north and south of Ras abu Sait in the south. For details see *Mediterranean Pilot*, 5 (1976), 117–26.

[6] See A. El-Hakim, *The Middle Eastern States and the Law of the Sea* (Manchester University Press, Manchester, 1979), 7.

[7] Decree-Law No. 138 of 16 Sept. 1983, Lebanon *Official Gazette*, 45 (10 Nov. 1983.)

[8] See *infra*, Ch. 8 s.3(*a*).

of the territorial sea of the Base Area is measured from the low-water line.

Gibraltar's baseline represents a particular problem. As the metropolitan State, the United Kingdom has promulgated no specific legislation regarding the question of the territorial sea of the Rock. This question, however, has been part of the larger dispute over sovereignty between Spain and the United Kingdom due to differences in the interpretation of Article 10 of the Treaty of Utrecht of 1713. The problem will be discussed elsewhere.[9]

The legislation of Libya does not expressly mention a specific baseline. Article 1 of the Libyan Act of 1959[10] fixes the territorial sea at 12 n.m. but does not mention the nature of the baseline from which this distance is measured. Only a small segment of the coast of Libya has been determined by a closing line along the mouth of the Gulf of Sert, a claim which will be considered in detail later.[11] For the rest of the coast, however, both conventional and customary laws, as mentioned earlier, consider the low-water line as the normal or standard rule of general application, as opposed to the other rules that may be invoked only in certain circumstances.[12] The application of this 'normal' rule on the coast of Libya seems also to be implied in Libyan law prior to the 1959 Act. Until then, the territorial sea of Libya extended 'six miles *from the coast*, as was the case before Libya's independence'.[13] The term 'coast' can at least be interpreted to mean some point between the low-water and high-water lines.

The legislation of Greece also represents a similar problem. Article 1 of the Law of 1936[14] gives an imprecise definition of the baseline. It merely provides that the breadth of the territorial sea of Greece is fixed at 6 n.m. 'from the coast'. Related laws and regulations also refer to certain distances measured from 'the coast'.[15] However, at the time of ratifying the 1958

[9] See *infra*, s.5(*c*)(ii).

[10] Act No. 2 of 18 Feb. 1959. Text in *UNLS* (ST/LEG/SER.B/16) (1974), 14.

[11] See next chapter.

[12] *Supra*, Ch. 1 s.2.

[13] Note from Libyan Ministry for Foreign Affairs, Nov. 1955. Text in *UNLS* (ST/LEG/SER.B/6) 1957 (emphasis added).

[14] Law No. 230 of 17 Sept. 1936. Text in *WEDLS* (1983) (Binder II) (Greece), 15. As for future development, the legal requirements for establishing straight and closing lines do exist for most of the Greek coast, which is highly indented and, in many localities, fringed by numerous islands. The utilization of these features by Greece to draw a system of straight baselines would include extensive areas of waters, especially in the Aegean Sea, where Greece and Turkey are currently involved in a dispute over the delimitation of maritime boundaries (see *infra*, Ch. 8 s.4(*b*)(i) and Ch. 10). Currently only a Neutrality Act, promulgated in 1913, provides for the enclosure (whenever necessary for reasons of security) of bays and gulfs the width of which does not exceed 20 miles. Text ibid. 1.

[15] See e.g. Law No. 4,141 of 26 Mar. 1913; Law No. 1165 of 17 Mar. 1918; Presidential Decree of 6/18 Sept. 1931. Text ibid.

Shelf Convention Greece made a reservation which stated that, for the purpose of Article 6 of the Convention, Greece 'will apply, in the absence of international agreement, the normal baseline system for the purpose of measuring the breadth of the territorial sea'.[16] It is thus reasonable to deduce, considering also the reasons just mentioned with regard to Libya, that the low-water line constitutes the *de facto* baseline for Greece.

3. COMBINATION OF THE LOW-WATER, BAY CLOSING, AND STRAIGHT BASELINES SYSTEMS

One of the serious difficulties in the study of straight baselines practices in the Mediterranean is the diversity of the relevant rules adopted by the twelve Mediterranean States which have claimed or enacted for straight baselines. This diversity is not only due to the physical variations in the formation of the States' respective coasts, but is also influenced by political choices as to whether or not to observe the relevant conventional rules laid down by the 1958 Territorial Sea Convention. Of all the twelve claimants, four are parties to this Convention. Another difficulty is that some of these claims have been promulgated without official charts being published to show the actual locations of the baselines along the coast. To overcome this difficulty, a hypothetical study of these claims is carried out herein, using available charts.[17]

In order therefore to assess the conformity of the practice of the Mediterranean States on baselines with the 'rules of law', it is thought necessary to deal with these practices one State at a time, starting from the coast of Spain and moving from there to the rest of the coasts of the Mediterranean basin in a clockwise direction.

(*a*) *Spain*

Spain has established, 'in accordance with prevailing international law and specifically with the Geneva Convention of 1958',[18] straight baselines along almost all its Mediterranean coasts. Article 1 of the Royal Decree of 1977[19] provides for drawing 'straight baselines' on, *inter*

[16] 847 *UNLS* (1972), 338.

[17] This study is conducted mostly in combination with British Admiralty Charts, published by the Hydrographer of the Navy (Taunton). Data regarding the physical characteristics of the Mediterranean coasts are mainly based on the studies conducted throughout the Mediterranean Sea in *Mediterranean Pilot*, 1–5.

[18] See Preamble to Act No. 10/77 of 4 Jan. 1977 concerning the Territorial Sea. Text in *New Directions*, vii. 62.

[19] Royal Decree No. 2510/1172 of 5 May 1977, text ibid. vii. 64. Art. 2 of this Decree states that Decree No. 627/1976 of 5 Mar. 1976 concerning the drawing of baselines, which by virtue of the 1977 Act was transitionally applied, is rescinded.

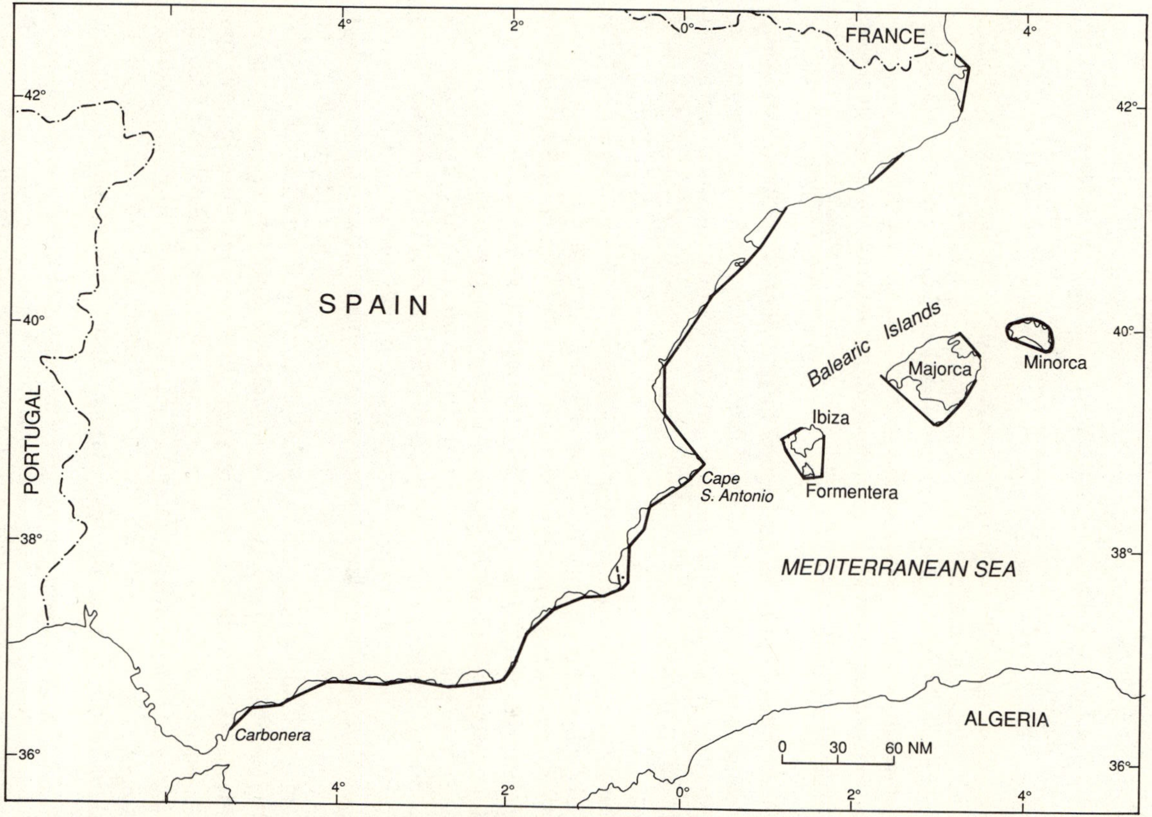

Source: D. Francalanci *et al*. (eds.), *Atlas of Baselines* (Giuffrè Editore, Milan, 1986).

MAP 2 Straight baselines of Spain (Mediterranean)

alia, the continental Mediterranean coast and the Balearic Islands. The Article states that those 'baselines' are drawn for the delimitation of Spanish jurisdictional waters referred to in Article 2 of the 1967 Act. The Act itself provides explicitly for the establishment of straight lines *and* closing lines joining the low-water lines of the natural points of entrance or opening of a bay provided that such an opening or entrance does not exceed 24 miles.[20]

The wording of the Decree, however, makes it extremely difficult to distinguish straight baselines from lines closing bays. Therefore the straight baselines which are determined by the Decree have to be examined in the light of the requisites of both Articles 4 and 7 of the 1958 Territorial Sea Convention to which Spain is a party.

It is convenient to divide the Spanish mainland coast into two segments (see Map 2). The first one extends for about 325 miles between Punta Carbonera and Cape S. Antonio. The coastline of this segment is gently indented and there are some islands in the vicinity of the coast. Along this segment thirty individual lines are drawn connecting points on the mainland as well as on the small islands of Homiga (2.2 miles offshore), Tabaraka (3 miles), and Portichol (lying north of Cape Nao). The lines do not deviate from the general direction of the coast, but in many places the coastline is too straight to justify the use of straight baselines.[21] The lines also close three major indentations: the Gulfs of Almeria, Valencia, and S. Jorge. The closure departs from the requirements of Article 7 of the Territorial Sea Convention, for although the closing line of each is less than 24 miles in length, none of them meets the semicircular test.

The second segment extends for about 270 miles from Cape S. Antonio to the French border. The coast along this segment possesses no indentations even for natural harbours.[22] There is only one group of volcanic islets and shoals: the Isoletes Columbretes, lying 27 miles offshore. Nevertheless most of the baselines of this segment are determined by ten straight baselines all drawn on points on the headlands, except for one turning-point situated on the islet of Maza de Oro. However, along the continental coast of Spain, the low-water line is still the baseline for only two segments: one from Cape Salou to Barcelona (the light) and the other from Arens de Mar (the end of the breakwater) to Cape Bagur.

Straight baselines are also drawn to determine the baseline of the Balearic Islands.[23] Two islands, Majorca and Cabrera, are tied together

[20] Art. 2 of Act No. 20 of 8 Apr. 1967. Text in *UNLS* (ST/LEG/SER.B/15) (1970), 668.

[21] Between Punta de Loma Palada and Monte, for example, the coast of the Gulf of Vera is almost regular; nevertheless, three legs of straight baselines are fixed along this segment.

[22] See *Mediterranean Pilot*, I (1978), 76.

[23] The enclosure of the Balearic Islands is of special interest. It is the only practice recorded in the Mediterranean that reflects the tendency among continental States to surround their offshore archipelagos with straight baselines. Art. 131 of the Informal

by four lines which surround the sea area around these islands almost entirely. On the north-eastern coast of Majorca there is a line of 16.8 n.m. This line does not only close the Bays of Pollensa and Alcudia, being the major indentations on this stretch, it also extends south-east to enclose a considerable part of the coast that ceases to exhibit the general characteristics required to justify a straight baseline system, a practice which seems to contradict the term 'in localities' of Article 4 of the 1958 Convention, which implies that a State would not be justified in making use of straight baselines along the whole of its coast, irrespective of the physical and geographical conditions, merely because those conditions warranted the use of baselines in one particular locality.

Another line of 14.2 n.m. is drawn along a minor curve on the eastern coast of Majorca. The curve is so gentle that the line runs almost parallel to the coast, thus having a minimum effect on the outer limit of the territorial sea. The third line extends for 19.4 n.m. along the eastern coast of the islands. The coast in this locality is neither deeply indented nor fringed by islands. The same remark is also applicable to the last line which ties the western coast of Cabrera with the north-western coast of Majorca. The line extends for about 34.8 n.m. and corresponds in direction with the south-western coast of Majorca, which possesses only one major indentation (the Bay of Palma). However, the low-water line is still the baseline along the north-eastern coast of Majorca (between Cape Formentera and Cape Llebeitx) where the coastline is regular.

Unlike the straight baselines of the islands of Majorca and Cabrera, most of the six baselines which unite the islands of Ibiza and Formentera seem to conform with the rules of Article 4. The only questionable line is that drawn between the islet of Tagomago north-east of Ibiza and the eastern coast of Formentera. Bearing in mind the relative narrowness of the Balearic Sea,[24] the line (drawn between two points so far out (22 n.m. apart)) does not truly reveal any relation with the coasts of the islands and, as a result, encloses a band of waters of about 200 sq.m. However, the other five lines are relatively short and are all connected with turning-

Negotiation Text of 1977 explicitly provided for such an enclosure. The Article was removed from subsequent drafts, despite the effort of some countries, including Canada, Ecuador, and India, to include it. However, it is beyond doubt that such a practice would still be legitimate under the provisions of Art. 121 of the 1982 Convention (Art. 10 of the 1958 Convention), which makes it clear that the territorial waters claimed from islands must be determined in accordance with the provisions of the Conventions which apply to other land territory. The rules for drawing straight baselines set out in Arts. 7 (1982 Convention) and 4 (1958 Convention) therefore apply to the islands which belong to mainland States.

[24] The Balearic Sea and its related area is about 29,664 sq. m. and borders three coastal States plus Gibraltar. See T. G. Carter, 'The Balearic Sea and the Algerian Basin', in D. Stanley (ed.), *The Mediterranean Sea: A Natural Laboratory* (Hutchinson & Ross, Stroudsburg, Pa., 1971), 9.

points on headlands as well as on fringing islets, such as Tagomago, Vedra, and Gleda Blana.

The baselines of Minorca, determined by six individual legs of straight baselines, comprise about 90 per cent of the island's coastline. The lines vary in length from 1.9 to 11.7 n.m. and are also drawn on headlands which alternate with smoothly curved localities, without departing appreciably from the general direction of the coast; and, as a result, have little effect on the extension of the outer limit of the territorial sea.

There are no special provisions either in the relevant Acts or the 1977 Decree concerning the 'Spanish enclaves' on or near the coast of Morocco. Reasons for this silence are perhaps related to the particular geographical position of these enclaves as well as the controversy with Morocco over their sovereignty. The problem will be discussed elsewhere.[25]

In conclusion, whilst certain extensions of the Spanish straight baselines system deviate from Article 4 of the 1958 Territorial Sea Convention, this deviation is effectively minimal, for the lines are generally short and drawn on relatively regular coasts, and as such have a very negligible effect on the extension of the internal waters of Spain.

(*b*) *France*

The French Decree of 1967[26] determines the straight baselines and the closing lines of bays from which the breadth of the territorial sea is measured. Article 1 of the Decree speaks of the combined use of 'straight baselines and . . . closing lines of bays', but the language of the Decree does not distinguish between the lines which are meant to close bays and those which are drawn on deeply indented coasts and around fringing islands. As a result, it is necessary to bear in mind the requisites of Articles 4 and 7 of the 1958 Convention on the Territorial Sea to which France is a party.

It is convenient, for the purpose of analysis, to divide the straight baselines of the continental Mediterranean and Corsican coasts into the following sectors (see Map 3):

Sector 1. A single line of 9.4 n.m. encloses the Gulf of Aigues-Mortes between the mouth of Grau de Palaves and the Espiguitte lighthouse. The Gulf is neither a juridical bay nor a deeply indented one. The baseline of the regular stretch eastward up to the border with Spain is given by the low-water line.

[25] See *infra*, s.3(*l*).

[26] Decree No. 255 of 19 Oct. 1967, repr. in 71 *RGDIP* (1967), 1165. Art. 1 of Law No. 71-1060 of 24 Dec. 1971 states that 'the low-water line, the straight baselines and the closing lines of bays, which are determined by Decree', constitute the baseline of France's territorial waters, *JORF* (30 Dec. 1971), 12899.

Sector 2. A single line of 8.25 n.m. which encloses the Gulf of Saintes Maries between the Grau d'Orgon and the Beauduc beacon. The Gulf is neither deeply indented nor does it meet the semicircular test.

Sector 3. Eight broken lines running from the western entrance to 'the Gulf of Fos' and finishing at the Toulon roadstead. The lines enclose embayments of an indented coast, utilizing islands such as Planior, Grand Rouveau, and Embiez, all in the immediate vicinity of the coast. The only line which seems to be questionable is the 9 n.m. line which encloses the Gulf of Fos, as the coast is neither deeply indented nor does it form a juridical bay.

Sector 4. Extends between east of the Toulon roadstead and the Cape of Camarat. The coast is deeply indented and fringed by islands, islets, and rocks. Seven lines have been constructed enclosing all the offshore islands. Further east, there are three major detached lines. One (10.25 n.m.) closes the Baie des Anges and the Baie de Villefranche. Calculated together, the two Bays are widely open with no off-lying islands and form a mere curvature of the coast. The second line further out closes the Baie de Beaulieu (about 3 n.m.). Further eastwards, the baseline is defined by the low-water line. The third line (2.6 n.m.) encloses the Baie de Roguebrune, a rather smooth indentation. Between these latter segments, the baseline is defined by the low-water line, apparently to prevent cutting off the territorial sea of the principality of Monaco (situated in between for about 1.69 miles) from the high seas or a prospective exclusive economic zone. This restriction is provided for in Article 4(5) of the Territorial Sea Convention (Article 7(6) of the 1982 Convention).[27]

Sector 5. Three segments of detached straight baselines are drawn along the western and south-western coasts of Corsica. The coast of the island where the lines are drawn is deeply indented with numerous bays alternating with rocky headlands and off-lying islands and islets. The first segment is a single 8.7 n.m. line which closes off the juridical bay of St Florent. The second segment is also a single line which closes the Gulf of Calvi, a juridical bay with an entrance of 3.45 n.m. The third segment comprises fourteen continuous lines drawn on twenty-five basepoints, and encloses the remainder of the western and south-western coast where there exist several bays and gulfs; some of them such as Porto and Ajaccio are juridical, but others are not, mainly because their enclosing lines deviate from the natural entrances to include other segments of the coast.[28] Along the east coast of Corsica the baseline coincides with the low-water line.

[27] See *supra*, Ch. 1 s.2(*b*).

[28] The line closing the Gulf of Volinco, for instance, ignores the natural entrance of the Gulf between Cape Pollo and Campa Maro Point.

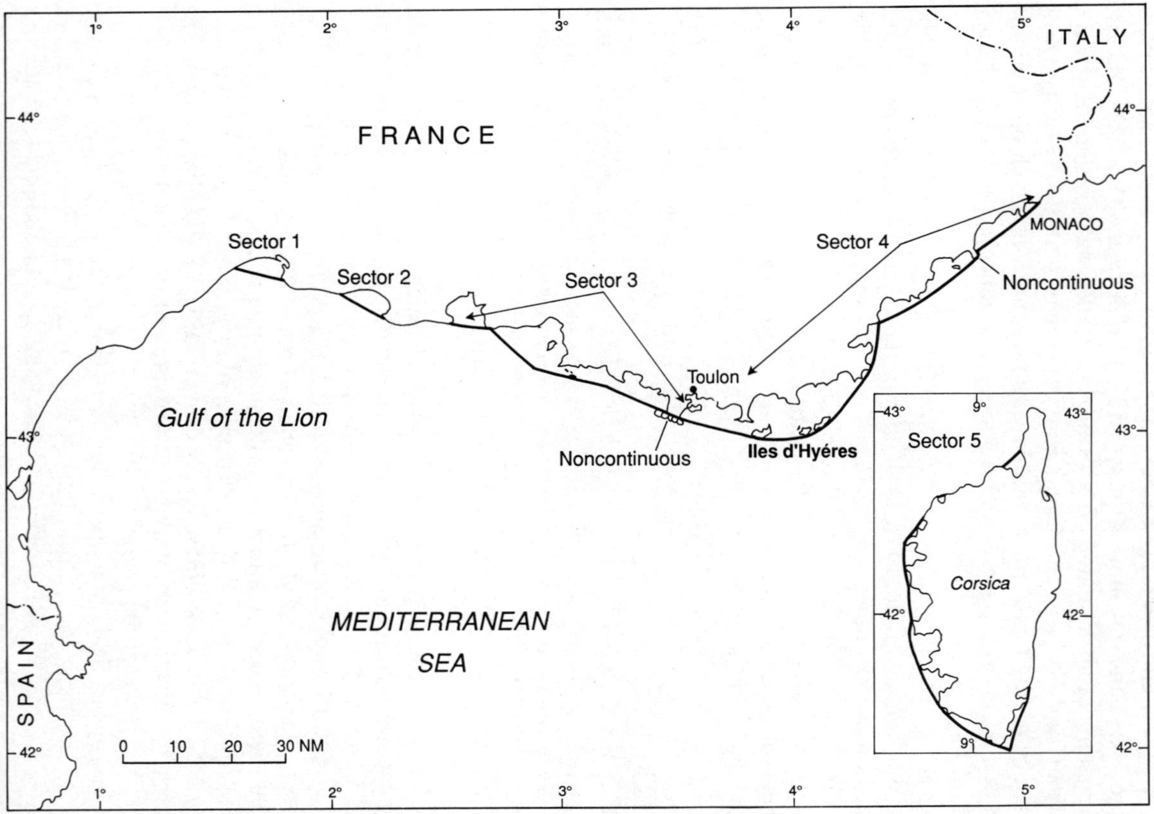

Source: Francalanci *et al.*, *Atlas of Baselines*.

MAP 3 Straight baselines of France (Mediterranean)

On the whole, the straight baselines of the Mediterranean coasts of France seem to be in line with the requisites of the closing and straight baselines systems. In a few cases, smooth curves are enclosed contrary to both systems. It remains, however, to mention that even with these latter cases the enclosures of small bays and gulfs merely reinforce an 1888 Decree which provided for closing, *inter alia*, most of the bays and gulfs mentioned above.[29] The long-standing enclosure by France of such curvatures with no apparent objection by other States[30] would seem to present a good claim on the basis of the concept of 'historic bays'.

(*c*) *Italy*

The Italian Decree of 1977[31] represents a major shift from the long-standing Italian practice of following the low-water line.[32] The Decree provides, *inter alia*, for the establishment of a system of straight baselines along specific localities on the peninsula as well as on the islands of Sardinia and Sicily (see Map 4). The system consists of 147 individual lines, situated on 185 turning-points. As the State consists of a peninsula, two main islands, and coastal archipelagos, Italy has used the system to enclose indentations, including bays and gulfs, as well as coastal archipelagos. It is the enclosure of these two distinct geographical features that seems to raise problems in conformity with the relevant rules of the 1958 Territorial Sea Convention, to which Italy is a party.

(i) Straight baselines closing coastal archipelagos

In the southern Adriatic, two straight baselines enclose the triangular area of sea between the port of Termoli, the point of M. Pucci on the mainland (the Gargano promontory) and the Tremiti group (four islands), the furthest lying about 15 n.m. from the coast of the mainland. These islands are so near to each other that they appear (to a navigator approaching the coast) as one island and, as such, cannot be easily regarded as fringing the coast, which itself is devoid of any indentation.

In the Tyrrhenian Sea, the closure of coastal archipelagos seems to be much more questionable. Straight baselines are drawn to contain the island of Capri and extend further west-north-west to close off the north-

[29] In execution of the Act of 1 Mar. 1888, which provided for the enclosure of bays the entrances of which do not exceed 10 miles, a Presidential Decree of 9 July 1888 listed twenty-one curvatures situated on the Mediterranean coast of France and Corsica, none of which exceeds 10 miles in width. Text in *UNLS* (ST/LEG/SER.B/6) (1957), 15–17.

[30] See L. Bouchez, *The Regime of Bays in International Law* (A. W. Sythoff, Leyden, 1964), 54–6.

[31] Presidential Decree No. 816 of 26 Apr. 1977. Text in 3 *IYIL* (1977), 570.

[32] The low-water line still prevails along most of the Adriatic coast. Few segments of the baseline along the eastern coast of Sicily and the south-western and north-western coast of the peninsula also coincide with the low-water line.

western group of the Pontine Islands, lying approximately 20 n.m. from the nearest point of the mainland coast. The group (three islands), seen from the nearest coast, is arranged like stepping-stones, almost at right angles to the general direction of the coast. Because of this distance and such an arrangement, it would be difficult to argue that the group constitutes a fringe of islands along the coast in its immediate vicinity. Moreover, the last segment, which unites the westernmost island of the group (Palmarola) with the mainland, extends in a north-eastern direction for a considerable distance and, as a result, enlarges the sea area of the internal waters significantly.

The last coastal archipelago to be considered is that of the Tuscan islands. These are eight islands and islets situated at various distances between the Tyrrhenian coast of the peninsula and the east coast of Corsica. Elba, the biggest island, is only 5 n.m. from the coast, whereas Sc. d'Africa and Capraria, the westernmost islands enclosed by the system, are respectively 40 and 25 n.m. from the coast. The archipelago is entirely enclosed by an almost continuous belt of straight baselines, which extend the outer limits of Italian internal waters to areas as far west as about 20 n.m. from the eastern coast of Corsica. This enclosure was challenged by France when the two States considered in the 1970s the delimitation of their maritime jurisdiction.[33] Seen in connection with the general direction of the coastline, the lines drawn around the Tuscan archipelago form a kind of wedge which stretches out towards the east coast of Corsica. It does not seem at all possible to argue that the drawing of such lines corresponds to the general direction of the coast. Nor is it evident that such a result is permissible under the general premiss of 'economic interests peculiar to the region', since there seem to exist no peculiar socio-economic circumstances associated with sea-related activity in these regions. For instance, it is reported that fishing near this Italian archipelago is poor and that there never has been evidence that the inhabitants of the region have substantial long-standing interests in fishing.[34]

(ii) Lines closing geographical bays and gulfs

Although Article 1 of the Italian Decree speaks of 'straight baselines and lines closing natural or historical bays', it does not indicate which system of the 1958 Territorial Sea Convention is used as a basis for drawing these lines, except for the Gulf of Taranto, which the Decree defines as a

[33] See *infra*, Ch. 9 s.2(*d*)(ii).

[34] Adam, commenting on this point and using data provided by the Italian Naval Ministry, concludes that 'the region in question [the Italian coastal archipelago] cannot . . . be made to benefit from the [economic interests] provision . . . either because of its character or due to the economic interests involved'. R. Adam, 'Un nuovo provvedimento in materia di linee de base del mare territoriale italiano', 61 *RDI* (1978), 469.

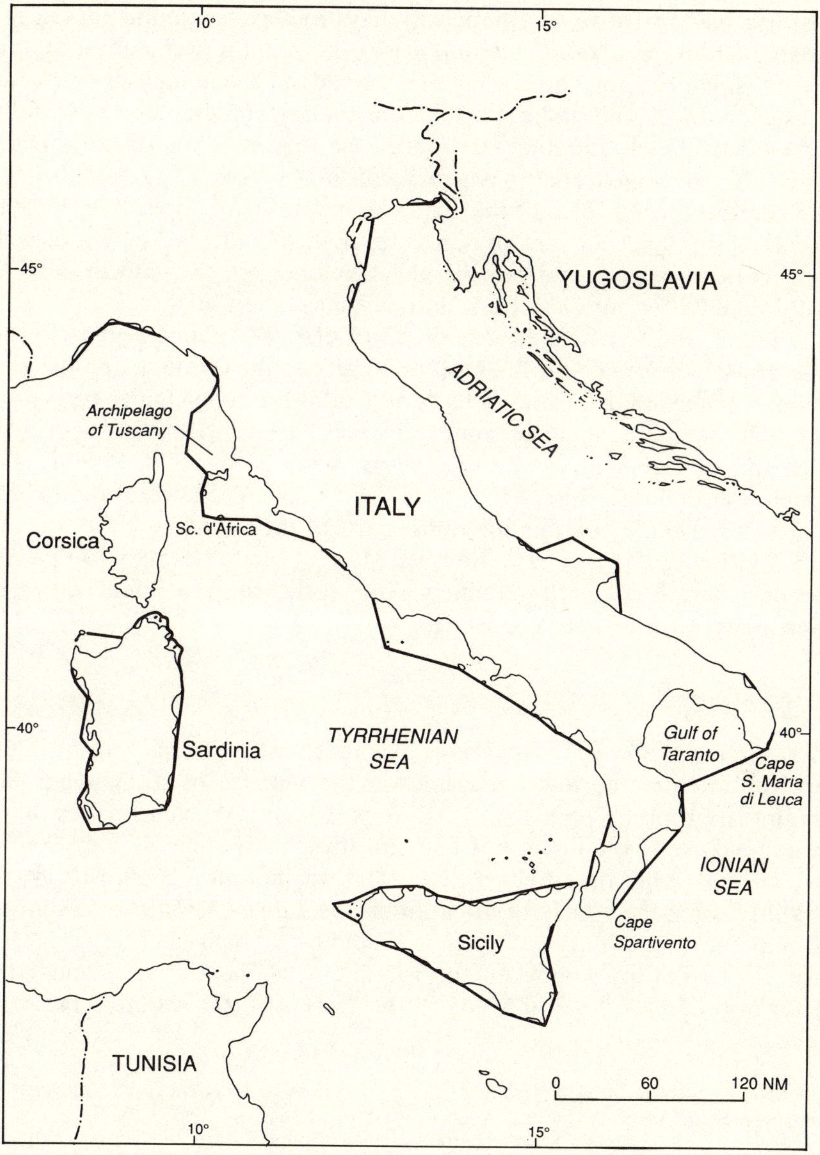

Source: Francalanci *et al.*, *Atlas of Baselines*.

MAP 4 Straight baselines of Italy

'historic bay'.[35] However, numerous bays and gulfs around the various coasts of Italy are closed. In some cases the natural entrance of the gulf or bay is ignored and the closing line extends to touch a basepoint which is further off.[36] This indicates that the method of the straight baseline rather than the closing line is primarily the reason behind this enclosure. Thus, the rules governing passage according to Article 5(2) of the 1958 Convention (Article 8(2) of the 1982 Convention) are applicable. In other cases, closing lines are drawn across the natural entrance points of some well-marked indentations which, nevertheless, fail to conform with the requirements for juridical bays. For instance, the Gulfs of Gioia and S. Eufemia in the Tyrrhenian Sea, the Gulf of Castellammare in Sicily, the Bay of Manfredonia, and the Gulf of Venice in the Adriatic are all closed by lines longer than 24 n.m. and do not fulfil the semicircular test.

Finally, the system closes some gulfs and bays which seem to be in line with the requirement for juridical bays, such as the Gulf of Policastro, which is closed by a line of 18.3 n.m. Two lines are drawn inside the Gulf of Trieste. These close the innermost parts of the Gulf where the coastline is not indented. Their effect, however, is very limited as they run parallel to the coast. Nor do they influence the territorial sea boundary in the Gulf between Italy and Yugoslavia.[37]

(*d*) *Malta*

Article 3(1) of the 1971 Act[38] provides that the territorial sea of Malta is 'measured from the low-water line on the method of straight baselines joining appropriate points'. It is obvious that the Article provides for the combined use of the low-water line and straight baselines systems, though it does this rather obscurely. However, it was not until 26 April 1983 that Malta, during the course of litigation in the *Libya–Malta* case, submitted a map showing twenty-six straight baselines[39] which enclose as internal waters the sea area between the islands of Malta, Comino, and Gozo. They also tie the island of Malta to the Rock of Filfla, where three of the

[35] The decision to close the Gulf of Taranto seems to be a typical case of enclosure for vital security interests. See next chapter, esp. s.3(*b*)(i).

[36] e.g. in the Tyrrhenian Sea the lines closing the Gulfs of Lerno and Naples extend far beyond their 'natural entrance' points.

[37] See *infra*, Ch. 8 s.3(*b*).

[38] Territorial Waters and Contiguous Zone Act, as amended in 1978. Text in *UNLS* (ST/LEG/SER.B/19) (1980), 62–3.

[39] *Memorial* submitted by the Republic of Malta (1983), vol. iii: *Maps*, map 2. The co-ordinate values of the turning-points of the straight baselines have not been officially published. It seems reasonable, however, to assume that these baselines as drawn on the map are regarded as presenting Malta's official position as to the determination of the baselines from which the breadth of its territorial sea is measured (see Map 5).

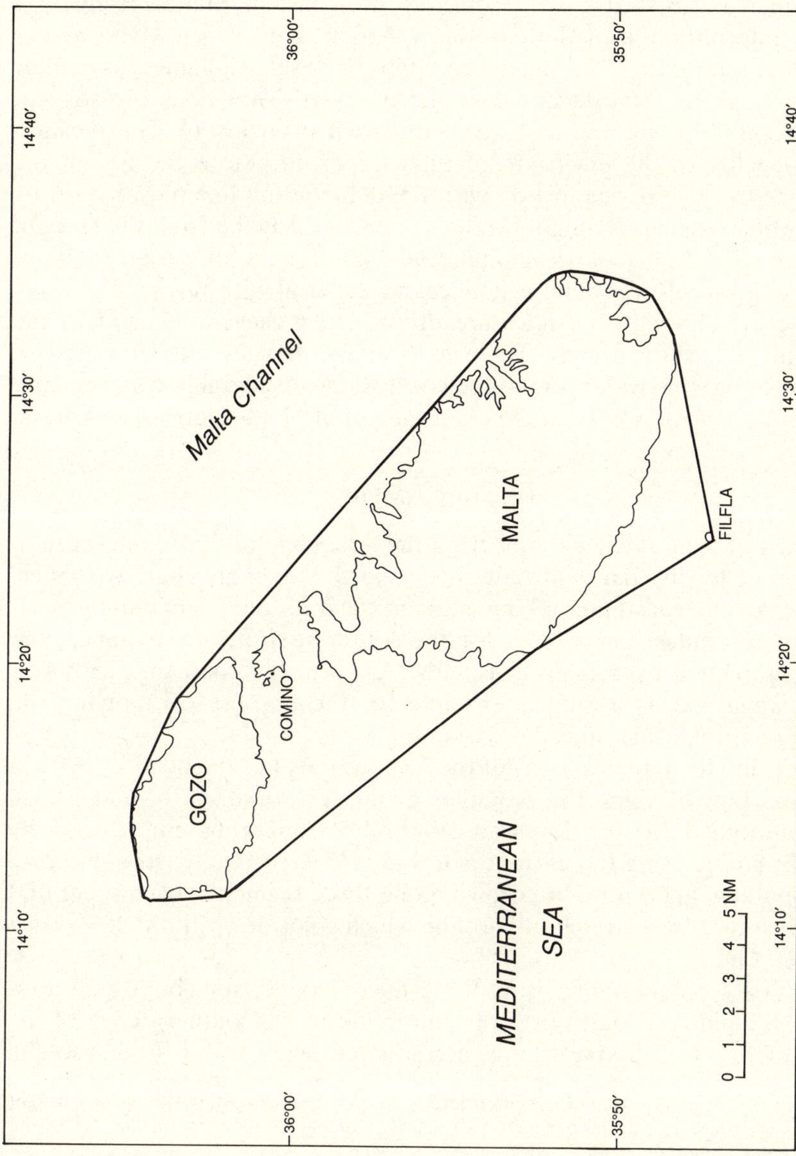

Source: Francalanci *et al.*, *Atlas of Baselines*.

MAP 5 Straight baselines of Malta

twenty-six turning-points are depicted. It is this last enclosure that seems to raise the question of conformity with Article 4 of the 1958 Convention on the Territorial Sea to which Malta is a party. Filfla is an uninhabited rock which is situated 3 n.m. south of the coast of Malta. Even by a flexible interpretation of the criteria of Article 4, to which Malta makes an explicit reference,[40] this enclosure seems to be questionable, for neither can the Rock be regarded as 'a fringe of islands' nor does the straight coast of southern Malta (the island) justify such an enclosure. The question of the legality of the enclosure of Filfla occurred before the ICJ in the *Libya–Malta* case in connection with the delimitation line put forward by Malta which was, according to Malta's view, constructed from the straight baselines. The Court, however, ignored Filfla for the purpose of delimitation and, therefore, found it unnecessary to deal with this point.[41]

However, although the lines are drawn along the entire coast of the Republic, in localities where the coast is regular they are interspersed with sections of the low-water line of the coast for a considerable distance and, as a result, have no effect on the external belt of Malta's territorial waters.

(*e*) *Yugoslavia*

The coast of Yugoslavia extends from the Albanian border in the south to the base of the Istrian peninsula in the north, for more than 250 miles; but the actual coastline, taking into account its many indentations, is nearly 2,500 miles, and except for the southern stretch of 60 miles, the coast is screened by a chain of islands. As such, the coast of Yugoslavia can be regarded as a unique example for the purpose of applying the system of straight baselines.

According to Article 2(1–3) of the Law of 1965 (as amended by Article 16 of the Law of 1987) the baselines of the territorial sea of Yugoslavia are determined by the low-water line, by straight baselines, and by 'straight lines closing the entrances to bays'.[42] Accordingly, straight baselines of about 344.7 n.m. are drawn along three segments, each separated from the other by a stretch of baseline which coincides with the low-water line (see Map 6).

The first segment, comprising three lines, extends for about 22.9 n.m. from the mainland coast north of Dubrovnik to the southern edge of the island of Mljet. Off-lying islands are situated landward of the lines. The

[40] A notice accompanying the map referred to *supra*, n. 39, states that the straight baselines are 'drawn . . . in accordance with article 4 of the Convention on the Territorial Sea and the Contiguous Zone'.

[41] See *supra*, Ch. 1 s.3(*b*)(ii).

[42] Law of 22 May 1965 on Yugoslavia's Marginal Seas, Contiguous Zone, and Continental Shelf. Text in *UNLS* (ST/LEG/SER.B/15) (1970), 133–4, as amended by the Law of 23 July 1987, *Sluzbeni list SFRJ* (*Official Gazette* of the SFRY), 99 (1987), 1211.

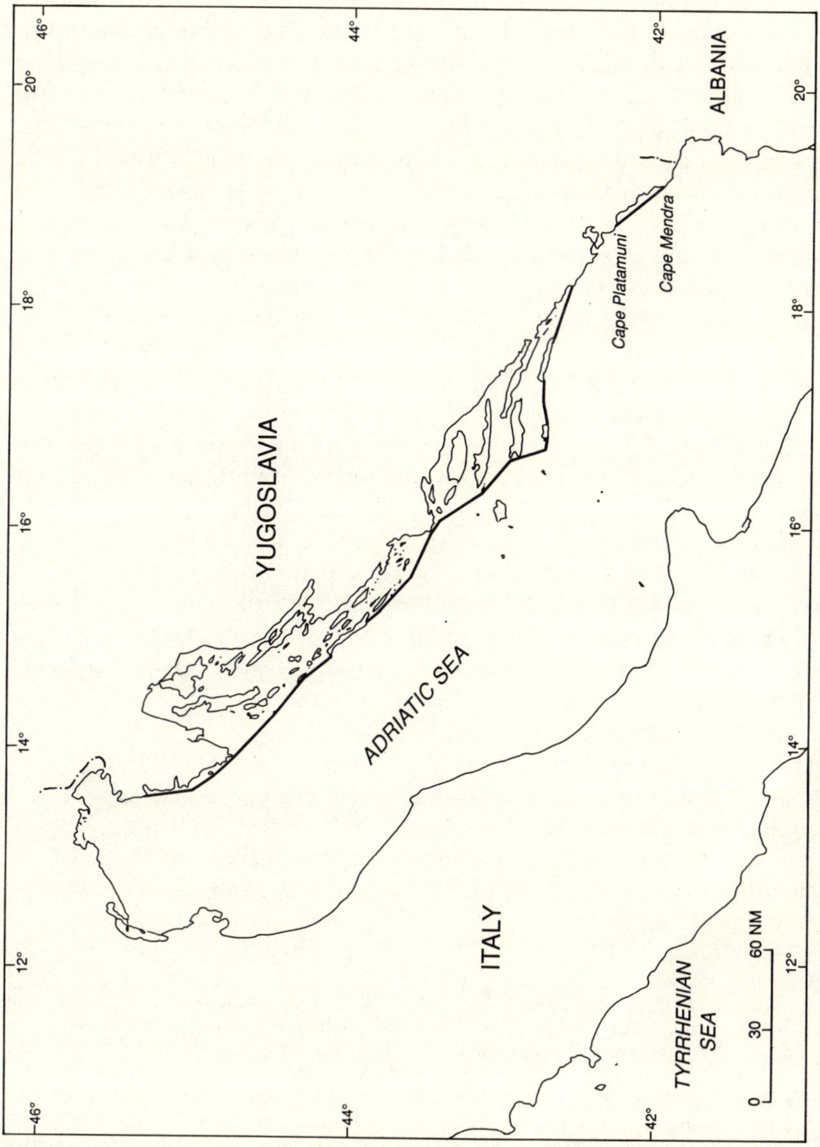

Source: Francalanci *et al.*, *Atlas of Baselines*.

MAP 6 Straight baselines of Yugoslavia

second segment has fifteen individual legs and extends for 129 n.m. from Cape Korizmeni on the foremost northern part of the island of Mljet to the island of Dugi Otok. All the seventeen turning-points are situated on either islands or low-tide deviations with lighthouses thereon. The presence of islands and other elevations in this area makes the considerable deviation of the straight baseline from the general direction of the mainland coast (nearly 45°) almost unnoticeable.[43] The final segment extends for 92.8 n.m. from Cape Veli Rat to Cape Kastanjia. Along this segment, the sea makes its widest and deepest penetration into the land, formed by the peninsula of Istria jutting into the sea. This area is also dotted with intersecting islands, the two biggest of which are Kar and Cres. All the nine turning-points are located on islands and low-tide elevations with lighthouses built thereon.

Four stretches of the coast of Yugoslavia still coincide with the low-water line. These are from the Albanian border to Zarubaca point, the low-water line of the Mljet island, the coast of Dugi Otok, and from Kastanjia to the Italian border.

The definition of 'bay' given by Article 3 mentions the semicircle test, though no maximum length of the closing lines is established.[44] Although the law does not provide for the closure of particular bays, Article 3 could be applied in the segments of the coast beyond Zarubaca Point, where no straight baselines have been drawn; there is at least one bay (near Boka Kotorsuka) to which the definition of Article 3 of the law is applicable. Finally, a Law of 30 March 1979 has added a new segment, which connects the Capes of Platamuni and Mendra. The area thus included within the internal waters cannot be regarded as a juridical bay.[45] The 1987 Act also retains the 1979 adjustments.

Apart from this latter enclosure, the straight baseline system of Yugoslavia reflects a noticeable conformity with even the most restrictive interpretation of the rules of law on straight baselines. It is a conformity that can be attributed to the geographical characteristics of its coast, as well as a tendency on the part of the State to take the relevant rules of law into consideration.[46]

[43] See *LIS* 6: 4.

[44] Art. 3 defines a bay as 'a distinctly limited inlet recessed into the land and of a sea area equal to or larger than the area of the semi-circle with a diameter equal to the length of the straight line closing the entrance into the inlet'. However, Yugoslavia is party to the 1958 Territorial Sea Convention, and signed the 1982 Convention.

[45] Law of 3 Mar. 1979, *Sluzbeni list SFRJ*, 13 (1979), 409. See T. Scovazzi, 'La linea di base del mare territoriale negli stati costieri del Mediterraneo', in B. Vukas (ed.), *The Legal Regime of Enclosed or Semi-enclosed Seas: The Particular Case of the Mediterranean* (Zagreb, 1988), 93, at 115.

[46] It would have been possible indeed for Yugoslavia to extend its baselines further seaward legitimately to enclose, e.g. the islands of Andrija, Susac, and Bisevo—lying outside the second segment of the straight baselines.

(*f*) *Albania*

Articles 1 and 4 of the 1976 amendment Decree[47] simply close off most of the Albanian coastline with sequences of straight baselines drawn on basepoints on the mainland with the exception of one basepoint on Sazani island. A small stretch of the coast, however, between the Cape of Gjuhezes and the Bay of Grama, seems to coincide with the low-water line (see Map 7).[48] The Albanian coast is not deeply indented with penetrations varying approximately between 3 to 4.5 n.m., and only one island (Sazani) exists in the immediate vicinity of the coast (about 3 miles off the nearest mainland). Nevertheless, the straight baselines utilize all the headlands along the coast and, as a result, close seven major indentations of which only two, i.e. the Bays of Drinit and Vlones, meet the legal definition of juridical bays. The 1976 Decree does not elaborate as to the basis according to which the lines are drawn. Albania, however, is not a party to the 1958 Convention on the Territorial Sea.

(*g*) *Turkey*

The baselines system of Turkey is not clear. Article 4 of the Law of 1964[49] states that the 'lowest ebb line' extending along the coast is the normal baseline from which the width of the territorial sea is measured. Paragraph 2 continues as follows: 'in indented coasts or in areas with islands located close to the shore, the method of the straight baselines connecting the foremost points of the shore and the islands does apply.' Article 5 also refers to the closing lines for gulfs with entrances not exceeding 24 n.m. The Article does not require the semicircular test, thus departing from the 1958 Convention rules for closing juridical bays. However, Turkey is not party to this Convention.

The 1964 Act, however, was repealed by the Act of 1982, which states in Article 3 that the breadth of the territorial waters of Turkey will be measured from basepoints to be delimited by the Council of Ministers.[50] On 29 May the same year the Council decreed that 'taking into account the power specified in [the 1982 Act] and taking into account the peculiar characteristics of the seas which surround Turkey and of the principle of equity, [it] has decided that the situation obtained before the promulgation of this law will be continued'.[51]

[47] Decree No. 4,650 of 1970, as amended by Decree No. 5,384 of 1976. Arts. 1 and 4 are repr. in *UNLS* (ST/LEG/SER.B/19) (1980), 3.

[48] This seems to be implicit in the language of both Arts. 1 and 4 of the Decree.

[49] Law No. 476 of 15 May 1964. English text in *LIS* 32.

[50] The Territorial Waters Act No. 2,674, 20 May 1982, *Resmi Gazete* (the *Official Gazette* of Turkey), 17,708 (29 May 1982).

[51] An official translation by the Turkish Embassy, Washington, DC, as quoted in

In stipulation of the application of the provisions of Articles 4 and 5 of the 1946 Act the Turkish Office of Navigation and Hydrography issued a map which depicts the Turkish coast with straight and bay-closing lines,[52] and this seems to be taken by the United States *LIS* series as representing the official baselines system of Turkey.[53] To facilitate analysis of the system drawn in this map, the coastline of Turkey may be divided into two sectors: the continental Mediterranean coast and the continental Aegean coast (see Map 8). The first sector extends from Kara Burun (36° 33′ W., 27° 59′ E.) to the Syrian border, some 400 miles east-souh-east. In localities where the baselines are drawn, the coast is deeply indented rather than fringed by islands. On this sector there seem[54] to be thirty-three continuous lines connecting points on headlands as well as points on some of the few islands lying in the vicinity of the coast, such as Strongili. The juridical bays of Iskanderun and Tasucu are closed by two lines measuring 23.5 and 7 n.m. respectively. On the whole, the drawing of baselines in this sector appears to be in line with the criteria provided in the 1958 Territorial Sea Convention. In fact, it is possible to argue that some segments of the lines could legitimately have been pushed further so as to enclose larger sea areas. This would be the case in the part of the coast where the islets of Yeovyios and Ketova could have been used as turning points.

The second sector appears to contain fifty-three continuous lines. Along this sector the coast is deeply indented and alternates with rugged peninsulas and bays. Rocks and small islands lying off entrances to bays are a common feature of this segment. At least six juridical bays are closed by closing lines.[55] The lines also include the three Turkish islands of Imyros, Bozacada, and the Rabbit Islands, and, as a result, close the entrance to the Straits of the Dardanelles.[56] Two segments of the Turkish baselines are shown on the map to coincide with the low-water line, one in the north between the boundary with Greece and Enez and the other in the south between Burun and the frontier with Syria.

The Office of Navigation map, however, cannot be regarded as representing the official coastline of Turkey as it did not emanate from the Council of Ministers, which is officially charged with the application of

G. Guttry, 'The Delimitation of Territorial Waters in the Mediterranean Sea', 11 *SJILC* (1984), 377, at 384.

52 *Turkiye, Karasulari esas ve duz Hatlari Haitasi*, scale 8,003, 1 : 100,000, 17 May 1965.

53 See *LIS* 32.

54 The copy of the Turkish Office map which is available to the author shows no more than 86 individual lines. The *LIS* series (No. 32), however, alludes to no less than 119 lines.

55 These are the Bays of Gokova, Kandalaya 14 n.m., İzmir 13 n.m., Comdarly 13 n.m., Edremit 11 n.m., and Saros.

56 Passage through the Straits is of course unaffected, as it is guaranteed by the Montreux Convention of 20 July 1936. Text in 31 *AJIL* (1937) (Supplement), 1.

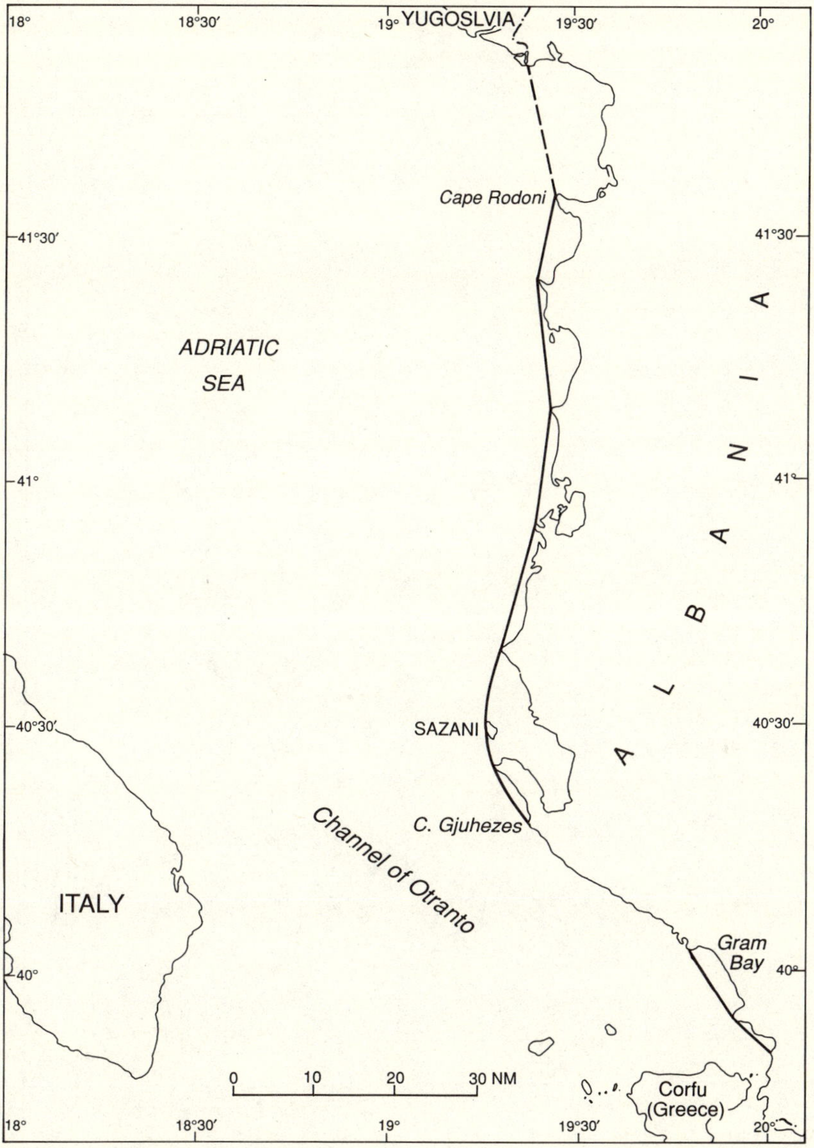

Source: Francalanci *et al.*, *Atlas of Baselines.*

MAP 7 Straight baselines of Albania

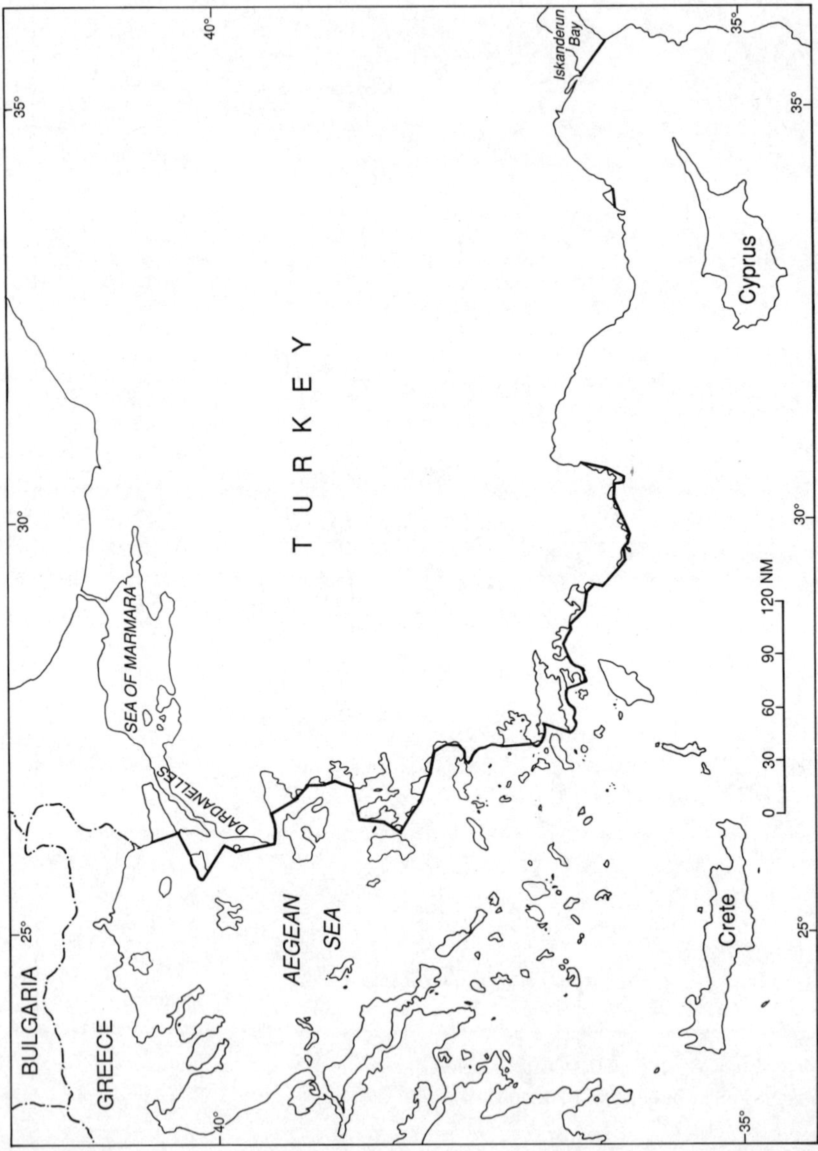

Source: Francalanci *et al.*, *Atlas of Baselines.*

MAP 8 Straight baselines of Turkey

the 1964 and 1982 Acts.[57] This may also find support in a Decree issued in 1987 by the Council of Ministers, which granted licences to the Turkish Petroleum Company to search for oil in the north Aegean continental shelf. A map appended to the Decree depicts the Turkish waters between the islands of Imyros and Tenedos, without showing any system of straight or closing lines.[58] The low-water line seems therefore still to represent the national coastline of Turkey.

(*h*) *Syria*

The legislation of Syria lays down the general guidelines according to which straight baselines may be drawn for the measurement of its territorial sea. These guidelines depart in many aspects from the conventional rules of the 1958 Territorial Sea Convention, although Syria is not a party thereto. Article 5 of the Syrian Legislative Decree of 1963[59] determined these general guidelines as follows:

1. In the case of the presence of a shoal not farther than 12 n.m. from the coast, lines shall be drawn from the coast all along 'the external edge of the shoal'. It is evident that the Decree permits the enclosure, by *straight baselines*, of shoals. They are defined in Article 1 of the Decree as 'every region within the territorial sea, covered with shallow water, part of which remains uncovered with water at the lowest level reached by the low tide'. This obviously departs from Article 4(3) of the Territorial Sea Convention which prohibits the use of low-tide elevations as basepoints unless they have a permanent structure built on them. While the Decree's definition implies that part of the 'shoal' may be a low-tide elevation, the requirement is not clear. However, none of the shoals lying off the Syrian coast seems to have such a structure.[60]

2. In the case of the presence of an island within 12 n.m. from the coast, lines shall be drawn from the mainland coast all along the external straight shores of the island.[61] The Syrian system therefore uses straight baselines from individual islands, a practice which goes beyond the Convention regime, which requires the presence of 'a fringe of islands' and not simply one island.

[57] Art. 5 of the 1982 Act states that the width of the Turkish territorial waters is measured from the baselines which are determined by the Council of Ministers, and displayed on large-scale nautical charts. A similar provision was contained in the 1964 Act.

[58] See *Resmi Gazete*, 19,412 (26. Mar. 1987).

[59] Legislative Decree No. 304 of 1963, repr. in *LIS* 53. The more recent Law No. 37 of 16 Aug. 1981 seems to maintain the system of 1963. See *LSB* 1 (1983), 61.

[60] See *Mediterranean Pilot*, 5 (1976), 115–19.

[61] Art. 1(c) of the Decree defines an island as 'land completely surrounded by the water and . . . always, in ordinary circumstances, above the level of high tide'.

3. In the case of the presence of a group of islands lying within 12 n.m. of one another with the nearest island not more than 12 n.m. away from the coast of the mainland, lines shall be drawn from the land all along the external shores of islands if they are in the form of a range, or lines shall be drawn along the external shores of the more prominent islands if they are not in the form of a range.

4. If the nearest island in the group is more than 12 n.m. from the mainland, lines shall be drawn along the external shores of the islands if they form a range or along the shores of the more prominent islands if they are not in the form of a range.

5. Pockets of high seas which may be left behind as a consequence of drawing the lines mentioned in (1)–(4) shall be included by a drawing of lines not exceeding 12 n.m. in length if the pockets are surrounded by the territorial sea on all sides.

6. In the case of the presence of a bay, lines shall be drawn at one point of the land from the entry of the bay to the other side. Article 1(8) of the Decree defines a bay as 'the pronounced curve which has a depth in relation to the width of its mouth so as to encompass water surrounded by land. The curve is not considered a bay unless its area is equal to or more than half that of a circle circumscribed with the mouth of that curve.'

Although Article 4 of the Decree talks of baselines 'shown on the large scale map of, and approved by the Syrian Arab Republic', the map has not been published. The precise locations of the straight baselines system cannot, as a result, be determined. However, a study of the geographical characteristics of the Syrian coast[62] reflects a rather limited application of the general guidelines identified above. For instance, the Syrian coast possesses no major indentations which correspond with the definition of a bay given in the Decree. Gentle curves, such as Khalij al Basit and the Ladhiqiyah coast, should have baselines given by the low-water line, which is qualified, under Article 5(1), by 'the lowest tide at ebb'. There are a few islets and shoals which the system can use, but they are all situated within 12 n.m. of the coast and of one another, a geographical factor that makes the treatment in detail by the Decree of some of the cases shown in (1)–(6) above superfluous. However, it is of interest to note that the Syrian Decree is the second of only two practices in the Mediterranean which stipulate a maximum distance between the coast and the offshore features and between the features themselves for the purpose of drawing straight baselines, a practice that is more restrictive than the Convention regime. Yet, on the other hand, by using straight baselines from single islands, the Syrian system goes beyond what the Convention allows.

[62] See *Mediterranean Pilot*, 5 (1976), 115–19; *LIS* 53: 7–9.

(*i*) *Egypt*

Unlike the Syrian coast, the Mediterranean coast of Egypt is almost featureless, without fringing shoals, reefs, or islands. In 1990 Egypt introduced its new system of straight baselines which effectively repeals the Royal Decree of 1951, which established general guidelines almost identical to those of Syria.[63]

The new system establishes a list of geographical co-ordinates (points of longitude and latitude) which connects fifty-three basepoints on the Mediterranean coast of Egypt with continuous straight baselines.[64]

Under the Decree of 1951, several indentations which did not conform to the requirements of juridical bays, such as the Gulfs of Salum, Abu Hashifa, El Hekma, and El Tina Bay, could be closed, in application of the Decree's wide interpretation of a 'bay' as 'any inlet, lagoon or other arm of the sea'.[65] In fact, in a comment on the Report of the ILC in 1955 the Egyptian *Revue de droit international* includes a map of closing lines for those features.[66] The same map shows a closing line for the Bay of El-Arab which Egypt claims as a 'historic bay'.[67]

Under the new system, the coast of Egypt is identified by continuous straight baselines which seem to have aimed at simplifying the coastal front of Egypt rather than having any effect on the extension of its territorial sea in the Mediterranean.

Remarkably, the system establishes straight baselines which follow the sinuosity of the gulfs and bays mentioned above (including the Bay of El-Arab) rather than enclosing their mouths, thus indicating Egypt's abandonment of its claim over El-Arab as a historic bay.

(*j*) *Tunisia*

Tunisia uses a combination of the low-water line and straight baseline systems. In application of the Law of 1973,[68] Article 1 of the 1973 Decree[69] establishes the baselines system of Tunisia which coincides with the low-water line between the Algerian borders and Cape Sidi Ali El Makki. From there three segments of straight lines are drawn, linking the entrance points of the Gulf of Tunis between Cape El Makki and Cape

[63] Royal Decree No. 180 of 15 Jan. 1951. For English text and analysis see *LIS* 22.

[64] Presidential Decree No. 27, dated 9 Jan. 1990, repr. in *LSB* 16 (1990), 5–11.

[65] Art. 1(b) of the 1951 Decree, *LIS* 22.

[66] See 'Report of the ILC on the Regime of the High Seas and of the Territorial Seas with Notes on the Egyptian Positions', 11 *REDI* (1955), 206.

[67] See *infra*, Ch. 7 s.3(*b*)(ii).

[68] 'Loi No. 73-49 du août 1973, portant délimitation des eaux territoriales', *Journal officiel de la République Tunisienne* (31 July/3–7 Aug. 1973), 1189; (7 Aug. 1973), 1189–90.

[69] Decree No. 73-527 of 3 Nov. 1973, *relatif aux lignes de base*, ibid. (Nov. 1973), 1697–8.

Bon. Proceeding eastwards up to Ras Kaboudia, the line once again coincides with the low-water line. The low-water line is also the baseline for the islets of Fratelli, Eastern Galitons, La Galite, Cani, and Pilau.

Straight baselines are also drawn from Ras Kaboudia to enclose the shoaly areas of Chebba and the sea area west and south-west of the Kerkennah Islands, and extend southward to Ras Marmor to enclose the Gulf of Gabes and the island of Jerba. From Ras Marmor to the border with Libya the baseline is measured from the low-water line. For practical purposes, the characteristics of the lines closing the Gulfs of Tunis and Gabes for 'historic' reasons are outlined elsewhere.[70] The rest of the straight baselines are depicted on fourteen turning-points, most of which are low-tide elevations. The Kerkennah Islands consist of two small islands, Chergui and Gharbi, as well as some islets and rocky patches. The average distance of the islands from the mainland is about 13 n.m.

The legality of the Tunisian baselines system, especially the lines drawn around the Kerkennahs and the Gulf of Gabes, was challenged by Libya in the *Tunisia–Libya* case.[71] Libya argued, *inter alia*,[72] that the Tunisian coastline did not conform to the conditions laid down in Article 4 of the 1958 Geneva Convention since, 'unlike the Norwegian coast, it is not deeply indented'.[73] It was also argued that the Kerkennah Islands are not 'part of the island fringe' but merely 'two localised and isolated islands'.[74] Tunisia contended that, as a non-party to the 1958 Convention, it was entitled to draw baselines with the same flexibility that is granted by customary international law,[75] especially in the light of the existence of economic interests, demonstrated by established fishery activities in the area landward of the baselines.[76] As mentioned earlier, the Court did not decide on this question, since the lawfulness of the Tunisian system was not considered relevant to a decision on the case.[77]

However, it would seem difficult to argue that the Kerkennahs do not constitute a fringe in the vicinity of the mainland coast. Compared to State practice, such an argument nowadays seems untenable.[78] It is

[70] See *infra*, Ch. 7 s.3(*b*)(ii).

[71] *Memorial* submitted by Libyan Arab Jamahiriya (1980), 24–6, 54–7; *Counter-Memorial* submitted by the Government of Tunisia (1980), 17–18. See also the Court's Judgment, ICJ Reports (1982), 71–7, 88.

[72] See also *supra*, Ch. 1 s.3(*b*)(ii).

[73] See Libyan Arab Jamahiriya, *Memorial* (1980), 55.

[74] Ibid.

[75] *Counter-Memorial* submitted by the Government of Tunisia (1980), 17–18.

[76] Neither the Law nor the Decree of 1973 expressly qualified the areas of waters within these baselines as internal waters. However, this is implied in the text of Art. 1 of the Decree, which reads: 'Les lignes de base, *à partir desquelles est mesurée la largeur de la mer territoriale tunisienne*' (emphasis added).

[77] See *supra*, Ch. 1 s.3(*b*)(ii).

[78] See e.g. *supra*, s.3(*a*),(*b*),(*f*). See also A. Gioia, 'Tunisia's Claims over Adjacent Seas and the Doctrine of "Historic Rights"', 11 *SJILCL* (1984), 327–75, at 350.

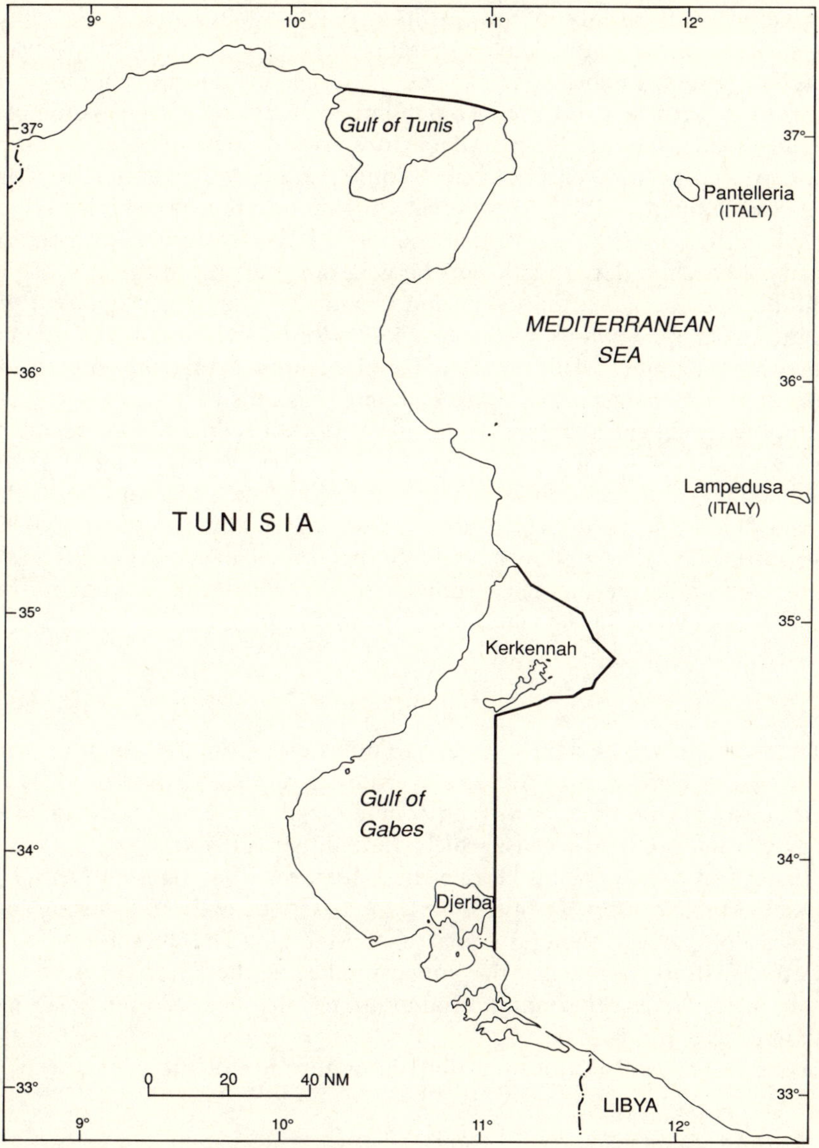

Source: Francalanci *et al.*, *Atlas of Baselines*.

MAP 9 Straight baselines of Tunisia

probably this flexibility in current State practice that led Judge Evensen to agree entirely with Tunisia's argument.[79] The legality of Tunisia's baseline system might, none the less, be questionable in relation to its marked departure from the mainland coast seaward of the Kerkennah Islands (see Map 9). Such a departure would stand against Tunisia's contention that there existed a close connection between its baselines and the land domain.[80] Tunisia, however, might still argue for the legality of this departure on the basis of the 'economic interests' proviso, which may justify a baseline departure from the general direction of the coast.[81] As will be shown later, there is sound evidence that the region of the Kerkennahs is known as a fishing area exploited by the local inhabitants since far-off times.[82] But whilst it could be argued that the 'protrusion' caused by the Tunisian baselines system could thus be justified by socio-economic realities, a different conclusion may be drawn with regard to the utilization of buoys. The Tunisian system of 1973 places eleven connecting points on 'balises', erected on low-tide elevations and other shallow areas some of which are always below water. Article 4 of the Territorial Sea Convention forbids the use of such features unless light-houses or similar permanent installations have been built upon them. It is very doubtful that buoys can qualify for this requirement.[83]

(*k*) *Algeria*

The Algerian coast extends about 550 miles east from the boundary with Morocco to Cape Roux. About eleven dispersed islands and islets lie off this coast, but none is more than 6 miles offshore. Fringing rocks are a common feature of the coast which, though generally straight, alternates with several narrow bights. Using these features, the Algerian Decree of 1984[84] establishes seventy-three straight baselines and lines closing bays drawn along almost the entire coast (see Map 10). The lines are drawn in relatively short segments; the longest, closing the Gulf of Arzew, is 23.15 n.m. Four other major indentations are closed, none of which qualifies as a juridical bay.[85]

However, it will be recalled that, in application of the French Law of

[79] See his Dissent in the *Tunisia–Libya* case, paras. 17–19.

[80] *Counter-Memorial* submitted by the Government of Tunisia (1980), 17–18.

[81] See *supra*, Ch. 1 s.2(*b*).

[82] See *infra*, Ch. 7 s.3(*b*)(ii).

[83] See Sep. Op. of Judge de Aréchaga in the *Tunisia–Libya* case, para. 120.

[84] Decree No. 84-181 of 4 Aug. 1984, *Journal Officiel de la République Algérienne* (7 Aug. 1984), 813.

[85] These are the Bays of Oran, Algiers, Annaba, and the Gulf of Timezrit with closing lines of about 16.3, 7.12, 21.96, 22.55 n.m., respectively. These enclosures do not conform with the semicircular test.

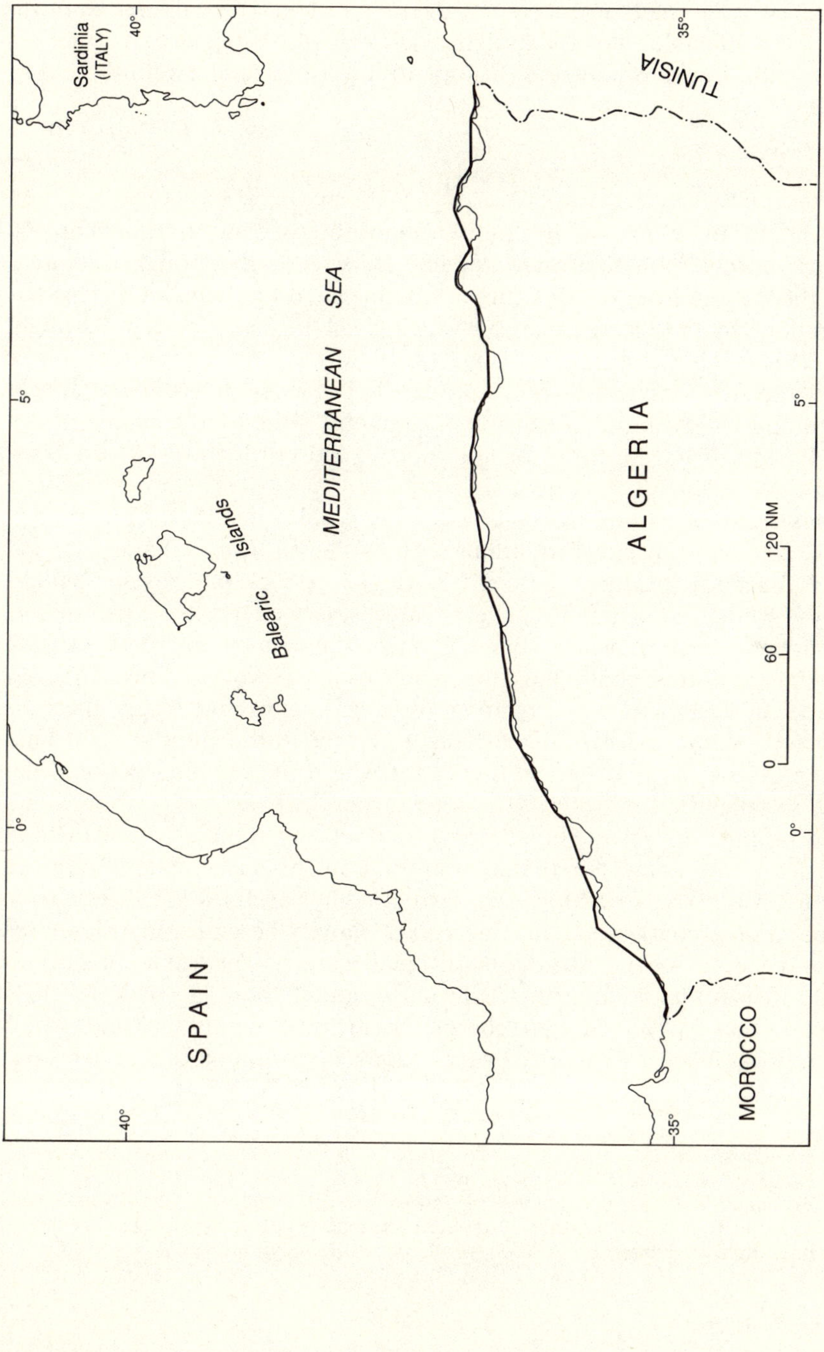

Source: Francalanci *et al.*, *Atlas of Baselines*.

Map 10 Straight baselines of Algeria

1888,[86] a Decree of 9 July 1888 was promulgated for the coast of Algeria and included seventeen bays, curvatures, and islands as part of the Algerian coast.[87] The same features are enclosed within the new Algerian system of 1984. Such enclosures by the new system which do not conform with the applicable rules for straight and closing lines might, nevertheless, be argued to have obtained a 'historical' title since their establishment in 1888.

(*l*) *Morocco*

Article 1 of the *Dahir* of 1973 provides that the baselines of the territorial sea of Morocco are the low-water line, as well as the straight baselines and the closing lines of bays that are determined by a decree.[88] This has been done by the Decree of 1975, which gives the co-ordinates of straight baselines closing a series of 'baies, rades, anses, îlots, rochers et autres anfractuosités des côtes Marocaines'.[89] The system encloses almost continuously—with only three minor segments—the Mediterranean coast of Morocco from the Strait of Gibraltar to the border with Algeria (see Map 11).

This relevant part of the coast is on the whole curved, with a few off-lying, separated insular formations such as Perejil, Mar, Chafarinas, and Leon. In many localities the coast is fringed by rock patches and shoals. On the whole the system uses these features, as well as eleven identifiable curvatures, none of which seems to meet the semicircular test. All the eleven curvatures could have been enclosed, however, since 1919. A Moroccan *Dahir* of 1919 relating to maritime fishing states that the territorial sea in the case of bays has to be measured from straight lines drawn across the bay, at the first point where the opening is not wider than 12 miles.[90]

Of particular interest, however, is the fact that the system encloses as Moroccan internal waters the areas around the 'Spanish enclaves' situated off or on the coast of Morocco. Two straight baselines are drawn from points on the coast of Ceuta which place almost the entire maritime area around Ceuta within Moroccan internal waters. The same practice is applicable to Melilla and the Chafarinas islands. The rocks of Velez de la Gomera and Al Hocaima are completely included within baselines which enclose water areas further seawards. Spain's position on these enclosures

[86] See *supra*, n. 29.
[87] *UNLS* (ST/LEG/SER.B/6) (1957), 16–17.
[88] *Bulletin officiel du royaume du Maroc* (7 Mar. 1973), 392.
[89] Art. 2 of Decree No. 2.75.311 of 21 July 1975, ibid. (13 Aug. 1975), 996.
[90] P. L. Rivière, *Traités, codes et lois du Maroc* (Paris, 1925), iii. 896.

does not seem to be publicly announced, although a Spanish writer recalls that Spain protested in this regard.[91]

This Moroccan system implies, of course, the deprivation of the 'Spanish enclaves' of any maritime jurisdiction, a practice that confirms Morocco's dispute with Spain over the political status of these enclaves. The implications of this dispute on a potential maritime boundary settlement are considered elsewhere.[92]

However, the utilization by the Moroccan system of these Spanish enclaves reveals a somewhat dubious stance. Whilst it on the one hand reaffirms Morocco's challenge to the Spanish sovereignty over, *inter alia*, the islands and Ceuta and Melilla, it seems, on the other hand, to give some implicit recognition of the rule which prohibits the extension of straight baselines in a way that would cut off the territorial sea of those enclaves (being foreign territories) from the high seas or an exclusive economic zone.[93] The Moroccan system leaves, as seen in Map 12, small areas of sea north of Ceuta, Melilla and the Chafarinas unconnected with baselines, thus allowing them to retain direct access to the area of sea northwards.

[91] Orihuela, 'La delimitación de espacios maritimes', ii, doctoral thesis (University of Murcia, 1986), 610.

[92] *Infra*, Ch. 8 s.4(*b*)(iii).

[93] See *supra*, Ch. 1 s.2(*b*).

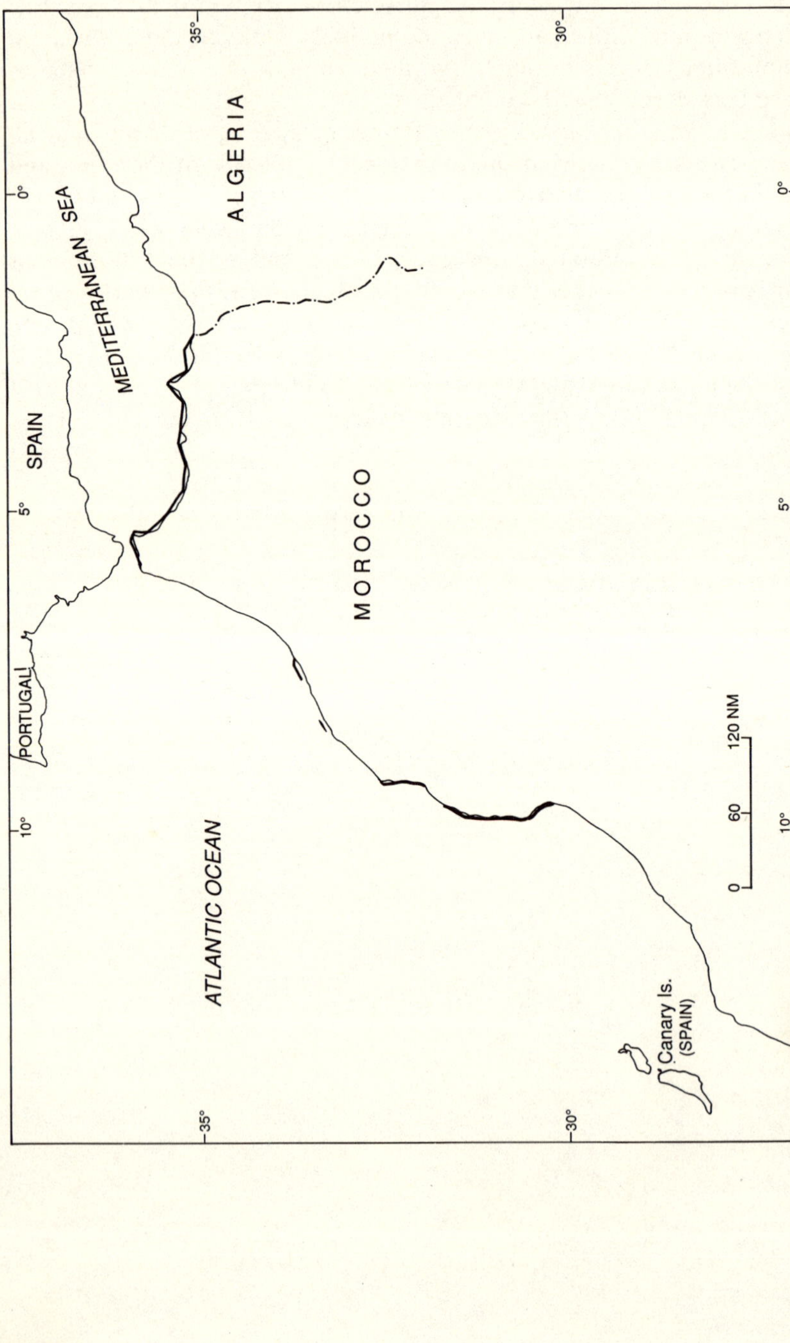

Source: Francalanci *et al.*, *Atlas of Baselines*.

MAP 11 Straight baselines of Morocco

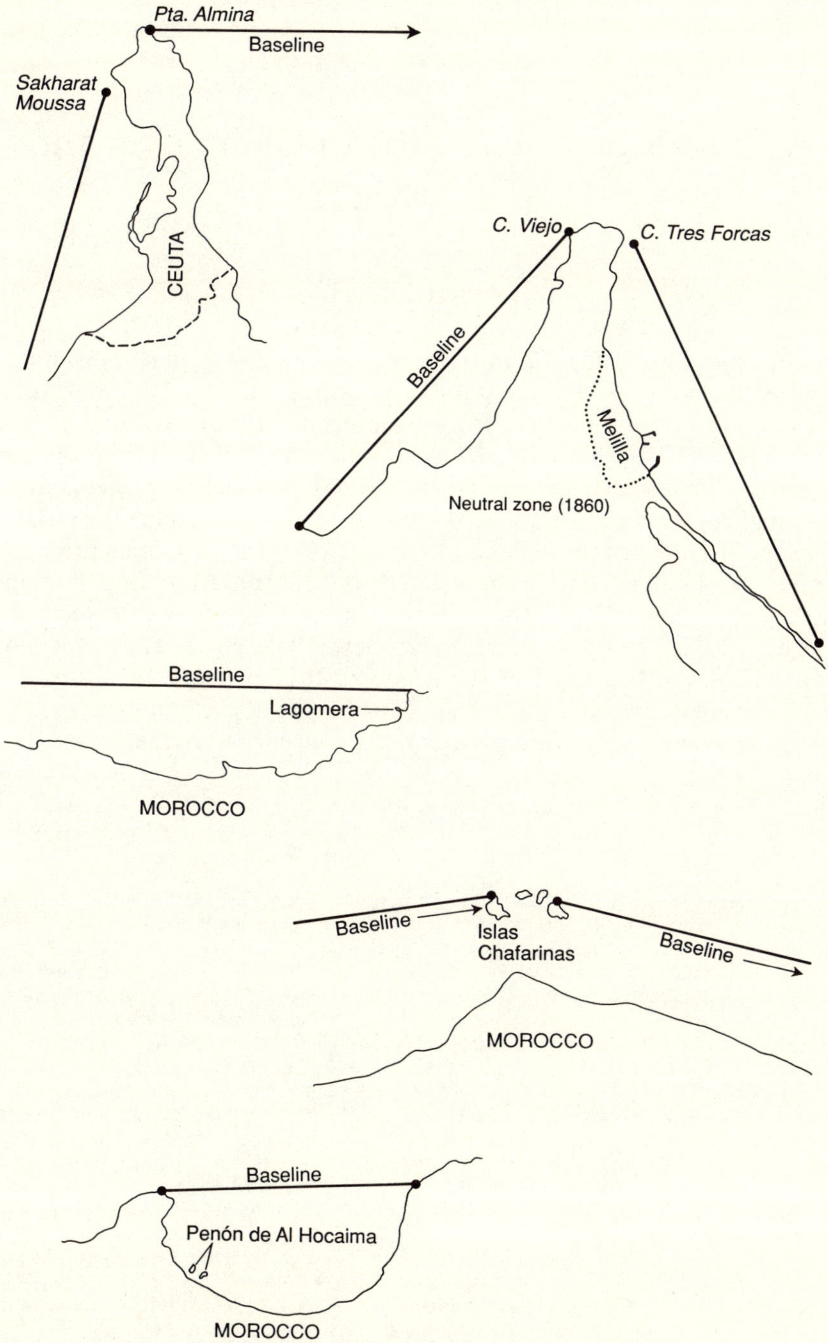

MAP 12 Morocco's baselines system around the 'Spanish enclaves'

7

A Case-Study: The Libyan Claim over the Gulf of Sert

1. INTRODUCTION

On 9 October 1973 the Libyan Government declared that the Gulf of Sert,[1] within an enclosed line parallel to latitude 32° 30′ N., 'constitutes an integral part of Libyan territory subject to its full sovereignty and is considered as internal waters from which the Libyan territorial sea begins'.[2] The Declaration gives two reasons for enclosing the Gulf of Sert, namely, (1) 'the undisputed exercise of Libyan sovereignty rights there and [treatment] throughout its long history as being Libyan property' and (2) the Gulf's 'vital importance for the Libyan Arab Republic and the necessity for the exercise of full control over it to safeguard the security and safety of the country in view of its geographical location commanding the heartland of the country'. It is evident, therefore, that Libya considers the Gulf of Sert an 'historic bay',[3] partly on the basis of exercising historic rights and partly on the concept of vital interests.

The doctrine of historic bays, recognized by adjudication as an established concept of international law in the *North Atlantic Fisheries* Arbitra-

[1] Alternatively known as Sidra, Sort, Surt, or Sirte. However, 'Sert' appears to be the nearest transliteration to the modern Arabic name given to the Gulf and its region, known also as the El Khalij region. Historically, the name goes back to the site of Punic, Roman, and early Islamic settlements known respectively as 'Chavax', 'Iscina', and 'Sort', now known as 'Medina Sultan', about 55 km. east of the modern city of Sert. The Arabic place-name 'Sort' was presumably derived from the ancient name of 'Syrtis', which was thought to have its origin in the Greek word 'suresthai' meaning 'to be drawn', in reference to the action of currents on ships trapped in the Gulf. After the decline and abandonment of 'Sort' its name came to be applied to the entire region. Around the 1840s the Turks erected a fortress at Marsa Zaafran (Modern Sert) and nominated a 'bey of Sert' to represent the Government. From then on, the name Sert rather than Sort or any other derivative has been used in modern Arabic. For a concise account of the origin of the name of the region and relevant sources see R. G. Goodchild, *Libyan Studies* (1976), 133–7.

[2] Declaration of 9 Oct. 1973, *Al-Jaridah Al-Rasmmiyah* (*Official Gazette*), Special Issue No. 5 (15 Oct. 1973).

[3] Whilst 'historic bays' are only, in the words of Bouchez, 'a *species* of the *genus* historic waters', for obvious reasons, the discussion in this chapter concentrates on historic bays. See e.g. C. Bouchez, *The Regime of Bays in International Law* (A. W. Sythoff, London, 1964), 199. Nor is any attempt made to examine the legality of the Libyan claim (and its consequences) under the relevant rules governing the system of straight baselines or the EEZ.

tion of 1910,[4] remains controversial. Although the doctrine has been extensively addressed by publicists[5] and is the subject of two United Nations studies,[6] two Law of the Sea Conferences have failed to codify a definition of the legal requirements of historic bays.[7] This failure underlines the absence of agreement in jurisprudence as well as between States to qualify certain norms which are often introduced as 'legal requirements' for the establishment of a historic bay.

Whilst this disagreement highlights the political complexities of the doctrine,[8] it also suggests, more significantly, the impossibility of formulating and applying general requirements to the unique and particular characteristics of each bay in question, in view of the rigid form in which the suggested 'legal requirements' are often presented. These legal requirements, often advanced by scholars, may be outlined as follows: firstly, there must be an effective exercise of sovereignty over the area claimed as historic; secondly, there must be a long continuous display of sovereignty; thirdly, the claim must be acquiesced in or, arguably, simply tolerated by other nations.

Historic claims, however, are basically asserted in response to economic necessity, national security, or another 'vital interest' of the coastal sovereign asserting control. Ideally, these interests gradually emerge when recognized by the community of nations after a long and open exercise of sovereignty. But these interests may also occur within a short period of time, raising the question of whether the vitality of such existing interests justifies the enclosure by the coastal State of a specific bay on the basis of the concept of 'historic bays'. A fourth element, often referred to as 'vital interests', is therefore relevant.

In the following section, the legality of the Libyan claim in the light of the first three 'requirements' will be discussed. For reasons related not only to the controversial nature of the fourth element but also to the special characteristic of the Libyan claim, a separate section will be dedicated to the issue of 'vital bays'. In a fourth section, the characteristics

[4] See *infra*, n. 117.

[5] e.g. Bouchez, *The Regime of Bays in International Law*; M. Strohl, *The International Law of Bays* (Martinus Nijhoff, The Hague, 1963). For a survey of international legal scholars, see UN Secretariat, *Judicial Regime of Historic Waters, Including Historic Bays*, UN Doc. A/CN.4/143, repr. in *Yearbook of the ILC* (1962), ii. 16–19 (hereinafter the UN Secretariat Report).

[6] UNCLOS I, *Official Records*, *Historic Bays: Memorandum by the Secretariat of the United Nations*, i (1958), 1–38 (hereinafter UN Memorandum); UN Secretariat Report.

[7] For a summary review of the codification efforts at the First and Third Conferences on the Law of the Sea, see L. Goldie, 'Historic Bays in International Law: An Impressionistic Overview', 11 *SJILC* (1984), 211, at 214 ff.

[8] A clear example is the Colombian draft article on historic bays, the only official proposal submitted at UNCLOS III. The sole purpose of the article's restrictive approach was to ensure difficulty for neighbouring Venezuela in establishing her historic claim over the Gulf of Venezuela. See ibid. 265–6; *infra*, s.3(*b*).

of the Gulf of Sert will be examined in the light of the concept of vital interests. The final section will attempt to draw some tentative conclusions regarding all the issues discussed.

2. LEGAL REQUIREMENTS FOR THE STATUS OF HISTORIC BAYS

(*a*) *Preliminary remarks*

In his argument for a realistic approach to the interpretation of the rules of international law, Charles de Visscher pertinently points out that

> Every rule of positive international law . . . presents two essential aspects for critical examination on different planes: the degree in which its content corresponds to social needs, and the accuracy of its formal expression compared with the practice of States. The rule of international law retains its full force in application only insofar as it satisfies this double requirement.[9]

A realistic analysis of the legal requirements of historic bays therefore demands knowledge of all social realities that determine these requirements as well as knowledge of the process of formal elaboration by which they take effect. Certain preliminary remarks are essential in order to put such knowledge into effect.

Firstly, it is important to bear in mind that 'a historic bay' is only one mechanism, amongst others, which serves to maintain a fair balance between the exclusive interests of the coastal State and the inclusive interests of the international community. This rationale underlies the whole idea, and indeed the overall necessity, of the law of the sea.[10] The concept of a historic bay gives effect to a presumed community policy that, prima facie, the intrusion of a sea area, even of relatively wide expanse, into the land territory of a State creates a usage that is more characterized by exclusive State control than by inclusive interests.[11] The

[9] C. de Visscher, *Theory and Reality in Public International Law*, trans. P. Corbett (Princeton University Press, Princeton, NJ, 1957), 133.

[10] For instance, the law of the territorial sea 'is essentially a compromise between the necessities and interests of the coastal State in the water off or near its shores, and the general concern of all states with the freedom of the seas—freedom of navigation, of communication and passage, of reasonable exploitation of the living and other resources, and of use for experiment and research'. G. Fitzmaurice, 'The Law and Procedure of the International Court of Justice, 1950–54: Points of Substantive Law', 31 *BYIL* (1954), 371–2. See also M. McDougal and W. Burke, *The Public Order of the Oceans* (Yale University Press, New Haven, Conn., 1962), 2 ff.

[11] 'The community policy in regard to internal waters has been based upon the recognition that the exclusive interests of states attach most fundamentally to waters which penetrate into the land mass. A coastal State has a pronounced interest in assuring that

decision-maker should be able to verify this prima-facie presumption by looking into the reality of the situation, i.e. by examining the conflicting interests. In other words, the peculiar geographical configuration of the bay provides the *material* and initial source for the claim of exclusive control; but it is the evidence of the pre-eminence of the exclusive interest in contrast to a minimum deprivation of community use that provides in a concrete case the *ratio legis* of a historic bay title. The approach of the ICJ in the *Anglo-Norwegian* case is illustrative.

The Court refused to restrict the use of straight baselines to certain situations designated as exceptional. It preferred a more general approach—the natural relationship between certain stretches of sea and the land. 'It is the land which confers upon the coastal State a right to the waters off its coasts.' It is the degree of intimacy between concrete and particular features of the land territory and certain parts of the sea that justifies extending the regime of inland waters to them.[12]

Secondly, also considering its social function in clarifying general community policy on internal waters, the concept of historic bays cannot be conceived as a departure from the principle of the freedom of the seas. A historic bay represents a feature in one of two parallels comprising this principle; for though representing a venerable and universally accepted rule of law, the principle of the freedom of the seas has never been considered so immutable as to preclude investigation into the possibilities of its adaptation to the legitimate interests of coastal States.[13]

The adaptation of the principle to, and indeed its coexistence with, basic coastal States' interests has been rooted in the development of the law of the sea: from the common law 'headland to headland' and *fauces*

water areas which are so intimately and strategically related to its land territory are not used to the disservice of the local population. Enclosure of these waters, therefore, is not perceived as an unreasonable infringement on the free use of the seas as long as the penetration of water area into the littoral is significant enough to evoke the reasonable concern of the coastal state and to suggest a balancing in favour of exclusive use.' G. Westerman, *The Juridical Bay* (Oxford University Press, New York, 1987), 23. For discussion of the development of international norms in regard to bays, see the same work, ch. 2, 'Fundamental Purposes, Principles and Policies'.

[12] This criterion, which serves for bays, should, the Court held, be broadly applied all along a coast where the geographic configuration is as peculiar as it is in Norway. 'Such a coast, viewed as a whole, calls for the application of a different method': that of baselines detached to a reasonable degree from the physical line of the coast. The qualification implicit in these last words is due to the same criterion, the close dependence of the sea area on the land territory. 'It follows that while such a State must be allowed the latitude necessary in order to be able to adapt its delimitation to practical needs and local requirements, the drawing of baselines must not depart to any appreciable extent from the general direction of the coast.' *Anglo-Norwegian Fisheries* case, *UK* v. *Norway*, ICJ Reports (1951), 129, 133.

[13] See e.g. H. Lauterpacht, 'Sovereignty over Submarine Areas', 27 *BYIL* (1950), 376, esp. 376–407.

terrae notions,[14] through Grotius's consent to 'a gulf or strait . . . occupied . . . with the intention to possess it as a private property',[15] down to the exercise of exclusive rights over more open areas of sea such as the contiguous zone, the continental shelf,[16] and the EEZ.[17] These conceded unilateral claims constitute, according to H. Lauterpacht, the 'second parallel stream' of the principle of the freedom of the seas, the first being insistence on the full right of freedom from any interference.[18] It is a parallel which springs out of the requirement of striking a fair balance between the exclusive and inclusive interests; for the 'evaluation of conflicting interests and the effective recognition of those which, on balance, are entitled to protection is a constant feature of juridical activity as, indeed, of law in general.'[19]

Neither—for the same reasons—can a historic bay be regarded as an exception to 'standard' rules applicable to the delimitation of maritime areas.[20] The regime of historic bays represents an autonomous rule (amongst others which may be called standard rules) according to which larger areas of sea—larger than that prescribed by the standard rules—may be claimed, the 'exception' relating to the frequency of use rather than to legalizing an otherwise illegal use. Both rules have the function of confirming a legal situation. This was advanced by the author of the United Nations study in the following words:

> general rules in this field . . . [assert] that, within the framework of customary international law, certain maximum limits for the territorial sea and the width of the opening of bays are—generally applicable and that in certain cases there exists an historic title to waters which do not come within these limits. The so-called general rules would then be 'general' in the sense only that they would be more generally applicable than the 'exceptional' title to 'historic waters'. But they would not be 'general' in the sense of having a superior validity in relation to the 'exceptional' historic title. Both the general rules and the historic title would be part of customary international law, and there would be no grounds for claiming a priori that the historic title is valid only if based on the acquiescence of the other States.[21]

[14] For the discussion of the influence of English law upon the rules of international law relating to bays, see D. O'Connell, *The International Law of the Sea* (Clarendon Press, Oxford, 1982–4), i. 338–49.

[15] Quoted ibid. i. 349.

[16] Lauterpacht, 'Sovereignty over Submarine Areas', 398–415.

[17] For a more comprehensive analysis see G. Smith, *Restricting the Concept of Free Seas: Modern Maritime Law Re-evaluated* (Robert Krieger Publishing Co., New York 1980).

[18] 'Sovereignty over Submarine Areas', 403–6.

[19] Ibid. 404.

[20] For authors who take this view and their relevant quotations, see the UN Secretariat Report, 7–10.

[21] Ibid. 10.

The above remarks lead, finally, to the sound propositions that (1) a framework of analysis of the legal requirements for a historic bay should not overlook the social function of this concept within the general aim of maintaining the necessary equilibrium between the inclusive (international community) and exclusive (coastal State) interests; that (2) the concept of a historic bay is considered neither a derogation of the free seas principle nor an exception to the rules applicable to the delimitation of baselines; (3) the analysis of the elements constituting title to a historic bay has therefore to be conducted independently and without being prejudiced by the a priori postulate that this is an exceptional claim, or by resorting to analogies from other modes of acquisition of title—such as 'prescription' or 'immemorial possession'—whose social function differs from that attributed to a maritime historic right.[22] With these propositions in mind, the following subsections attempt to seek some plausible suggestions for the clarification of these 'legal requirements'.

(*b*) *Effective exercise of authority*

It is generally agreed that, since a State's claim to a historic bay means a claim to sovereignty, then the authority exercised by the claimant State 'which is the basis for the claim, must also be sovereignty'.[23] Writers insist that the manifestation of sovereignty requires two elements. One is intentional, that is, the authority exercised should display the intention to act as a sovereign. Secondly, the claimant State has actually to display its authority.[24] In *US* v. *Louisiana*, the Supreme Court of the United States held that the Inland Waters Line defined by Congress in 1895 to govern navigation in the inland waters of the United States could not be the basis of a historic claim because it was, on the evidence, not intended to correspond with the national boundary.[25]

The element of intention seems, however, to assume an orderly and consistent reasoning on the part of the State similar to that which can be attributed to individuals. It involves, as one authority notes, 'the imputation of a state of mind, involving a *legal* assessment and "judgment", to those ordering state activity'. It is an approach which 'expects too much and is unrealistic in seeking a particular and coherent intention in a mass activity by numerous individuals'.[26]

Furthermore, the excessiveness of the element stems from its abstract

[22] See *infra*, s.2(*c*), (*d*).
[23] The UN Secretariat Report, 13.
[24] Ibid.
[25] 394 US 11, 22 L. Ed. 44 (1969), 46–7.
[26] I. Brownlie, *Principles of Public International Law* (Clarendan Press, Oxford, 1979), 143.

treatment, which is unrelated to any contingent situation. The decision to claim a historic bay usually springs out of various acts of 'exclusiveness' (exclusion of foreign fishermen, effective use of the area for local communication, control of navigation, etc.). Each of such acts may exhibit a historic or traditional right over some activity, but not a historic title to the sea. It is the combined effect of such acts which may generate an attitude of exclusiveness over the maritime area and finally inspire the State to embody its intention in a piece of legislation or proclamation. This latter act merely gives an expressive value to an attitude of exclusiveness already in existence.

More important for legal analysis, however, is the second element of effective display of authority; for it is only through a concrete display of State activity consonant with sovereignty that it can clearly be shown that the State considers the area involved as an integral part of its territory.[27]

Although it is not possible to specify categorically what kind of acts produce sufficient evidence of sovereignty, the exclusion of foreign vessels (or the attempt to do so) from entering the bay and prohibition of fishing by foreigners stand as classic examples of acts which are commensurate with the claim of sovereignty.[28]

What amounts to an effective display of authority varies, however, in accordance with the limitations of geography. A priori analogies from arbitral decisions on land territories indicate that the sovereign rights which are exercised might be limited in scope and frequency in accordance with the accessability, inhabitability, and climatology of the area involved. This will be further emphasized elsewhere.[29] Nor does manifestation of sovereignty possess in all cases an identical intensity. For example, if the waters claimed give access to maritime routes used by international navigation the exercise of sovereignty is required to be demonstrated more clearly than in the case of similar claims to a water area navigated mostly by a limited number of ships passing to or from the ports of the coastal State.

In fact, there may appear to be no need at all for the coastal State actually to take a specific action to enforce its claim; for 'it is not impossible that [such a claim] was respected without the State having to resort to particular acts of enforcement'.[30]

Thus although a coastal State has, for example, the privilege of excluding from its internal waters foreign vessels in innocent passage, the need to exercise that privilege may never arise.

[27] The UN Secretariat Report, 15.

[28] G. Gidel, *Le Droit international public de la mer* (Sirey, Paris, 1934), vol. iii. 633; Bourquin, *Les Baies historiques*: *Mélanges Georges Sauser-Hall* (Paris, 1952), 43.

[29] See *infra*, s.2(*c*).

[30] The UN Secretariat Report, 15.

Indeed, in the present case, the records do not indicate that, until recently, there was ever any occasion to exclude foreign vessels from navigating in the Gulf of Sert. This is not surprising since, as will be seen later, foreign vessels have little interest in entering the Gulf except to reach the Gulf's ports.[31] This lack of interest in the Gulf's waters for international navigation seems, therefore, to support rather than disprove the claim of historic title to the Gulf (or at least to part thereof) as inland waters.[32]

As mentioned earlier, it is the concurrence of special circumstances and facts (which are not alike in all cases) which would finally push a State to make specific a claim to a certain water area, through the enacting of legislation. The absence of incidents or disputes over the area considered by the coastal State as under its sovereignty may delay the necessity of declaring *de jure* what has been already considered *de facto* part of the territorial sovereignty of that State. The necessity of displaying some positive acts of sovereignty may therefore never arise.

Thus it was only in 1981 and afterwards that Libya found it necessary to undertake concrete action to enforce its 1973 Proclamation, manifested in its continuous and costly opposition to the United States Sixth Fleet sailing into the Gulf of Sert.[33]

The above is not meant to suggest that the lack of necessity in the past to exercise actual sovereign powers over the maritime area should always rectify any defect regarding the 'historicity' of the claim. Such an argument will turn the whole idea of 'legal requirements' for establishing a historic bay into absurdity. All it calls for is the necessity for a reasonable legal analysis to link this requirement to the reality of each concrete case in order to avoid unrealistic over-conceptualization.

(*c*) *The time factor*

The time factor implies an exercise of authority over a period of time sufficient to indicate a pre-eminent exclusive usage.[34] In this context, the term 'usage' means a uniform course of conduct, or a repeated activity, by the claimant State in the area claimed as historic. However, the period of time necessary for usage to be manifested is a matter of subjective

[31] See *infra*, s.2(*d*)(iii).

[32] The US Supreme Court employed this reasoning in deciding that the Mississippi Sound constitutes a historic bay. See *infra*, text at n. 136.

[33] Between 1977 and 1986 the US Navy conducted nine large-scale manœuvres in the Gulf of Sert. Within this period the Libyan and US military forces were reported to have confronted one another at least four times. For a detailed account see S. Ratner, 'The Gulf of Sidra Incident of 1981: A Study of the Lawfulness of Peacetime Aerial Engagements', 10 *YJIL* (1984), 59; *Time* (31 Aug. 1981); *Time* (21 Apr. 1986).

[34] The UN Secretariat Report, 15.

judgment. Traditional codification projects and writers on the subject frequently mention imprecise notions of perpetuity, including 'usage of long standing', 'established usage', 'immemorial usage', and 'continuous and immemorial usage'.[35] These notions are said to be 'wholly impractical'[36] and contemporary jurisprudence suggests that the length of time during which exclusive usage exists is highly flexible and should vary according to the facts of each case.[37]

The identification of the time factor with 'usage' seems to qualify the absolute notion of 'long-standing time'. It seems to suggest that there should be an adequate period of time sufficient (1) to admit the existence of the factor of usage, i.e. the commencement by the State of the exercise of authority; (2) to indicate the scope of this authority, i.e. to show that the authority displayed is sovereignty; and (3) to ascertain that a peaceful and undisturbed display of authority has existed.

The problem, however, is how to determine the length of this 'adequate period'. The United Nations Study admits that this will be a matter of judgment that takes into account the circumstances of each particular case.[38] Although this answer reflects the complexities of codification referred to earlier, it also emphasizes the basic rationale for the requirement of historic use. As Visscher pertinently noted,

> Proven long usage, which is [the historic title] foundation, merely represents a complex of interests and relations which in themselves have the effect of attaching a territory or an expanse of sea to a given State. It is these interests and relations, varying from one case to another, and not the passage of a fixed term, unknown in any event to international law, that are taken into direct account by the judge to decide *in concreto* on the existence or nonexistence of a consolidation by historic titles.[39]

The time factor should therefore be investigated in the light of this underlying rationale. Three categories of circumstances are submitted to be relevant to such an investigation. The first is geographical/environmental, relating to the bay under consideration, since the regularity of usage of the bay is determined largely by the location of the bay, the degree of accessibility of its waters, and the quality and availability of its resources. For instance, the regularity of usage in an Arctic bay is much less than that expected in a bay located, say, in the North Sea. The

[35] See draft articles of international codifications and opinion of authors in the UN Memorandum.

[36] The UN Secretariat Report, 15.

[37] Scelle, for instance, stated that the period of prescription 'is indeterminate and must in each case be submitted to the test of reasonableness'. G. Scelle, *Droit international public*, ii (Paris, 1946–7), 435.

[38] The UN Secretariat Report, 15.

[39] Visscher, *Theory and Reality*, 200.

second category pertains to the special socio-economic conditions of the claimant State, that is, the power that the State can display over the bay in accordance with its maritime experience (as a major, average, or small maritime power), the size of its population, its technological ability, etc. The third category is strategic, relating to the special vulnerability of the coastal State's security interests. It might be that the intrusion of the bay into the mainland, coupled with vital water-related establishments, would entail a display by the coastal State of certain precautionary acts including surveillance and police and military patrols for the protection of its security. These acts, if they are regular, are submitted to be sufficient to prove the State's continuous authority over the bay.

The importance of the above categories in determining the time factor is submitted to conform not only with authoritative decisions[40] but also with the basic community policy underlying the law on historic waters. The rationale for the requirement of 'long usage' is to ensure that certain interests exist which are sufficiently vital to have motivated the coastal State to display its sovereignty over an indentation regularly, and in such a noticeable way that other States have acquiesced in or tolerated this sovereignty. But this attitude of exclusivity emerges out of flexible cultural, psychological, economic, and even pure nationalistic tendencies so monotypic in each individual case that they cannot be given an overall, comparative definition.[41] This confirms not only the difficulties encountered by codification efforts in reaching agreement on defining (in contrast to other coastal State jurisdictions) the concept of historic bays, but also the importance for legal analysis of recognizing the continuing quality of international law to adapt to changing economic and political factors within a diverse 'community' of nations.

The monotypic nature of these elements provides the proper environ-

[40] e.g. the following statement of Judge Huber provides an *a fortiori* analogy to historic bays: 'Manifestations of territorial sovereignty assume . . . different forms, according to conditions of time and place. Although continuous in principle, sovereignty cannot be exercised . . . at every moment on every point of a territory. The intermittence and discontinuity compatible with the maintenance of the right necessarily differ according as to whether inhabited or uninhabited regions are involved, or regions enclosed within territories in which sovereignty is uncontestably displayed or again regions accessible from, for instance, the high seas.' *Palmas Island* Arbitration Award, 22 AJIL (1928), 867, at 877. Similarly, in the *Rann of Kutch* Arbitration it was held that because of the inhospitable nature of the terrain which until recently made the Rann virtually 'incapable of permanent occupation, the requirement of possession cannot play the same important role in determining sovereign rights therein as it would have done otherwise'. Text in 7 ILM (1968), 633, at 679. See also statement to the same effect in the *Eastern Greenland* case, PCIJ Rep. (1933), Ser. A/B, No. 53, at 69; the *Minquiers and Ecrohos* case, ICJ Reports (1953), 53.

[41] In the *Tunisia–Libya* case the Court pointed out that international law did not 'provide for a *single* "regime" for "historic waters" or "historic bays", but for a particular regime for each of the concrete, recognized cases of "historic waters" or "historic bays"'. ICJ Reports (1982), para. 100.

ment for the 'attitude of exclusivity', not necessarily only through a historic pattern of life, but also 'from emerging and anticipated conceptions of new needs, caused by changes in technology and customs, whose satisfaction is expected to be partially derived from uses of maritime resources'.[42] The insistence on a 'long-standing historic use' in such a situation would, it is submitted, extend beyond the social function attributed to the factor of time. Under the concept of historic bays the factor of 'history' is not the underlying reason for accommodating a given community-specific interest. It is the urgency of that interest that matters. The 'history' factor only represents objective, incontrovertible evidence of the emergence of such interest. When a State's alleged needs are not supported by 'long usage' the burden of proof on that State is heavier. In such a case the State has to demonstrate that its particular interests are of the kind that have always exercised a decisive role in favouring a State's exclusive authority, in the light of the general community policy on the law of the sea, namely, the formulation of a set of rules aiming to achieve an appropriate balance between the exclusive interests of coastal States and the inclusive interests of the whole community of States.[43]

The task of identifying the pre-eminence of exclusive interests is by no means formidable. The principle of good faith is often introduced by courts for examining a State's conduct and can legitimately play an extended role in ensuring that a fair balance is kept between the alleged rights exercised by, and the duties imposed on, the claimant State.

The principle of good faith is in the first place a basic rule of the law of treaties, already essential for treaty implementation and interpretation.[44] In this context parties to the 1958 Territorial Sea Convention are under an obligation to 'perform' the right to claim a historic bay provided for in Article 7 'in good faith'.[45] This obligation is explicitly stressed in Article 300 of the 1982 Convention. It reads:

> States Parties shall fulfil in good faith the obligations assumed under this Convention and shall exercise the rights, jurisdiction and freedoms recognized in this Convention in a manner which would not constitute an abuse of right.

Of particular interest is the Article's express inclusion of 'abuse of rights' as the negative form of the principle of good faith. But both the principle and its negative form[46] apply also to the general performance of

[42] McDougal and Burke, *The Public Order of the Oceans*, 308.

[43] See *supra*, s.2(*a*).

[44] See Art. 31(1) of the Vienna Convention on the Law of Treaties.

[45] Art. 26 of the Vienna Convention reads, 'Every treaty in force is binding upon the parties to it and must be performed by them in good faith.'

[46] There are few writers who consider the concept of abuse of rights redundant when the criteria of good faith and reasonableness are provided by an existing legal rule. See e.g. Brownlie, *Principles of Public International Law*, 443–5.

a State's rights and obligations under international law.[47] As the ICJ puts it, good faith is 'one of the basic principles governing the creation and performance of legal obligations, whatever their source'.[48]

The prohibition of abuse of rights entails *inter alia* that a State is under obligation not to use its *formal* rights in the interest of its own policy, to the detriment of the general aim of the international community.[49] In the case under investigation a coastal State's abuse of the right to claim a historic bay may occur when it uses the claim for ulterior purposes unconnected with the maintenance or protection of a pre-eminent exclusive interest; for instance to interfere with and restrict the right of a neighbouring State in the freedoms of the high seas,[50] or to evade a treaty obligation regarding pollution control or supervision.[51] The use of historic bays is extended in these instances to cover the commission of what in fact is a fictitious, and therefore abusive, exercise of a right.[52]

It is therefore possible by subjecting a historic bay claim to the test of good faith and reasonableness to ascertain the existence or otherwise of a pre-eminent exclusive interest. A historic claim which is not supported by a historic use which is 'clear and beyond doubt' has to be made in pursuit of genuine exclusive interest of the kind that the concept of

[47] This general influence of the principle of good faith owes its growing importance mainly to the Charter of the UN. According to Art. 2(2) of the Charter States are obliged to fulfil in good faith the obligations assumed by them in accordance with the Charter.

[48] *Nuclear Tests* cases (*Australia* v. *France*), ICJ Reports (1974), 253, para. 46. It was largely on the basis of this principle that the Court gave binding effect to France's unilateral declarations concerning the French intention to abstain from future atmospheric nuclear tests in the area of the South Pacific. See *Nuclear Tests* cases (*New Zealand* v. *France*), ICJ Reports (1974), 457, para. 49.

[49] The *Walter F. Smith* claim (1929) relating to expropriation proceedings may present a good analogy. The Arbitrator considered it contrary to the principle of good faith and unlawful for a State to go through the forms of an expropriation procedure in order to seize private property not for public purpose, but for the use of some individuals for private profit (*USA* v. *Cuba*, 2 RIAA, 913, at 917–18). For a more detailed discussion see B. Cheng; *General Principles of International Law as Applied by International Courts and Tribunals* (Grotius Publications, Cambridge, 1987), 105–36; H. Lauterpacht, *The Function of Law in the International Community* (Clarendon Press, Oxford, 1933), 286–306; B. Iluyomade, 'The Scope and Content of a Complaint of Abuse of Rights in International Law', 16 *HILJ* (1975), 47–92.

[50] See e.g. freedoms of high seas mentioned in Art. 2 of the 1958 Convention on the High Seas.

[51] e.g. Art. 4 of the first Protocol to the Barcelona Convention for the Protection of the Mediterranean Sea against Pollution provides that the 'dumping into the Mediterranean Sea Area of wastes or other matter listed in annex I to this Protocol is prohibited'. A State which might wish to allocate 'a convenient' site near its coast to dump such materials might evade such a prohibition by declaring the area 'internal waters'. Art. 1(2) of the Convention, to which the said Protocol refers in Art. 2, reads that (for the purposes of the Convention and its Protocols) '[e]xcept as may be otherwise provided in any Protocol to this Convention, the Mediterranean Sea Area shall not include internal waters of the contracting Parties'.

[52] For other forms of fictitious exercise of a right, see Cheng, *General Principles of International Law*, 122–3.

historic bays is destined to protect and so overriding that it was not calculated merely to cause any unfair prejudice to the inclusive interest of the international community.[53]

However, in respect of the claim over the Gulf of Sert, the Libyan Declaration of 1973 mentions, as seen earlier, Libya's exercise of 'sovereignty over the Gulf . . . throughout its long history'. Although Libya's history reveals very little on ancient Libyan interest and activity along its Mediterranean coast the following account of its known maritime history should enable a reasonable conclusion to be reached.

Whilst Libya's masterly role in the Mediterranean in the seventeenth and eighteenth centuries (identified then with the powerful Karamanli Dynasty) is well documented,[54] there seems to be no evidence suggesting particular Libyan interest in the Gulf of Sert. The resources of the then powerful Libyan Fleet (named by Christian nations as the Fleet of the Barbary State of Tripoli) were at the time concentrated on maintaining a predominant role, amongst other maritime powers, in the then prevailing politics of the Mediterranean's high seas, rather than being deployed offshore.[55]

Libya's interest in the offshore areas, however, appeared to have increased when Italy occupied Libya in 1911. At the time foreign fishing activity in Libyan waters (especially by neighbouring States, such as Greece, Malta, and Italy) was specially related to sponges, for which the coasts of Tunisia and Tripolitania were famous. In this respect, the Gulf of Sert is not reported to have had a more significant economic value than the rest of the Libyan coast.[56]

The Italian authorities in Libya were relatively active in fishing activities. They undertook a number of scientific surveys of Libyan waters and sought to regulate and control the exploitation of offshore resources,

[53] See also the application of the test of 'reasonableness' and 'moderation' by the ICJ in the *Anglo-Norwegian* case in determining whether Norway committed a 'manifest abuse' in delimiting the baseline of the Lapphavet basin, 141–2, 153, 156ff.

[54] See e.g. E. Rossi, *Storia di Tripoli e della Tripolitania della Conquista Araba al 1911* (Arabic trans.) (Dar Al-Thakhaffa, Beirut, 1974); L. Wright and J. Macleod, *The First Americans in North Africa* (Arabic trans.) (Al Ferjani Bookshop, Tripoli, 1974).

[55] The American and European ships were engaging in rivalries in the Mediterranean for the purpose of maritime blockades, prizes, and the protection of their merchant ships from piracy, which was a thriving trade in the Mediterranean. These powers had to come to terms with the then strong Libyan Fleet. For instance, in the period between 1662 and 1816 Mediterranean as well as non-Mediterranean powers had concluded at least seventeen treaties of peace and commerce with the bashaw of Tripoli. The main theme of these treaties was to allow foreign ships to navigate without hindrance by the Libyan Fleet and, occasionally, to provide protection for ships of the contracting power against pirates. See e.g. a series of treaties in *BFSP*, i (pt. 1) (1812–14), 710–33; iii (1815–16), 513; H. Miller (ed.), *Treaties and Other International Acts of the United States of America* ii (Government Printing Office, Washington, DC, 1931), 349, 529.

[56] See e.g. L. Anderson and G. Blake, 'The Libyan Fishing Industry', in J. Allan (ed.) *Libya since Independence* (Croom Helm, London, 1982), 73ff.

particularly the sponge banks. Regulations were promulgated and illegal fishing was actively dealt with.[57] Although these regulations continued in force well up to the independence of Libya in 1951, there seems to be no evidence of special regulatory or protective measures being taken with regard to the Gulf of Sert or any part of it.

During its occupation of Libya, Italy did not establish a consistent distance for measuring the breadth of the territorial waters of Libya. The extension of the 'territorial sea' varied according to the terms of legislation in effect at any given time.[58] Of particular interest are two Italian Decrees. The Royal Ordinance of 1931 relating to sponge fishery provided that sponge banks stretching continuously from the coast up to any distance beyond the territorial sea were deemed to be part of the said sea. Some of these banks were reported to extend 25 miles from the coast.[59]

Royal Decree No. 595 of 6 June 1940 forbade national ships, warships, and neutral merchant ships to approach, *inter alia*, the territorial waters of Libya. The Decree mentioned certain areas considered dangerous for navigation, including 'the 12-mile-wide band along the coasts from the Tunisian border to the Egyptian border'.[60] Nothing in these decrees or in any other similar regulation[61] appears to embody any particular rule for the Gulf of Sert.

It can, therefore, be concluded that, well into the years leading to Libyan independence, the Gulf of Sert constituted no special economic or strategic value to Libya, apart from its projection southwards into the Libyan heartland, thus dividing the major population centres of Libya into two regions.[62]

This picture, however, changed radically when the discovery of oil in the Sert basin in the 1960s turned Libya from one of the poorest countries

[57] e.g. under Italian Instructions of 16 Apr. 1919, permits were required for sponge or coral fishing beyond territorial waters, declared then at 'three marine miles from the coast'.

[58] e.g. Royal Decree No. 85 of 4 Feb. 1913 fixed the breadth of the sea for the purposes of customs surveillance at 'twelve miles . . . from the shore along the coast of Tripolitania and Cyrenaica'. The same distance was adopted in Royal Decree No. 402 of 18 Mar. 1915, also on customs regulations for Tripoli and Cyrenaica; whereas according to the Italian Instructions of 1918 the limits of the territorial waters 'are to be intended as established at three marine miles from the coast'. For an English trans. see the *Tunisia–Libya* case, *Documentary Annexes* submitted by the Libyan Arab Jamahiriya, vol. ii (1983), annexes 23, 24, 26.

[59] See G. Panayotacos, 'Un cas d'extension arbitraire des limites de la mer territoriale', 6 *Revue hellénique de droit international* (1953), 253, at 255.

[60] *Tunisia–Libya* case, *Documentany Annexes*, Annexe 27.

[61] e.g. the Declaration of 1911 established 'an effective blockade from September 29th [1911] by Italian ships of war of the coasts of Cyrenaica and Tripoli'. Text in L. Hertslet (ed.), *Commercial Treaties*, xxvi (Butterworth, London, 1913), 708.

[62] The Gulf's intrusion into the centre of Libya's most arid coastline was the major factor in dividing the State for administrative as well as political purposes (when Libya was a Federal State between 1952 and 1963) into, *inter alia*, two provinces: Cyrenaica and Tripolitania.

in the world into one of the world's major producers of oil. Thus to consider Libya's long-standing interest and activity in the Gulf (see below s.4(*b*) and (*c*)) is a matter of dealing not with a distant past, but with a span of less than three decades. Therefore, to argue the Libyan case on the basis of a rigid standard of 'historic usage' appears to be unfounded.

(*d*) *The attitude of other States*

The attitude of foreign States is usually discussed under the imprecise heading of 'acquiescence',[63] which may be defined as 'the inaction of a State which is faced with a situation constituting a threat to or infringement of its rights'.[64] In the *Gulf of Maine* case, the Chamber defined acquiescence as 'equivalent to tacit recognition manifested by unilateral conduct which the other party may interpret as consent'.[65]

Acquiescence is, therefore, an essentially negative concept, which simply implies that foreign States have been inactive; and it is with respect to historic claims in particular that acquiescence merely indicates that there has been no opposition by other States to the State's peaceful display of authority over the historic waters.[66] The view that 'something more than the mere failure to object must be shown'[67] is regarded as rigid, for it is widely accepted that the 'general toleration' by other States of a well-defined and uniform claim of a State would suffice as the 'basis of a historical consolidation which would make [that claim] enforceable as against all States'.[68]

But although acknowledging this qualification in principle, the discussion in literature seems to pay little attention in the elaboration of this requirement to its specific function in the law on historic bays. 'Toleration' represents an essential element whereby presumed community policy towards a particular State claim can be identified. It is a function which not only reflects the requirement's association with the rationale for maritime historic claims[69] but also draws attention to the necessity for its consideration within the context of that rationale.

The urgency of this realization is especially highlighted in view of the misleading association in literature of title to historic bays with claims

[63] e.g. P. Jessup, *The Law of Territorial Waters and Maritime Jurisdiction* (G. A. Jennings, New York, 1927), 382; C. Colombos, *The International Law of the Sea* (Longman, London, 1967), 186; Bouchez, *The Regime of Bays in International Law*, 203.

[64] I. MacGibbon, 'The Scope of Acquiescence in International Law', 31 *BYIL* (1954), 143.

[65] *Gulf of Maine* case, ICJ Reports (1984), para. 130.

[66] The UN Secretariat Report, 17.

[67] *US* v. *Alaska*, 422 US 184, 45 L. Ed. 2d 109 (1975), at 122.

[68] *Anglo-Norwegian* case, 138. See also MacGibbon, 'The Scope of Acquiescence', 143; Bourquin, *Les Baies historiques*, 46; Y. Blum, *Historic Titles in International Law* (Martinus Nijhoff, The Hague, 1965), 132.

[69] See *supra*, s.2(*a*).

based on acquisitive prescription.[70] Acquisitive prescription assumes, even when dissociated from immemorial possession,[71] the possession by one State of a title which was *originally invalid* as a result of the consent or acquiescence of the *dispossessed State*. It is a concept which involves an active and a passive element: a proprietary action by the offending State and passivity and inaction on the part of the original sovereign.[72]

The role of acquiescence in the creation of a prescriptive title is therefore evident. The acts leading to possession are, in their inception, illegal; the resulting possession is adverse and unlawful, and it is only on the basis of the gradual change in the quality of those acts, produced by the *inaction or silence by the original sovereign*, that the resulting possession is given legality. It is this 'tacit abandonment, surrender, or acquiescence, that constitutes the creative factor in the acquisition of a title by prescription'.[73]

Acquiescence in this case attributes validity to the enjoyment of a certain right, originally belonging to, or being exercised by, another State, without effective protest on the part of the other State. The underlying principle is that a State which has 'slept upon its rights' should not be allowed to revive them against a State that has been in constant enjoyment of those rights. Acquisition of title in this manner shares with extinctive prescription the legal effect of creating an estoppel against the losing State whose claims have become obsolete.[74]

In order to establish a title by prescription, the claimant State must therefore show an acquiescence on the part of the other State clear and beyond doubt. It is a requirement which reflects the principle's development from the *usucapio* of Roman law,[75] whose later adaptation to international law did not face major difficulties, as it was considered

[70] See e.g. Gidel, *Le Droit international public de la mer*, 621–5; D. Johnson, 'Acquisitive Prescription in International Law', 27 *BYIL* (1950), 332, at 349; G. Fitzmaurice, 'The Law and Procedure of the International Court of Justice, 1951–54: General Principles and Sources of Law', 30 *BYIL* (1953), 1, at 27–8.

[71] See e.g. D. Johnson, 'Consolidation as a Root of Title in International Law', *Cambridge Law Journal* (1955), 215, at 219–22. He points out the tendency among writers to regard acquisitive prescription as a means of acquiring title through 'an adverse' holding, that is, 'prescription properly so called', and to distinguish it from 'straightforward', 'immemorial possession'.

[72] G. Fitzmaurice, 'The Law and Procedure of the International Court of Justice, 1951–54: Points of Substantive Law', 32 *BYIL* (1955–6), 20, at 31.

[73] Ibid. n. 1. For a similar view see G. Verykios, *La Prescription en droit international public* (Paris, 1934), 75; C. Hyde, *International Law*, 2nd rev. edn. (Little, Brown & Co., Boston, 1945), i. 386–90.

[74] D. Greig, *International Law* (Butterworths, London, 1976), 163.

[75] Interruption of possession, known as *usurpatio*, was fatal to the acquisition of title by *usucapio*. Johnson, 'Acquisitive Prescription', 334–5; Brownlie, *Principles of Public International Law*, 156–8.

largely within the framework of *adverse* (State v. State) claims over land areas[76] which are physically capable of appropriation.

Such conditions make reasonable the argument for a strict proof of possession and a very long period of possession in order to show beyond any doubt the acquiescence of the losing State in the adverse holding of the other. The importance of acquiescence here relates, furthermore, to the exceptional character of acquisition by prescription, for it is only through the acceptance of the previous holder that the change of sovereignty becomes lawful.

Such conditions and factors are extraneous to the concept of a historic bay. As stated earlier, a maritime exclusive title is not regarded as derogation of the principle of the freedom of the high seas. It is a 'functional adaptation' inherent in the flexible virtue of this principle in order to strike the necessary balance between the exclusive interests of coastal States and those of the international community as a whole.[77] It is very artificial to argue that a State making an exclusive claim to a water area geographically within its land territory implies that it is engaging in wrongful conduct.[78] The configuration of the bay in relation to the land territory of the claimant State constitutes, prima facie, a bona fide test of the original quality of the claim. The function of acquiescence here is to make (through the reaction of other States) objective the subjective assessment of the claimant State regarding the territoriality of the bay. This function stems from the realization that a historic bay conveys no 'adverse title' against an alleged right of one specific State.[79] Rather, it is a claim to an area as part of a State's territory, with the possibility that one or more States contend that it is part of the high seas. It is because of this latter possibility, i.e. to safeguard the *res communis* principle, that acquiescence (or toleration) is advanced as a legal requirement for a maritime historic title.

The artificiality inherent in the assimilation of a historic bay claim to acquisitive prescription occurs most clearly in the discussion of the character of the 'effective protest' which ousts the presumption of toleration.

[76] It suffices to mention that, in all arbitral decisions which dealt, though in passing, with prescription, the dispute was related to two *contesting claims* over *land* areas. See e.g. relevant abstracts of the *Island of Palmas*, the *Clipperton Island*, *Minquiers and Ecrehos*, and the *Chamizal* cases, in D. J. Harris, *Cases and Materials on International Law* (Sweet & Maxwell, London, 1983), 151–68.

[77] See *supra*, s.2(*a*).

[78] This, for instance, seems to be argued by Fitzmaurice, who stated that an exclusive right to waters 'which would otherwise, in law, be high seas, can only be acquired by prescription, i.e. by a consistent assertion of jurisdiction over them acquiesced in . . . by other States in such a way as to amount, in effect, to an abandonment, actual or constructive, of the communal rights in those waters they would otherwise have retained'. 'The Law and Procedure of the ICJ', 30 *BYIL* (1953), 27–8.

[79] The UN Secretariat Report, 11.

Envisaged within the context of 'adverse interests between two parties' a historic claim is said to be defected by the protest of the States 'which are directly interested'. This category involves, it is suggested, 'neighbouring States'[80] or 'other States whose maritime interests are affected such as maritime powers of the world'.[81] A right will not come into existence if the majority of the directly interested States expressly reject the claim.[82] In such selective categories protest is envisaged as a sanction by single States to prevent encroachment on their individual rights.

Such a selective approach fails to take into account the necessary link between the nature of the legal obligation of the claimant State and the requirements which reflect that obligation. In the *South-West Africa* case Judge Alvarez suggested that there are cases in which States may be under an obligation 'without the beneficiary of the rights relating to these obligations being known. The beneficiary is the international community.'[83] In one sense this could be regarded as true of all obligations arising out of general rules of international law, but there is perhaps no better relevant example than the corresponding obligation of a coastal State claiming a historic bay. Its obligation, that is, not to inflict undue encroachment on the *res communis* principle, is owed to the international community at large, not to any particular category of States as beneficiary. This conclusion would not be affected by the fact that there might be, as is often the case, a particular State or States who benefit most from the restriction imposed on the claiming State. Their particular interest, however, is not the legal cause which gives rise to the corresponding obligation.

An appropriate criterion of an effective protest must therefore reflect its function to keep operative the necessity of establishing a balance between the *res communis* and the individual interests of the coastal States. It should provide a reasonably objective test of the point at which the one set of interests must be regarded as being displaced by the other. The action of protest *per se* possesses no conclusive and decisive value in establishing such a process.[84] Its conclusiveness is established only

[80] Gidel, *Le Droit international public de la mer*, 634; Bouchez, *The Regime of Bays in International Law*, 267.

[81] Johnson, 'Acquisitive Prescription', 351, 353–4.

[82] Bouchez, *The Regime of Bays in International Law*, 267–8.

[83] *International Status of South-West Africa* case, ICJ Reports (1950), 177.

[84] Even in the case of rival claims to a territory, protest is only given the function of ascertaining which of the rival claims would be capable of support on another variety of grounds. This was applied by the ICJ in the *Minquiers and Ecrehos* case, over a French protest against a British legislative Act including the 'Ecrehou Rocks' within the limits of the Channel Islands. This protest, the Court said, 'could not deprive the Act of its character as a manifestation of sovereignty'. ICJ Reports (1953), 66. This approach was also taken by Judge Carneiro. Citing the *Eastern Greenland* case, he said that 'the character of the acts of the British Government is not altered by the protests which, from time to time, were made by the French Government'. Ibid. 106.

through its ability to demonstrate that an overriding community interest is violated.[85] This in turn requires the identification in a concrete case of the interests at stake and the consideration of the impact that a particular claim may make upon the interests involved. It is an approach which not only reflects the social function of the law on bays but also embodies the principle of good faith as an objective element, governing States' conduct. The authority and continued usefulness of 'protest' can be safeguarded (like any other principle of law) only if it is acted upon in accordance with its true purpose and the ever valid test of reasonableness. The true purpose of 'protest' is to ensure freedom of navigation and other bona fide community interests from unnecessary prejudice by individual States. A protest is effective, and therefore has a legal power in defecting a historic claim, if it pursues such a purpose, that is, to serve to protect some bona fide community interest.[86]

What constitutes a pre-eminent community interest is obviously not a static notion, but one which evolves according to the development of international life. The inclusive use of the sea for navigation, flying, and fishing is the major instance which international law has accepted as a clear case of inclusive community uses. But as the international community becomes more integrated, more concerns are laid upon its members which could reflect issues of common interests, such as protection and preservation of marine environment,[87] or adoption of conservation and management measures for fishing resources.[88]

These are the major concerns leading States to protest against a historic claim. The very fact that States protesting against or 'reserving' their position regarding the Libyan claim over the Gulf of Sert variously appealed to some of these factors[89] in justification illustrates the real function of the requirement of acquiescence. The impact of the Libyan

[85] e.g. if a water area outside the territorial sea is of great interest as an international sailing route for the whole community of States, such waters may not fall under the sovereignty of the coastal State. In this case, the principle of freedom of the seas must prevail over the interests of the coastal State. The establishment of rights to historic bays is in law almost impossible, and the protest of only few States would suffice to defect such a claim.

[86] A State may protest for purely political motives, e.g. to act in accordance with the aspirations of a political partner. It is also possible that a State may protest in order to obtain some individual facility not shared by other States (e.g. the use of the empty waters of the Gulf of Sert (see *supra*, n. 33, and *infra*, s.2(*d*)(ii), (iii)) by the US Fleet for military manœuvres). The ineffectiveness of such and other similar protests, as a result of their failure to reflect a genuine interest of the international community, is reinforced by the requirement of good faith. International law assumes that every right is the legal protection of a legitimate interest. An alleged exercise of a right not in furtherance of such interest, but for a different one (for which the right has not been created), can no longer claim the protection of the law which prohibits the abusive use of rights. See *supra*, text at nn. 45–52.

[87] See e.g. pt. XII of the 1982 Convention, esp. Arts. 192–208.

[88] See e.g. Arts. 61–7 of the 1982 Convention.

[89] See *infra*, n. 102.

claim on three of these major freedoms (fishing, aviation, and navigation) should therefore be considered.[90]

(i) Fishing

The wealth of fishing stocks depends on many factors, all connected with an abundance of phytoplankton. The waters of the Gulf of Sert appear to be one of the lowest areas in the Mediterranean in phytoplankton productivity. This seems to be related to two factors. First, the surface currents in the Gulf are not strong enough to cause upwelling and mixing

[90] The challenge by the US Fleet to the Libyan decision to close the Gulf (see *supra*, n. 33) raises the question of the military use of the high seas as an allegedly fourth high-seas freedom. Whilst lack of space does not permit elaboration, the question which seems to arise in this regard is whether the term 'navigation' mentioned specifically in Art. 58(1) of the 1982 Convention would embrace vessels and aircraft performing military exercises within a coastal State's EEZ. The compromise language of Art. 58(1) reflects disagreement between the major maritime powers and coastal States at UNCLOS III (see A. Lowe, 'Some Legal Problems Arising from the Use of the Seas for Military Purposes', 10 *MP* (1986), 171, at 178 ff.) and the issue still remains unsettled in jurisprudence. One prevailing view, however, seems to admit that military exercises might be regarded as within the general meaning of 'navigation', so long as they are confined within the normal acts of traversal or transit within the EEZ, but not when they appear to be extensively and repeatedly performed (see *supra*, n. 33, for the US Sixth Fleet manœuvres within the Gulf of Sert) within a certain area of the EEZ. In this regard attention is drawn to Art. 301 of the 1982 Convention which embodies the principle of peaceful use of the waters of the EEZ and therefore obliges States not to pursue aggressive policies inconsistent with the UN Charter against the territorial integrity of any State. (See e.g. P. Gralindo, 'The Exclusive Economic Zone in the Light of Negotiations of the Third United Nations Conference on the Law of the Sea', in F. Vicuna (ed.), *The Exclusive Economic Zone: A Latin American Perspective* (Westview Press, Boulder, Colo., 1984), 31, at 55; F. Francioni, 'Peacetime Use of Force, Military Activities and the New Law of the Sea', 18 *CILJ* (1985), 203, at 216; but see also E. L. Richardson, 'Power, Mobility and the Law of the Sea', 58 *Foreign Affairs* (1980), 902). However, assuming the legality of conducting manœuvres within an EEZ in principle, the legality of engaging in large-scale exercises such as those conducted by the US Sixth Fleet still has to be considered in the light of general community policy. In other words, does the alleged right of the USA to conduct such military exercises prevail over the interest of Libya in protecting its security by prohibiting warships in the Gulf? Whilst the matter has not been judicially tested, State practice seems to leave open in this case the possibility of favouring the pre-eminence of the State's exclusive interest. Exclusion zones promulgated by various States in different circumstances (such as 'security zones', 'air and identification zones', 'quarantine' zones, 'embargo' zones, and 'nuclear tests and danger zones') all seem to serve 'as a prescription against a reasonable restriction on the free use of the sea by all States where this restriction is grounded on a valid peace, security, or self-defence basis and the responsive action is necessary and proportional to the threat'. C. Q. Christol and Davies, 'Maritime Quarantine: The Naval Interdiction of Offensive Weapons and Associated Material to Cuba', 57 *AJIL* (1963), 525, at 540. Art. 59 of the 1982 Convention can be said to reflect a general acceptance of this trend. It states that 'in cases where this Convention does not attribute rights or jurisdiction to the coastal State or to other States within the exclusive economic zone, and a conflict arises between the interests of the coastal State and any other State or States, the conflict should be resolved on the *basis of equity and in the light of all the relevant circumstances, taking into account the respective importance of the interests involved to the parties as well as to the international community as a whole*' (emphasis added).

of more nutrient-rich intermediate water layers. This is coupled with the absence of any permanent rivers to bring in nutrients or cause upturning of waters.[91] This explains why, throughout history, fishing in the Gulf has been exercised mainly by local inhabitants,[92] except for a limited sponge bank situated near the north-western arm of the Gulf which used to be fished by foreign fishermen.[93] However, the impact of the closing line on international fishing will soon be an irrelevant argument. A Libyan EEZ claim will practically exclude all foreign fishing within areas well beyond the closing line.

(ii) International aviation

Ironically, the absence of regular civil aviation over the Gulf was one of the 'factors', according to United States officials, that made the Gulf of Sert 'an ideal location of safe conduct of [US manœuvres] in the Mediterranean'.[94] Civil aviation routes, except for domestic flights, do not cross the Gulf of Sert but fly directly over or into Tripoli and Benghazi. Thus the airspace above the Gulf is not used for international air traffic by non-Libyan aircraft.[95]

(iii) International navigation

Perhaps the most important factor in the determination of the legality of a coastal State's exclusive claim is the impact of such claims upon international navigation.[96] The Mediterranean basin's role in international trade is significant. It accounts for approximately 15 per cent of total world trade.[97] The figures are even higher for petroleum cargo, with 24.2 per

[91] See UN, FAO, *Atlas of Living Resources of the Seas* (Rome, 1981), sheets 1.1, G.1.

[92] See *infra*, s.4(B)(i).

[93] See the *Tunisia–Libya* case, *Map Annexes* submitted by the Libyan Arab Jamahiriya (1981), map 13 (showing 'Sponge Grounds in Libya').

[94] D. Neutze, 'The Gulf of Sidra Incident: A Legal Perspective', 108 *Proc. US Naval Inst.* (1982), 26.

[95] See International Civil Aviation Organization, 'Regional Manual: European-Mediterranean', Doc. 4600, Amendment No. 60 (15 Oct. 1952), EU-RE 27, 28.

[96] The insignificance of a bay for international navigation has been one of the important factors brought up by States in support of their 'historic' claims. See e.g. the US Attorney-General's opinion in connection with the status of Delaware Bay, J. Moore, *A Digest of International Law*, i (Goverment Printing Office, Washington, DC, 1906), 738; the opinion of the Second Court of Commissioners of *Alabama Claims* (1885) regarding the status of Chesapeake Bay, ibid. 741; the UK argument in the *Fisheries* case regarding the Norwegian enclosure of, *inter alia*, the Inderleia and the Court's answer, the *Anglo-Norwegian* case, 132 and 142; Conversely, navigation has been the major factor prompting States to protest against claims to historic bays. See e.g. the protests of Japan, the UK, and the USA against the Soviet Union's claim over Peter the Great Bay in 1957, in Strohl, *The International Law of Bays*, 350–5.

[97] S. Truver, *The Strait of Gibraltar and the Mediterranean* (Sijhoff & Noordhoff, The Hague, 1980), 41.

cent of the world total being unloaded in Mediterranean ports and 16 per cent of the world total transported in international shipping loaded in Mediterranean ports.[98]

Against this background it is striking that the contribution of the Gulf's waters (south of the closing line) is almost non-existent. This, however, is explained by the fact that, owing to the Gulf's huge incision into the Libyan mainland, ships not bound for the coasts of the Gulf have no reason to enter its waters. To illustrate this factor further the relevance of two major routes in the central Mediterranean is considered.[99]

1. *Inter-oceanic routes*

These comprise two major trades, namely, east–west trans-Mediterranean cargo and oil routes. Combining the two together, one of the world's busiest and most important shipping routes lies some 150 n.m. north of the line of latitude 32° 30′ N., linking the Atlantic and Indian Oceans via the Strait of Gibraltar and the Suez Canal. There are estimated to be between 50 and 150 vessels passing particular points on the route in a single day, possibly 13 per cent of which will be oil tankers.[100] The route passes to the north of Malta, through the Sicily channel, emphasizing the non-existent effect of the Libyan enclosure of the Gulf on the inter-oceanic routes of the Mediterranean.

2. *Intra-Mediterranean routes*

The intra-Mediterranean shipping includes general cargoes, scheduled ferry and passenger services, and oil routes. The relevance of the first two activities to the navigation in the Gulf is almost entirely absent; for although Libya is heavily dependent on imports from a variety of neighbouring States, the most important destinations of these imports are Tripoli in the west and Benghazi in the east.

In the Gulf of Sert there are no regular shipments of general cargoes. While small ships carry goods to the oil-exporting terminals in the Gulf (such as supplies and equipment for the oilfields and terminals), port facilities for general cargo purposes are limited.

Scheduled ferry and passenger services in Libyan ports have almost certainly been far below the levels of activity in most other ports of the Mediterranean. This is largely because of the absence of a large resident population and no transit trade, especially in the southernmost part of the Gulf's hinterland which is mostly unpopulous.

The Gulf contains a relatively large number of oil shipment routes from

[98] Ibid. 44–5.

[99] See also IMO, *Ships Routing* (London, 1984), pt. B: Traffic Separation Schemes, s.iii (Mediterranean and Black Seas).

[100] Suez Canal Authority, *Yearly Report 1983* (Ismailiya, 1983).

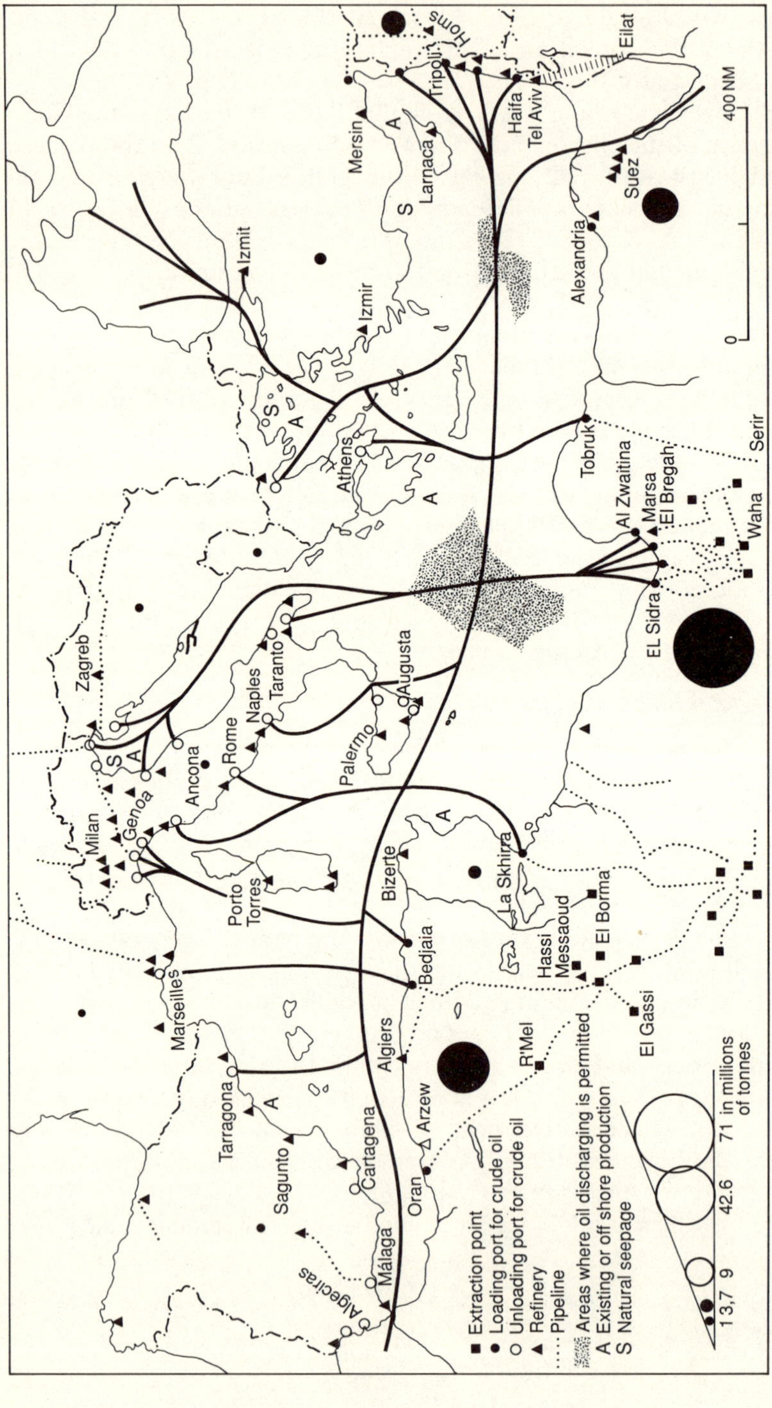

Source: P. Lourd, 'Oil Production in the Mediterranean Sea', 6 *Ambio* (1977), 317.

MAP 13 Production and transport of oil in the Mediterranean Sea

Libya, partly because of the variety of destinations within and outside the Mediterranean but primarily because almost all Libyan terminals are located within the Gulf.[101] This is the only major contribution of the Gulf to international navigation.

What emerges from this background is the fact that the Gulf's economic contribution to the world community at large is almost non-existent. It is this factor which seems to frustrate the argument that the closure of the Gulf would cause a genuine deprivation of community interests. In this regard it is indeed difficult to substantiate that the protests held against the Libyan Declaration identify any bona fide community interests. Their legal validity reflecting the absence of 'acquiescence' is questioned, furthermore, for their failure to comply with the concept of good faith, especially in the face of the 'silence' or 'indifference' of eleven Mediterranean States, against one (Malta) who protested expressly and four (Greece, France, Tunisia, and Turkey) who reserved their rights to express their position regarding the enclosure.[102]

3. THE RELEVANCE OF THE CONCEPT OF VITAL INTERESTS

(*a*) *Discussion in the jurisprudence*

The concept of vital interests is traditionally linked with the doctrine of absolute sovereignty and used as a plea in defence of action taken in response to any kind of act or situation whether it be of a lawful or unlawful character. In the words of Brierly,

> On the whole it remains that States in general regard some of their interests and policies as so 'vital' that they mean to remain free to assert them, if need be, whatever the law may say about them, and moreover to decide for themselves which among their national interests they will regard in this way.[103]

[101] See *infra*, s.4(*b*)(ii), (iii) and Map 13.

[102] The dispatches of thirteen governments expressing their attitude towards the Libyan decision to close the Gulf were provided by courtesy of the Legal Adviser's Office, the People's Bureau for Foreign Liaison, Tripoli. In addition to the Mediterranean States the position of the other States is as follows: Belgium, Bulgaria, the UK, the USA, and the USSR protested expressly; whereas Austria, Denmark, and the Federal Republic of Germany reserved their right to express their position. Although Italy appears to oppose the Libyan claim, it has not done so publicly. It was reported that, in a meeting held in Tripoli on 23 Apr. 1974 with a Libyan official, the Counsellor of the Italian Embassy stated that his Government objected to the Libyan claim though it did not wish to do so in writing. Ibid. To the same effect, see a written reply by the Italian Under-Secretary for Foreign Affairs to a member of the Italian Parliament, in 2 *IYIL* (1976), 422–3.

[103] J. Brierly, 'Vital Interests and the Law', 21 *BYIL* (1944), 51. In that sense Brierly's definition is similar to Brown's definition of the concept of necessity or self-preservation

Under contemporary international law, however, such a plea has lost much of its logic.[104] Politico-economic interactions between States have resulted in factual and legal constraints under which competing interests are tested by the standards of 'good faith' and 'reasonableness'. Obviously, international law allows a State a wide discretion in determining the nature, extent, and duration of the interest qualified as vital. This discretion extends also to the methods best calculated to meet such vitality of interests.[105] But this discretion must be exercised in good faith, and the law will intervene in cases where it is abused.[106] In the law on maritime boundaries, for instance, it no longer carries conviction for a State to consider the enclosure of an area of the high seas as a non-negotiable national interest, completely withdrawn from the dominion of the law. As stated by the ICJ in the *Anglo-Norwegian* case, it is incumbent upon the coastal State to conduct its policy 'within the bounds of what is moderate and reasonable'.[107] The Court emphasized the principle that 'the delimitation of sea areas has always an international aspect; it cannot be dependent merely upon the will of the coastal State as expressed in its municipal law'.[108] Having stated this, the Court introduced the parallel principle that a State 'must be allowed the latitude necessary in order to be able to adapt its delimitation to practical needs and local requirements'.[109]

However, the specific question which has recently arisen in the doctrine is whether the principle of vital interests, thus subjected to the bounds of 'reasonableness' and 'good faith', may allow a State's appeal to its vital needs to justify an 'historic title' which might not be recognized if the element of 'long usage' was strictly applied.

Some commentators take a theoretical approach and argue that whilst 'vital interests' can be used as an element to fortify a certain claim which is usually based on other well-established criteria, they can play no determinative role in the acquisition of a historic title. The premiss of this argument is that 'the historic bay concept has always been connected with past events. By definition it had no reference at all to emerging needs and

which 'merely excuses what would otherwise have been unlawful'. 'The Lessons of the Torrey Canyon: International Law Aspect', 21 *CLP* (1968), 113, at 127.

[104] For a concise account of the trend in legal discussions to redefine the concept of absolute vital interests see K. Partsch, 'Vital Interests', *Max Planck Encyclopedia* (Elsevier Science Publishers, Amsterdam, 1987), x. 526–8.

[105] See Cheng, *General Principles of International Law*, 132–3.

[106] While, e.g. the right of a State in war to determine what are military necessities is left to the discretion of that State, international tribunals are entitled to intervene 'in cases of manifest abuse of this freedom of judgment'. *Spanish Zone of Morocco Claims* (1923): Rapport III (1924), 2 *RIAA*, 615, at 645 (trans.). See also *US Nationals in Morocco* case, ICJ Reports (1952), 212.

[107] *Anglo-Norwegian* case, 142.

[108] Ibid. 132.

[109] Ibid. 133.

aspirations.'[110] Some proponents of this view, however, also postulate a category of 'vital bays' by which newly urgent needs might be recognized in exceptional circumstances and States 'should not be forced to rely on an escape valve, not easily open to them, in the form of historic title'.[111] Bourquin, for example, argues that 'there seems little doubt the vital interests of the State in the possession of a bay constitutes side by side with historical tradition one of the bases on which it may rely in claiming sovereignty therein', but, he concludes, the two institutions must not be confused.[112]

An opposite view, however, reflecting State practice and judicial decisions, suggests that vital interests can only be relevant as part of the doctrine of historic bays. The suggestion according to this view is that vital interests play an equitable role, that is, moderating the requirements traditionally suggested for historic bays. An example of this liberalizing role is that suggested by O'Connell, whereby vital interests can be used 'in the evaluation of long usage'.[113] According to this argument the traditional requirement of 'long usage' should be moderated so as to take into account the interests of newly emergent States who have not been sovereign for long enough to identify their national interests and, consequently, to establish a historic claim on the basis of their own interest.[114]

Beside this 'special circumstance' of newly independent States, it has been further argued that vital interests constitute the *de facto* element in the claim to a historic bay. Gidel, for instance, acknowledged the fact that 'the description "vital bays" . . . sums up in one word the conditions of substance to be fulfilled by the areas in question [i.e. historic bays], whereas the expression "historic bays" suggested conditions of formality'.[115]

This argument has been reinforced in leading municipal and international decisions. The often quoted Judgment of the Court of Commissioners of *Alabama Claims* regarding Chesapeake Bay is significant:

> can we permit belligerent operations between foreign nations within the shores of this bay? What injustice can be done to any other nation by the United States

[110] See McDougal and Burke, *The Public Order of the Oceans*, 358. See also the UN Secretariat Report, 20; Bouchez, *The Regime of Bays in International Law*, 300.

[111] McDougal and Burke, *The Public Order of the Oceans*, 341.

[112] Bourquin, *Les Baies historiques*, 51, quoted and translated in UN Memorandum, 30. See also Goldie, 'Historic Bays in International Law', 226–9; the UN Secretariat Report, 20.

[113] O'Connell, *The International Law of the Sea*, i. 438.

[114] The rationale which has persistently been put forward for this view is that the traditional concept of historic bays had been elaborated in response to a former need that does not necessarily comply with the position of the newly independent States. See e.g. comment of García Amador, *Yearbook of the ILC* (1955), i. 211–12, paras. 59 and 61; statement of the Venezuelan delegate at the 1958 Conference, Doc. A/CONF. 13/39, p. 23; the Resolution of the African States' Regional Seminar on the Law of the Sea, Yaoundé, 2–30 June 1972, circulated as a Document of UN Seabed Committee II, Doc. A/AC. 138/79.

[115] Gidel, *Le Droit international public de la mer*, 629.

exercising sovereign control over these waters? What annoyance and what injury may not come to the United States through a failure to do so?[116]

In the *North Atlantic Coast Fisheries* Arbitration, the Permanent Court of Arbitration referred to the 'conditions of national and territorial integrity, of defence, of commerce and of industry' as all being 'vitally concerned with the control of the bays penetrating the national coast line'.[117]

The actual expression 'vital interests' enhanced this line of thought in the *Gulf of Fonseca* case, where the tribunal said: '[i]t is necessary to specify the characteristics proper thereto from the threefold point of view of history, geography and the vital interests of the surrounding States.'[118] Ruling on the relevance of these factors in the establishment of a historic claim, the Tribunal emphasized the supremacy of the vital interests factor. After reviewing the factors of history and geography related to the Gulf of Fonseca the Tribunal concluded that these factors 'give an exact idea of how vital are the interests guarded by the Gulf of Fonseca, and [how] those interests are of incalculable value in making up the characteristics of an "historic bay"'.[119] The Tribunal then called attention to certain economic and defence interests in order to support the decision that the waters of the Gulf were national waters: commercial and industrial centres were situated in the area around the Gulf; railways had been constructed connecting these waters with the hinterland; and the configuration of the Gulf facilitated the enforcement of the fiscal and other essential regulations of the littoral States. The interest of defence of this region was a necessary consequence of the aforementioned economic interests.[120]

In the *Anglo-Norwegian Fisheries* case, the ICJ found that 'certain economic interests peculiar to a region, the reality and importance of which are clearly evidenced by a long usage',[121] may be related to the basis by which a State advances a historic claim. In this connection, the Court took notice of the view of the Counsel for Norway:

> The Norwegian Government does not rely upon history to justify exceptional rights, to claim areas of seas which the general law would deny; it invokes history, together with other factors, to justify the way in which it applies the general law. This conception of an historic title is in consonance with the Norwegian Government's understanding of the general rules of international law.[122]

[116] Moore, *Digest of International Law*, 741.
[117] *North Atlantic Coast Fisheries* Arbitration, 4 AJIL (1910), 948, at 979.
[118] *Gulf of Fonseca* case, 11 AJIL (1917), 647, at 700.
[119] Ibid. 704.
[120] Ibid. 704–5.
[121] *Anglo-Norwegian Fisheries* case, 133.
[122] Ibid.

Of particular interest is the Court's emphasis that history and vital interests are interdependent factors. Referring to the fishing grounds of Lapphavet, the Court said:

> Such rights, founded on the vital needs of the population and attested by very ancient and peaceful usage, may legitimately be taken into account in drawing a line which, moreover, appears to the Court to have been kept within the bounds of what is moderate and reasonable.[123]

Of interest in this respect is the most recent decision of the Supreme Court of the United States involving a claim to historic waters.[124] It represents a major retreat from its previous decisions which had taken an extremely narrow view of the requirements concerning the establishment of a historic bay. It will be clear from the following quotations that this has been possible as a result of the Court's recognition of 'vital interest' as an essential factor in determining whether a body of water is a historic bay.

In the previous cases the Court had consistently pointed out that a historic bay 'is . . . essentially a bay over which . . . a coastal nation has *traditionally* asserted and *maintained* dominion with the *acquiescence* of foreign nations'.[125] With regard to the scope of the authority required, the Court had maintained that a historic claim should be 'clear and beyond doubt' in order to overpower the Federal Government disclaimer;[126] and, consequently, the Court had declared that the enforcement of regulations aiming at the exclusion of foreigners from fishing was not legally sufficient 'to demonstrate the type of authority necessary to establish historic title' and that 'the exercise of authority necessary to establish historic title must be commensurate in scope with the nature of the title claimed'.[127]

In respect of the position of other nations, the Court had again taken the narrowest view: 'we feel that something more than the mere failure to object must be shown.'[128]

In the 1985 Judgment, however, the Court took a more relaxed approach. Having mentioned the three 'traditional' requirements for historic claims,[129] the Court added that

> In addition, there is substantial agreement that a fourth factor to be taken into consideration is the vital interests of the coastal nation, including elements such as

[123] Ibid. 142.

[124] *US* v. *Louisiana et al.* (*Alabama and Mississippi Boundary* case), 470 US 93, 84 L. Ed. 2d 73 (1985).

[125] See *US* v. *California*, 381 US 139, 14 L. Ed. 2d 296 (1965); *US* v. *Louisiana* (*Louisiana Boundary* case), 394 US 11, 22 L. Ed. 2d 44 (1969); *US* v. *Alaska*, 422 US 184, 45 L. Ed. 2d 109 (1975); *US* v. *Maine et al.* (*Rhode Island and New York Boundary* case), 469 US 504, 83 L. Ed. 2d 998 (1985).

[126] e.g. *US* v. *California*, 318; *Louisiana Boundary* case, 91.

[127] *US* v. *Alaska*, 117–18, 120.

[128] Ibid. 122.

[129] *Supra*, s.2(*b*)–(*d*).

geographical configuration, economic interests, and the requirements of self-defense.[130]

On the requirement of acquiescence the Court observed that

There is substantial agreement that when foreign governments do know or have reason to know of the effective and continual exercise of sovereignty over a maritime area, inaction or toleration on the part of the foreign governments is sufficient to permit a historic title to arise. Moreover, it is necessary to prove only open and public exercise of sovereignty, not actual knowledge by the Foreign Government.[131]

With respect to its newly identified requirement of vital interests, the Court was unequivocal in rejecting the argument of the Federal Government that such a factor was irrelevant for the claim to a historic bay:

The United States argues that this official recognition of Mississippi Sound as an internal waterway of commercial and strategic importance has no relevance to the Sound's status as a historic bay. It would support this argument with a citation to the 1962 United Nations study of historic waters . . . at 56–58. The cited pages of the study discuss the view taken by some authors and governments that such circumstances as geographic configuration, requirements of self-defense, or other vital interests of the coastal state may justify a claim to historic bay status without the necessity of establishing long usage. The study notes, *ibid.* at 58, that '[t]here is undoubtedly some justification for this view', but ultimately suggests that it does not make sense for 'historic title' to be claimed in circumstances where the historic element is wholly absent . . . The study, however, does not suggest that such circumstances as geographic configuration and vital interests are irrelevant to the question whether a body of water is a historic bay and, indeed, it affirmatively indicates that such circumstances can fortify a claim to 'historic bay' status that is based on usage.[132]

The Court went on to conclude that such vital interests, including the recognition of commercial and strategic waterways, were very relevant in the present case:

the evidence demonstrates that the United States historically and expressly has recognized Mississippi Sound as an important internal waterway and has exercised sovereignty over the Sound on that basis throughout much of the 19th century.[133]

Arguing on the basis of the previous Court cases, the Federal Government attempted to introduce a rigid test whereby the requirement of sovereignty, manifest in the exclusion from the area of all foreign navigation in innocent passage, 'is . . . the only conduct that conclusively demon-

[130] *Alabama and Mississippi Boundary* case, 80–1.
[131] Ibid. 86.
[132] Ibid. 83.
[133] Ibid.

strates that the nation exercises authority over the waters in question as inland waters and not merely as territorial sea'.[134] The Court dismissed this argument as 'rigid' and 'unrealistic' and added

> To the contrary, in advocating a flexible approach to appraisal of the factors necessary to a valid claim of historic inland-water status, two leading commentators have stated: 'A relatively relaxed interpretation of the evidence of historic assertion and of the general acquiescence of other states seems more consonant with the frequently amorphous character of the facts available to support these claims than a rigidly imposed requirement of certainty of proof, which must inevitably demand more than the realities of international life could ever yield.'[135]

After quoting from the United Nations Secretariat Report which gave evidence to the same effect, the Court concluded,

> Thus, although a coastal nation has the privilege to exclude from its inland waters foreign vessels in innocent passage, the need to exercise that privilege may never arise. Indeed, in the present case, as the United States seems to concede, the record does not indicate that there ever was any occasion to exclude from Mississippi Sound foreign vessels in innocent passage. This is not surprising since, as noted above, foreign nations have little interest in Mississippi Sound and have acquiesced willingly in the United States' express assertions of sovereignty over the Sound as inland waters. We conclude that the absence in the record of evidence of any occasion for the United States to have exercised its privilege to exclude foreign navigation in innocent passage from Mississippi Sound supports rather than disproves the claim of historic title to the Sound as inland waters.[136]

It was not surprising, therefore, that the Court ruled that the development of Mississippi Sound for the purpose of intra-coastal waterways, construction of a military fortification (abandoned fifteen years after construction), erection of lighthouses, and a statement in a 1906 case that the Sound was 'an enclosed arm of the sea, wholly within the United States' were activities sufficient to assert exclusive jurisdiction for the purpose of a historic claim.[137]

(*b*) *State practice*

The need to have recourse to 'vital interests' to justify a State's action is reaffirmed in recent State practice. Perhaps the most recent example is the statement of reasons given by the United States State Department on its withdrawal from the proceedings in *Nicaragua* v. *United States*.[138] The

134 Ibid. 88.
135 Ibid.
136 Ibid. 89.
137 Ibid. 82–4.
138 *Military and Paramilitary Activities in and against Nicaragua* (*Nicaragua* v. *US*) (Merits), ICJ Reports (1986), para. 10.

statement explained: 'we have acted in the exercise of the inherent right to collective self-defense . . . We have done so in defense of the vital national security interests of the United States.'[139]

However, the relevance of 'vital interests' to the concept of historic bays[140] has been thought to be self-evident upon a careful analysis of contemporary State practice. Whilst States frequently close certain areas primarily on security grounds they naturally desire to fit such claims within the established and widely recognized concept of historic waters. This has been postulated by a recent commentator who advises that the

> number and frequency of coastal States' claims . . . shows that the old concept of an historic bay is currently evolving into a more flexible notion whose crucial elements are the *bona fide* assertion of State interests and the recognition of and acquiescence of third States, rather than immemorial usage and the long passage of time.[141]

On the basis of this background the same commentator expresses his doubt that 'the doctrine of "vital" bays may be labelled as a radical and isolated Libyan doctrine'.[142] Whilst the practice of States includes numerous assertions of such claims, dating even from very early times when there was no particular insistence that baselines had to be limited,[143] there has been no single claim which is wholly and exclusively supported by historical assertions of title. The following selective analysis of recent State practice, however, proposes to address the basic factors motivating States to claim 'historic bays'. These factors are either proclaimed by the claimant State or, as is often the case, deduced from the factual circumstances accompanying the decision to close the bay. In this regard, two factors appear to be most relevant.

(i) Enclosures rooted in vital security interests

There is no doubt that the national defence factor has exercised special influence upon the law of bays. From the strategic point of view, an

[139] Quoted in *Max Planck Encyclopedia*, x. 528. See also the statement of the Canadian Prime Minister in the House of Commons in Apr. 1970 defending the Canadian reservation concerning the jurisdiction of the ICJ in matters of the law of the sea, ibid.; a letter from the Minister for Foreign Affairs of Iceland to the ICJ on 29 May 1972 expressing his Government's unwillingness to confer jurisdiction on the Court 'considering that the vital interests of the people of Iceland are involved.' *Fisheries Jurisdiction* (*UK* v. *Iceland*), ICJ Reports (1973), 7.

[140] Judge Drago is believed to have been the first one who emphasized this relevance. See extract from his Diss. Op. in the *North Atlantic Sea Coast Fisheries* Arbitration, quoted in the UN Secretariat Report, 17–18.

[141] F. Francioni, 'The Status of the Gulf of Sirte in International Law', 11 *SJILC* (1984), 311, at 325.

[142] Ibid. 323.

[143] See e.g. Bouchez, *The Regime of Bays in International Law*, 216–37.

indentation in a country's coast usually commands special attention. It is generally within this indentation that ports are located. It is through these ports that the country's domestic and international commerce takes place. Consequently, communications, industrial areas, and big cities are established and extended around these ports. The depression in the coastline created by the bay is of itself a geographical advantage. It may 'shorten' the distance over which crude oil or other raw materials have to be transported from the hinterland to the coast, for export or industrial purposes. The ports also serve as sheltered areas for naval bases and naval facilities. The bay as a whole may offer the advantage of enclosed areas for the manœuvring and operation of seaborne or airborne forces. Bays in relatively warm and deep waters provide excellent conditions for submarines to operate comfortably. In response to these vital interests, the coastal State usually concentrates its defence in such areas; for, in times of crisis, such areas become a prime target for foreign offensives.

For these reasons, it is common practice that States in peacetime have closed certain bays to protect their security interests. The earliest indication usually cited as part of State practice is the 1793 opinion of the Attorney-General of the United States given in connection with the 'territoriality' of Delaware Bay. In affirming that all of the Bay was within the United States, the opinion rested only in part upon the fact that prior governments had claimed exclusive jurisdiction over the Bay. There was, further, insistence that no community interest would be affected by honouring the United States claim, whereas the United States might be gravely injured if the Bay were not regarded as part of its exclusive domain.[144]

In 1957 the Soviet Union closed Peter the Great Bay as internal waters.[145] The Soviet Union's justification for the enclosure was that the Bay 'historically belongs to Soviet internal waters, owing to the special geographical conditions of the Bay and to its significance in the economy and national defence'.

However, beside the 'historic rights' argument brought by the Soviet Union,[146] security considerations were certainly the vital motive behind the Soviet annexation of Peter the Great Bay. Vladivostok is the area

[144] The fact that a British ship (which had been captured by the French in the waters of the Bay) was returned to Britain was 'generally regarded as manifesting the acquiescence of both States to the United States claim over the Bay, and this whole transaction is thought to provide an instance of a claim which ought to be generally accepted irrespective of any particular prescriptions relating to baselines'. Jessup, *The Law of Territorial Waters*, 395–7.

[145] Decree of the Council of Ministers of 4 July 1957. This claim was confirmed by the Soviet Union in a 'notice to seafarers' published in 1986. The notice also includes four other 'historic bays', namely, the White Sea, the Bay of Chesk, the Bay of Bajdarack, and the Bay of Penzinck. For details see T. Scovazzi, 'Developments Concerning Soviet Straight Baselines', 3 *IJECL* (1988), 37–43.

[146] For a detailed account of the Soviet argument see Strohl, *The International Law of Bays*, 332 ff.

located in the middle of the Bay. The Bay's intrusion into the Soviet Union hinterland facilitated means of communication with the Asiatic interior including the Soviet Union. The port of Vladivostok played an important role as a rear base for Russian troops during the Russo-Japanese war. The port has been an important naval base and the entire maritime district of Vladivostok is heavily militarized.[147]

On similar grounds China in 1958 claimed the Pohai Bay as internal waters.[148] Whilst there is some evidence that China regards Pohai Bay as an 'historic bay',[149] the genuine motive of the enclosure is related to strategic and economic factors. In 1969 Cheng wrote:

> There can be little doubt that Pohai Bay is vital to the security of Communist China. Access to its waters would permit an attack supported by naval forces against any part of an area within a gigantic circle from Manchurian industrial centres to the heart of the North China plain. It would have been extremely difficult, if not impossible, for the Western Powers to bring the rulers of the Ching Dynasty to their knees in the nineteenth century without the use of Pohai Bay.[150]

However, since 1955 China has claimed the whole bay as a 'Military Warning Zone' and reiterated this position in its 1975 Fisheries Agreement with Japan.[151]

The Italian claim to the Gulf of Taranto is also rooted in security concerns. The Italian Decree of 1977 established a closing line, about 60 miles long, across the entrance of the Gulf between Cape S. Maria di Leuca and Alice Point. The decree justified the line by proclaiming the Gulf of Taranto as an 'historic bay'.[152] Ronzitti, in a critical analysis of the Italian claim, finds no evidence to suggest that Italy claimed or exercised exclusive authority over the Gulf before 1977.[153] However, as Francioni observes, the main reason behind the Italian action was purely strategic: 'to secure the Gulf from the unwelcome visits of non-NATO

[147] See ibid. 332, n. 9. For these reasons the Soviet Union was adamant that it should secure sovereignty over the Bay, despite strong protests from Japan and several Western powers. See Bouchez, *The Regime of Bays in International Law*, 224–5.

[148] See a 'hypothetical study' of the Chinese straight baselines system in *LIS*, 45.

[149] See J. A. Cohen and H. Chiu, *People's China and International Law: A Documentary Study* (Princeton University Press, Princeton, NJ, 1974), who gave the full text of Fu Chu's article 'Concerning the Question of Our Country's Territorial Sea' (Peking, 1959). Fu Chu specifically argues that Pohai Bay is historic. Further evidence of the Chinese attitude is found in B. Cheng, 'Communist China and the Law of the Sea', 63 *AJIL* (1969), 47, at 60–1.

[150] Cheng, 'Communist China and the Law of the Sea', 61.

[151] In the Agreement both parties maintained their position with regard to this military zone. See formal letters by China and Japan appended to the 1975 Agreement in *LIS* 70: 20–1.

[152] Art. 1 of the 1977 Decree, *supra*, Ch. 6 n. 31.

[153] N. Ronzitti, 'Is the Gulf of Taranto an Historic Bay?', 11 *SJILC* (1984), 275. Most Italian lawyers who commented on the subject arrived at the same conclusion. See e.g. B. Conforti, *Lezioni de dirrito internazionale* (Giuffrè Editore, Milan, 1983), 193; M. Fusillo, 'Baselines for Delimiting the Territorial Sea', 3 *IYIL* (1977), 570, at 575.

naval units.'[154] The Gulf contains important Italian naval installations, constituting part of NATO's military garrisons in the Mediterranean. This probably explains why the traditional opponents of extended maritime claims such as the United Kingdom and the United States did not challenge or question the Italian decision.[155] Fusillo, commenting on this decision in the light of the current practice of States, concludes: '[w]hilst the classification of the Gulf of Taranto as a historic bay cannot be considered as lawful on the basis of the trends prevailing when the Geneva Convention was adopted, nevertheless it appears to form part of the current development in practice.'[156]

Similar 'historic' claims rooted in national security interests are pressed by several other States such as Argentina and Uruguay in relation to the Río de la Plata,[157] Panama with respect to the Gulf of Panama,[158] and the Socialist Republic of Vietnam with respect to the Gulf of Bac Bo[159] (Tunkin) and the south-eastern sector of the Gulf of Thailand.[160]

(ii) Enclosures for vital economic interests

Fishing is the 'traditional' economic interest which has been associated with, and indeed influenced the development of, the law on bays.[161] The Norwegians' successful recourse to their vital economic interests in the

[154] Francioni, 'The Status of the Gulf of Sirte', 323. It was reported that in Aug. 1982 Soviet submarines entered the Gulf of Taranto. Italy protested and the Soviet Union denied the incident. See N. Ronzitti, 'Sommergibi non identificati, Pretese Baie Storiche e contromisure dello Stato Costiero', 66 *RDI* (1983), 5.

[155] The only reported comment of a NATO member is that made by the then British Foreign Secretary, Lord Carrington, in 1981. Asked whether the UK or NATO recognized the Italian claim to the Gulf of Taranto, he replied: 'Italy claims the Gulf of Taranto as internal waters. This is not consistent with our interpretation of the 1958 Geneva Convention on the Territorial Sea. NATO does not take a position on the territorial limits of its members.' 424 *Parliamentary Debates* (House of Lords Official Reports), Fifth Series (1981), 368. This cannot be interpreted in any way as a protest.

[156] Fusillo, 'Baselines for Delimiting the Territorial Sea', 575.

[157] Joint Declaration of Argentina and Uruguay of 30 Jan. 1961, in *LIS* 44.

[158] Panama claimed the Gulf of Panama as historic waters in 1956. Colombia and Costa Rica, which concluded maritime boundary agreements with Panama in 1976 and 1980 respectively, agreed they would not object to the Panamanian claim. See *LIS* 79, 97.

[159] According to the Vietnamese Declaration of 1977, the area of the Bac Bo (within a line connecting the Chinese–Vietnamese border with a point at the entrance of the Gulf—to be agreed upon with China) is claimed as internal waters. This was reiterated in 1982. A statement by the Government provided that '[t]he part of the Gulf appertaining to Vietnam constitutes the historic waters and is subjected to the juridical regime of internal waters of [Vietnam]'. *LSB* 1 (1983), 74. China disputes this claim and the dispute is expected to escalate following recent Chinese hydrocarbon exploration activity in the Gulf's seabed. See Ying-jeou Ma, *Legal Problems of Seabed Boundary Delimitation in the East China Sea* (University of Maryland, School of Law, Baltimore, 1984), 67.

[160] *LIS* 99.

[161] See e.g. C. Hurst, 'The Territoriality of Bays', 3 *BYIL* (1922–3), 42.

1951 *Fisheries* case has made it clear that the whole doctrine of the maritime domain 'is suffused with the notion of vital interests'.[162]

The special value of the waters of the Gulf of Concale for the fishing of oysters was the primary reason prompting France to conclude the 1839 Fishery Agreement with Great Britain providing for the exclusive right of French subjects to fish in the waters of the bay.[163] In 1929, France, invoking this treaty, claimed the Gulf of Concale as a historic bay.[164] Reservation of fishing for the nationals of Sweden was also the basic element upon which Sweden claimed Laholm as a historic bay.[165]

The existence of a local salmon fishing industry vital to the inhabitants led the Polish authorities in 1956 to claim as internal waters the western sector of the Gulf of Gdansk, landward of a closing line connecting Cape Hel with the Polish–Soviet border.[166] In accordance with the 1926 Decree-Law on Fishing and Sponge-Fishing, Egypt claimed the Bay of El-Arab as part of Egyptian 'territorial waters'.[167] It was the existence of traditional sponge-fishing activity in this shallow bight, situated west of Alexandria, which led to the enclosure of the Bay, which is 70 miles wide and 18 miles in depth.

The interests of coastal communities in fishery resources have been invoked by many other States to justify claims to 'historic bays'. Examples include the claims of Thailand over the Bay of Thailand,[168] Burma over the Gulf of Martaban,[169] and Bulgaria over Varna and Burgas Bays.[170]

162 O'Connell, *The International Law of the Sea*, i. 437.

163 27 *BFSP* (1838–39), 983, at 988.

164 See the French reply to the enquiries advanced by the Preparatory Committee of the 1930 Hague Conference, quoted in the the UN Memorandum, 3.

165 Bouchez, *The Regime of Bays in International Law*, 218–19.

166 The Polish Decree of 1956 does not designate the waters of the bay as historic, although the latter solution is probably the easiest for Poland to follow, as it removes the problem of conformity with the 1958 Convention rules on juridical bays (the closing line of the Gulf is 30 miles). However, Sebek (*EESDLS* (1979) (Binder I), 175–6) mentions that in order not to prejudice its claim over the bay, Poland opted not to ratify the 1958 Convention on the Territorial Sea.

167 Quoted in the UN Memorandum. The expression 'territorial waters' seems to mean internal waters. According to the Royal Decree of 1951 (*supra*, Ch. 6 n. 63), the inland waters of Egypt included 'the waters of the bays along the United Kingdom of Egypt', including, *inter alia*, the Bay of El-Arab. See also the Egyptian Government reply to question No. 2 of the Committee of Experts for the Progressive Codification of International Law, quoted in UN Memorandum, para. 24.

168 Thailand claimed the Bay of Thailand as an historic bay in 1959. However, in 1970 Thailand proclaimed its straight baseline system comprising three segments of baselines along its coast, two of which are in the Gulf of Thailand. See *LIS* 31. See the statement of the Thai delegate to UNCLOS I in UNCLOS I, *Official Records*, iii. 17.

169 Although the Burmese Declaration of 1968 (which closed, *inter alia*, the Gulf of Martaban within a closing line of 222.3 n.m.) did not make any reference to historic factors, it justified the claim by reference to the 'geographic conditions prevailing along' the coast and to the desire to 'safeguard vital economic interests'. Text in *LIS* 14.

170 According to the Bulgarian Decree of 1935, the Bulgarian Bays of Varna (or Stalin) and Burgas are regarded as Bulgarian internal waters. They have closing lines of 16 and

Vital economic interests are not, however, confined to fishing. The present practice of States shows that they invoke their alleged 'historic rights' in order to secure effective possession of and a bigger share of seabed areas of hydrocarbon potential. China's vigorous maintenance of its 'historic' claim over the Pohai Bay is understood in the light of the recent oil exploration activity in, *inter alia*, the Pohai basin and the adjacent seabed. Since 1978, Sino-foreign joint ventures have engaged in such activity and oil discovery in the basin has already been registered. China's policy has been to exclude the Pohai waters from any potential continental shelf agreement with the other neighbouring States.[171] Another similar example is the Colombian–Venezuelan territorial sea/continental shelf dispute in the Gulf of Venezuela. Venezuela claims most of the Gulf—bordered also by Colombia—south of a line drawn between Castillete (at the two States' border) and Salinas as historic waters.[172] Hydrocarbon resources are predicted in the continental shelf area north of this closing line. Venezuela's policy has been not to yield any part of the Gulf south of the closing line.[173] It argues for a continuation of the land frontier to a point 12 miles off the coast. Colombia, on the other hand, argues for a normal, equidistant boundary.[174] Each position with regard to this initial boundary would produce a different median line for the continental shelf boundary between the two States.[175]

In 1973, five years after the Libya–Tunisia continental shelf dispute arose, Tunisia established its straight baseline system enclosing, *inter alia*, the Gulf of Gabes.[176] In 1980, during the proceedings of the *Tunisia–Libya* case, Tunisia argued that the enclosure of the Gulf of Gabes could be based on the doctrine of historic bays.[177] Tunisia appears to have a strong case for the enclosure of the Gulfs of Tunis and Gabes on historic

23.5 n.m., respectively. The Decree does not make explicit reference to historic bays, but, according to Nikolaev (cited by Sebek, *EESDLS* (Binder I), 172), the enclosure was made to protect the vital economic interests of Bulgaria.

171 See Ying-jeou Ma, *Legal Problems of Seabed Boundary Delimitation*, 49 and map 11, p. 274.

172 The line was drawn by the Venezuelan Government in 1939 and confirmed in 1940. See K. Nweihed, 'EZ (Uneasy) Delimitation in the Semi-enclosed Caribbean Sea: Recent Agreements between Venezuela and Her Neighbours', 8 *ODIL* (1980), 1, at 9.

173 Ibid. 10.

174 See Bowett, *The Legal Regime of Islands in International Law* (Oceana Publications, New York, 1979), 313 ff.

175 Ibid. esp. fig. on p. 316.

176 Decree-Law of 1973, *supra*, Ch. 6 n. 69. The closing line of the Gulf of Gabes joins a buoy situated off the Kerkennah Islands and the north-eastern extremity of the island of Jerba. The choice of these basepoints rather than other points on the mainland is explained by the fact that this line is a continuation of the straight baselines drawn, *inter alia*, around the Kerkennah Islands. See *supra*, Ch. 6, text at nn. 69–76.

177 See *Memorial* submitted by the Government of Tunisia (1980), 92.

grounds.[178] The fact, however, remains that the Tunisian historic claim, and indeed the whole straight baselines system of 1973, was, as manifested in the *Tunisia–Libya* case,[179] to exclude all the enclosed areas from any proportionality test applied by the Court for delimitation purposes. This would have allocated to Tunisia large areas of the continental shelf within which Libyan and Tunisian oil concessions had overlapped. But, for reasons shown earlier,[180] the Court rejected the relevance of Tunisia's historic rights to the shelf boundary problem.

(*c*) *Conclusion*

The above review of contemporary State practice reveals that the old concept of historic bays is currently evolving into a more flexible regime whereby history plays a less prominent role than was traditionally assumed and strategic and economic factors a much greater one. While States invoke abstract and imperfectly formulated requirements such as 'long-standing history' and 'acquiescence' to support their claims, the real issue at stake in such claims is the existence of strategic and/or economic interests emanating from the inevitable interaction between two geographical and political phenomena, namely the penetration of the bay into the mainland on the one hand and the existence of a coastal community on the other. It is often the case that this interaction has existed for a long time, for it is not the State which confers or attaches such a great value to the interests, but rather that the economic, social, and security conditions of a given period are the ones which give birth to such vital interests and thus indicate their great value.

However, as the reality of the situations shows, such an importance may crystallize as a result of contemporary strategic and economic needs, so much so that the time required for this crystallization may be short and incidental. The recent opinions of authors seem to recognize such development and tend to acknowledge its relevance for the theory of historic

[178] Tunisia's regulations over sedentary and mobile fisheries in the banks and shoals off its coasts appear to date as far back as 1117. See A. Moussa, *La Tunisie et le droit de la mer* (Paris, 1981), 25–9. In 1871, the Law Officer of the Crown, advising the British Government not to protest about the assertion of rights over sponges and coral by the bey of Tunis in that year, stated, 'there is no objection in principle to the Bey of Tunis asserting an exclusive right to the *fructus* of the banks off the Coast of Tunis, to which Sponge and Polypi attach themselves, although the banks in question are at a greater distance than three miles from the coast, provided the Bey can show a prescriptive enjoyment of such *fructus*'. (A. McNair (ed.), *International Law Opinions* (Cambridge University Press, Cambridge, 1956), i. 259.) The historic claim over the Gulfs of Tunis and Gabes, enclosed by closing lines 23 and 45 n.m. long, respectively, was examined recently by A. Gioia, 'Tunisia's Claims over Adjacent Seas and the Doctrine of "Historic Rights"', 11 *SJILC* (1984), 327.

[179] See the *Libya–Tunisia* case, paras. 98–105.

[180] See *supra*, Ch. 6, text at n. 77. See also *infra*, Ch. 8 s.4(*b*)(ii)(3).

bays. In the following section the merits of the Libyan claim in the light of this doctrinal and practical development will be discussed.

4. ASSESSMENT OF LIBYA'S CLAIM ON THE BASIS OF THE CONCEPT OF VITAL INTERESTS

As noted earlier, Libya bases its exclusive jurisdiction over the Gulf of Sert partly on the principle of vital interests. Unfortunately, Libya has not so far made public any particular factors which may indicate any security or economic needs in the Gulf which are different from those generally attributed to huge coastal indentations such as the Gulf of Sert. It is proposed in this section to assess this particular Libyan argument in the light of the Gulf's geographical/economic/strategic characteristics.

(*a*) *The relevant physical characteristics of the Gulf of Sert*

The Gulf of Sert intrudes into the land territory of Libya from roughly the central point along its coast, dividing the country's coastline into two close to the parallel of 30° 30′ N. latitude. From its outer limit along the closing line of latitude 32° 30′ N. the Gulf encompasses a water surface area of 23,531 sq.n.m. This surface area includes the most southern and, consequently, the warmest waters in the Mediterranean. This southerly position results, in addition, in the passage of fewer depressions than anywhere else except the Algerian coast. On average, the Gulf is slightly more sheltered than other such locations in the southern Mediterranean, and it enjoys one of the lowest percentage frequencies of winter gales.[181]

In contrast to other major indentations in the Mediterranean, the Gulf's actual intrusion into the land mass of Libya is obscured by the unique projection of its arms. Unlike, for instance, the Gulfs of Taranto and Gabes, the Gulf of Sert's arms do not correspond to each other in their penetrating trends landwards. Whilst the eastern, shorter arm has a horizontal, south–north trend, the larger western arm has a gradual, south-east–north-west direction, which has consequently exhausted what would otherwise be a much deeper south–north coastal penetration. A relevant factor which further affects the overall picture of the Gulf is that a considerable portion of its coast is smooth and plain and surrounded by a great plateau, known as the Sert basin. The Sert coastal plain begins

[181] This geophysical description of the Gulf of Sert and the Sert region is a specific synthesis of information found in the following materials: *Mediterranean Pilot*, 5 (1976), 46 ff.; *Al Atlas Al Watani* (the National Atlas) (Secretariat of Planning, Tripoli, 1978); Hydrographer of the Navy, *Admiralty Chart No. 3354* (Taunton, Oct. 1965).

east of the city of Misratah and runs along the Gulf until reaching the Gabal Al-Akhdar mountain range, east of Benghazi.

Because of this peculiarity in the Gulf's coastal projection and topography it may appear at first sight to be more dominated by the open sea than by the land mass of Libya.[182] A more careful insight, however, would demonstrate the striking land-dominated nature of the Gulf of Sert. Two major factors determine the politico-geographical characteristic of an indentation: the degree of its intrusion into the land mass and the percentage of its coastal length in the entire coastline of a country. The total 'encroachment' of the Gulf of Sert into the central land territory of Libya may be seen from the fact that as compared to its closing line of approximately 304 n.m. in length its greatest penetration landward is 134.6 n.m. at longitude 19° 11′ E., a ratio of slightly over 2:1. As to the Gulf's total 'participation' in Libya's coastline, the following statistics are of interest:

1. entire Libyan coastline: 944 n.m.;
2. outside the closing line: 466 n.m. (49.4 per cent);
3. inside the closing line: 478 n.m. (50.6 per cent).

In other words, the Gulf comprises slightly over one-half of Libya's entire coastline.

Morphologically, the seabed of the Gulf, beyond a narrow coastal water strip, comprises part of a larger rise, known as the 'Sert Rise'.[183] In many seabed areas within the Gulf, however, there exist a number of well-defined shelves and ledges with numerous rock patches in between. The 200-metre isobath is located in the Gulf from about 18 to 25 miles offshore. The seabed dips fairly gently but steadily; the 1,000-metre isobath lies some 37 to 50 miles offshore. In the centre of the northern part of the Gulf (within the closing line) depths of 1,800 metres are reached. The possible strategic implication of such morphology will be identified below.

(*b*) *Economics*[184]

(i) Fishing

Although, as noted earlier, fish stocks in the Gulf of Sert are small, the coastal communities in the Mediterranean depend in varying degrees on

[182] See e.g. J. Spinnato, 'Historic and Vital Bays: An Analysis of Libya's Claim to the Gulf of Sidra', 13 *ODIL* (1983), 65, at 80; W. Burnett, 'Mediterranean Mare Clausum in the Year 2000?', 34 *Naval Law Review* (1985), 75, at 100.

[183] F. L. Marchant, 'Physiography of the Central Mediterranean, Ionian Sea', in D. Stanley (ed.), *The Mediterranean Sea: A Natural Laboratory* (Hutchinson & Ross, Stroudsburg, 1971), 14.

[184] Unless otherwise indicated, all figures and statistics provided in this subsection are

fishing as an element of their economy[185]—and this is an area of the Libyan economy that has received increased attention.[186] In the Gulf of Sert reports on a masterplan to develop Libya's fishing industry have pointed out the possibility of an increase in fish-landing in the Gulf's local ports.[187] Whilst no confirmed data and statistics are available, annual fish catches within the Gulf of Sert were estimated in 1980 at about 4,411 tonnes,[188] approximately 60 per cent of Libya's entire fish catch in 1982.[189] Most of the catch is landed in the Gulf's seven local ports, where fishing boats are located for at least part of the year.[190]

The above indicates that fishing in the Gulf is not a significant element in the economic importance of the Gulf. Its prospects are inevitably limited by the noticeably small stocks of the Gulf and shortage of manpower and local markets in the limited coastal settlements. Control of the Gulf against over-fishing does not seem to be a factor that affects the status of the Gulf of Sert. At the same time its paucity of fish marks it as purely of interest to Libyan local economy from this point of view.

(ii) Oil and petrochemical industry

A preliminary remark that should be emphasized with regard to this sector is the dominant position of the oil sector in the Libyan economy. Before the discovery of oil in 1961 Libya was one of the poorest countries in the world.[191] At present, although Libyan oil exports have declined as

taken from the following documents. Secretariat of Planning, *Socio-Economic Transportation Plan, 1981–1985*, pt. 1 (Tripoli) (in Arabic); C. Babinski, 'Report on the Export Refinery and Ethylene Projects in Libya' (Secretariat of Planning, Tripoli, Oct. 1977); Speerplan/Finnmap (Consulting Engineers), 'El Khalij Region: Master Plan (Sirt)' (Secretariat of Utilities, Tripoli, May 1981).

185 G. Blake, 'Mediterranean Non-energy Resources: Scope for Cooperation and Dangers of Conflict', in G. Luciani (ed.), *The Mediterranean Region* (Croom Helm, London, 1984), at 47 ff.

186 See e.g. expenditures on 'marine living resources' and related industries in Secretariat of Planning, *Socio-economic Transportation Plan*, 40, 75.

187 See e.g. Sogreah (Consulting Engineers), 'Study for a General Masterplan for the Development of Fishing Ports in the Libyan Arab Republic: Final Report', (Grenoble, 1975), 33. The *Instrupa Report* (Final Report on Results of Test Fishing Program, Gulf of Sirte (Tennenwalallec, Federal Republic of Germany, July 1977) concluded that prospects for commercial fisheries in the Gulf of Sert were good and drew attention to the potential for inshore fishing of shrimps, a marked concentration of which was reported along the western coast of the Gulf.

188 *Al Atlas Al Watani*, 107–8.

189 FAO, General Fisheries Council for the Mediterranean, *Statistical Bulletin No. 5: Nominal Catches 1972–1982* (Rome, 1984). However, it should be noted that this percentage is not accurate, for there is a discrepancy between the FAO's statistics of 1982 and the *Atlas*'s figures for catches in 1980.

190 These are from west to east Marsa Draiza, Qasr Ahmed, Sert, Harauh, Ben Jawad, Azzuwaytinah, and Benghazi. Of these, only Sert and Benghazi are important. See *Al Atlas Al Watani*, 41–2.

191 See e.g. J. Allan, *Libya: The Experience of Oil* (Croom Helm, London, 1981).

a result of both Libya's oil reserve conservation policy and the prevailing conditions of the world oil market, Libya produces about one million barrels a day. In 1986, crude oil exports alone totalled 96 per cent of Libya's total exports. If one adds that the proportion of government revenue generated by the oil sector has been over 95 per cent for almost two decades, the vitality of oil in the Libyan economy is self-evident.

Whilst Libya's offshore exploration has been active since 1974 Libyan offshore oil reserves seem to be less promising than those on land.[192] On land, all of the principal producing fields are located in the Sert basin, south of the Gulf of Sert coast. Because of the Gulf's intrusion into the land it was natural that pipelines were linked to and tanker terminals were constructed on the Gulf of Sert. Crude oil is pumped to the export terminals on the coast via thirty-one pipelines. With the exception of one major line to Marsa El Harigah near Tobruk all of them exit on the coast of the Gulf of Sert. There are four export terminals on the coast, together with a fifth at Marsa El Harigah for the eastern fields.

Reserves of Libyan natural gas were estimated in 1985 at 600,000 million cubic metres. Gas is principally produced in the Zelten and Ragubah fields in the Sert basin. It is transported to the Gulf of Sert via a 175 km. pipeline.

The country's two largest refineries, at Marsa El Bregah and Ras Lanuf, are both on the coast of the Gulf. In fact, the biggest oil-related industrial scheme in the country is the refinery and petrochemical complex at Ras Lanuf. In this respect, it is pertinent to mention that the majority of Libya's investment in the development of heavy industry has been focused on petrochemicals. It is, therefore, natural that over 97 per cent of the country's investment in this field has been concentrated on the coast of the Gulf. In addition to the Ras Lanuf complex, Libya is establishing a urea-based fertilizer and ammonia production centre at Marsa El Bregah, which is also the site of a liquidation of natural gas plant. Plans have also been made for the construction of a major petrochemical complex at the city of Sert.

(iii) International trade and export terminals

The bulk of Libyan crude oil exports leave the country through terminals located on the Gulf of Sert. In the first six months of 1986 Libya was estimated to have exported approximately 897,000 barrels a day. Beside crude oil, petroleum products and natural gas are exported from the Gulf

192 'Status of Offshore Oil Concessions in Libya', a restricted diagram on active and relinquished offshore operations in Libya in 1986, shown to author, Tripoli, Secretary of Heavy Industries, Dec. 1986.

of Sert coast. In 1983, Libya exported 2,047,000 tonnes of natural gas and 1,160,000 tonnes of petroleum products. Libya's petrochemical exports were targeted to fetch 114 million Libyan Dinars in 1985.[193]

It is obvious from this that the Gulf of Sert is by far the most important 'export terminal' in the country. In terms of both volume and value its exports certainly far exceed any other depot. In 1981, the Speerplan/Finnmap Report concluded that, for decades to come,

> The El Khalij [Region] will still be the main exporting zone of Libya. Foreign exchange earnings by oil exports, natural gas exports, and exports of petrochemical items, including refined oil, will be responsible for more than 95 per cent of exports by value.[194]

The above clearly indicates that the Gulf of Sert constitutes an *economic unity* with the mainland, for the dependence on its intrusion into the mainland especially for oil exportation and petrochemical industry is of the first importance.

(iv) Population settlement and communication

The above account illustrates how the discovery of oil in the 1960s has changed the Gulf and basin of Sert from a no man's land, situated between Tripolitania and Cyrenaica, into Libya's natural resource base and vital industrial centre. This has led to the development of the infrastructure to support the oil and petrochemical industry. In addition to Benghazi (the second largest city in the country) and Misratah, new towns at Ras Lanuf and Bregah have been established and are projected to have populations of 124,000. All the major settlements will be linked to the national electricity grid, with power stations being built at Ras Lanuf, Bregah, and Al Zwaitina. In addition to the country's principal, east–west, roadway, new roadways were constructed to link the new settlements with each other and with other towns and newly developed projects in the south. A major railway line of 999 km. is planned to bring iron ore from the south to the Misratah iron and steel complex, but the start of work on the scheme has been delayed. The complex constitutes Libya's largest industrial project outside the oil and petrochemical sector.[195] To complement the infrastructure necessary for the industrial development of the region (falling, like most of the country, in the arid

[193] Against 1980 prices; see table 3 in Secretariat of Planning, *Socio-economic Transformation Plan*, 117.

[194] *Supra*, n. 184.

[195] The complex is planned to produce 7 million tonnes of iron and steel when the third phase of the project is completed by the year 2005. The cost of the first phase was estimated at $3,500 million.

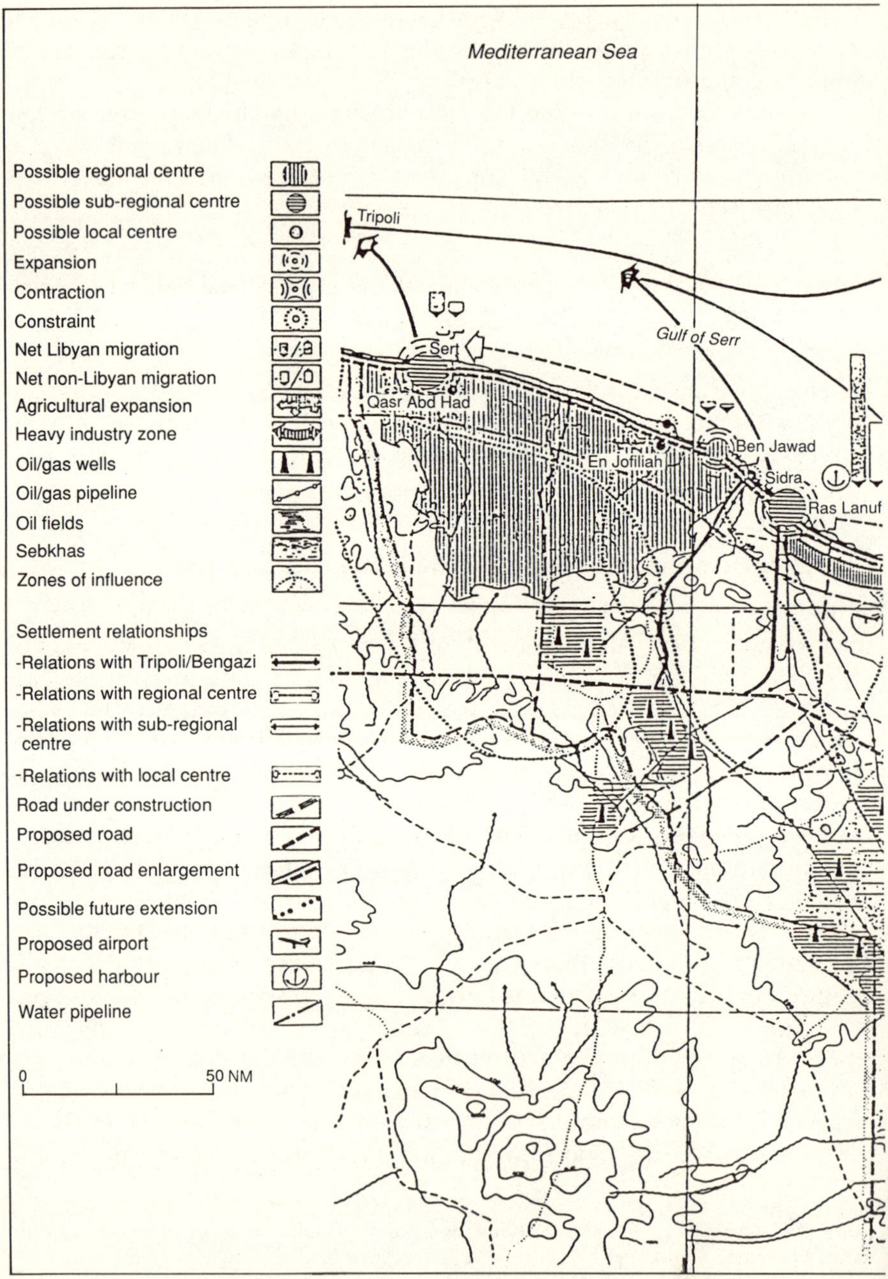
Mediterranean Sea
Possible regional centre
Possible sub-regional centre
Possible local centre
Expansion
Contraction
Constraint
Net Libyan migration
Net non-Libyan migration
Agricultural expansion
Heavy industry zone
Oil/gas wells
Oil/gas pipeline
Oil fields
Sebkhas
Zones of influence
Settlement relationships
-Relations with Tripoli/Bengazi
-Relations with regional centre
-Relations with sub-regional centre
-Relations with local centre
Road under construction
Proposed road
Proposed road enlargement
Possible future extension
Proposed airport
Proposed harbour
Water pipeline
0
50 NM
Tripoli
Sert
Gulf of Serr
Qasr Abd Had
Ben Jawad
En Jofiliah
Sidra
Ras Lanuf

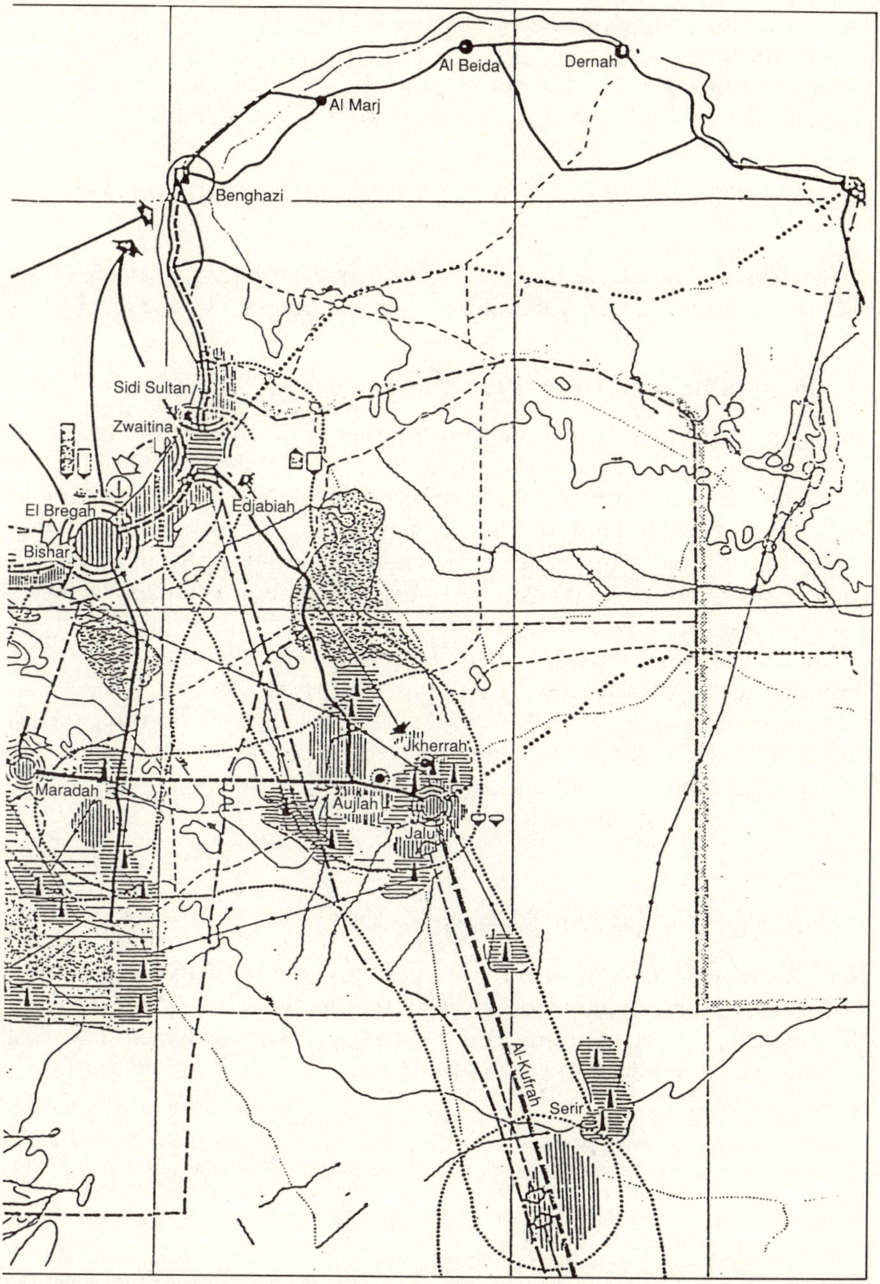

Source: Speerplan/Finnmap OY, 'El Khalij Region: Regional and Urban Planning: Summary Report' (Secretariat of Utilities, Tripoli, 1982).

MAP 14 Regional strategy: El Khalij region 1980–2000

zone) a major project has been under construction to bring water from the underground aquifers at Serir, Tazirbu, and Al-Kufrah to the coast of Libya. The first phase of this huge project[196] will link these aquifers by pipeline to a holding tank at Ejdabiah near the coast of the Gulf of Sert.

(*c*) *The strategic implications of the Gulf's physiography and economy*

In the light of the above factors, the principal grounds on which Libya may argue for its alleged vital interests are as follows.

(i) The strategic importance of the Gulf's resources

The land region into which the Gulf of Sert projects has become, since the early 1960s, the economic heartland of Libya—its major source of oil, the centre of its petrochemical and other heavy industry, and a key in the land communication between the two populous parts of Libya, and from this industrial centre to the sea. It is striking that a country of as large a land mass as Libya[197] has just one principal centre of economic activity. This characteristic (owing to geological, climatic, and geographical factors) of the Libyan economy will become more pre-eminent in the next two or three decades. It is submitted that Map 14 shows clearly the intensity of Libya's present and future investment in the Gulf of Sert region. It illustrates a detailed plan on 'the preferred course of development in the El Khalij region' well into the year 2000. The plan was prepared by Speerplan/Finnmap and was finalized and agreed upon by the Libyan Government in April 1981.[198]

(ii) The Gulf's vulnerability to pollution

Because of its relatively small size, the narrowness of its opening, the slow rate of rejuvenation of its waters (taking almost a century)[199] and the intensity of its maritime traffic, the growing volume of industrial activity and oil production and transportation, and the high concentration

[196] The project involved the construction of a massive water pipeline called the Great Man-Made River. It is the most expensive single project in Libya today. The total cost for the three-stage project (extending the pipeline to the west and east of the country) is expected to be $18,000 million, with annual foreign exchange costs to Libya running at $500 million until completion towards the end of the 1990s.

[197] Approximately 1,750,000 sq. km.

[198] The map is prepared by Speerplan/Finnmap OY, 'El Khalij Region: Regional and Urban Planning: Summary Report' (Secretariat of Utilities, Tripoli, 1982).

[199] International Maritime Consultative Organization Doc. MP/Conf.1/C.2/WP.28, Oct. 1983, at 3.

of coastal populations, the Mediterranean is especially vulnerable to the risks of pollution. By the 1970s, the Mediterranean was declared one of the most polluted seas in the world,[200] prompting its littorals to engage in concerted efforts to reduce pollution, starting with the 1976 Barcelona Convention and its related Protocols on pollution abatement in the Mediterranean.[201]

The potential danger in the Gulf of Sert relates more to oil transportation than land-based pollution, which is the main source of pollution in other parts of the Mediterranean.[202] However, the high concentration of oil transportation in the central Mediterranean combined with the clockwise surface currents in the Gulf[203] creates an area of risk for Libya, especially along the eastern and southern shores of the Gulf. Whilst these particular circumstances may create a special reason for treating much of the Gulf as a specially protected area—thus consistent with Protocol 4 of the Barcelona Convention[204]—the total enclosure of the Gulf on the basis of this special circumstance alone does not seem to be warranted in international law. However, the Canadian claim in 1970 to exclusive jurisdiction to regulate all shipping in zones within 100 miles of its Arctic coasts in order to guard against pollution is of interest. In response to a strong objection by the United States, Canada replied that its action was taken on the basis of 'the overriding right of self-defence of coastal States to protect themselves against great threats to their environment' and that 'such extensions of jurisdiction for limited protective purposes outside the territorial sea had ample precedent, particularly in American practice'.[205]

[200] For a detailed survey in the 1970s see UN, FAO, GFCM, *The State of Marine Pollution in the Mediterranean and Legislative Controls*, No. 51 (Rome, 1972).

[201] By Sept. 1985, the Convention and two of the four Protocols were in force among all the Mediterranean States, except Albania, plus the EEC. See B. Vukas, 'The Protection of the Mediterranean Sea against Pollution', paper presented at Conference on the International Juridical Regime of the Mediterranean Sea (Rome, 18–19 Oct. 1985). For the text of the Convention and the Protocols see UN Environment Programme, *Convention for the Protection of the Mediterranean Sea against Pollution and Its Related Protocols* (UN, New York, 1982).

[202] See S. Kuwabara, *The Legal Regime of the Protection of the Mediterranean against Pollution from Land-Based Sources* (Tycoolly Int. Publishing, Dublin, 1984).

[203] Principally due to the intrusion of the Gulf's eastern coast into the land, the surface movement of currents in the Gulf takes the form of a separate clockwise flow in contrast to the eastward circulation pattern of currents in the north, which is not 'trapped' within the Gulf's curvature. See *Mediterranean Pilot*, 5 (1976) 47. For a figure showing the surface currents of the Mediterranean see Truver, *The Strait of Gibraltar*, 32.

[204] The Protocol on 'Specially Protected Areas', signed in Geneva on 3 Apr. 1982, is designed to preserve and protect natural resources and natural environment in the Mediterranean Sea through establishing specifically protected areas encompassing marine areas and their natural environment. The contracting parties are entitled, under Art. 4, to establish the aforementioned areas on the basis of common rules and criteria.

[205] R. S. Reid, 'The Canadian Claim to Sovereignty over the Waters of the Arctic', 12 *CYIL* (1974), 111, at 128.

(iii) The Gulf's divisive impact

The Gulf's dominant position in the geography of Libya was shown earlier. This factor can further be explained by the fact that the water area of the Gulf literally separates the two major population centres of Libya, namely, the coastal regions of Tripolitania in the west and Cyrenaica in the east. Historically, this isolation was highlighted by the fact that, due to its arid nature, the hinterland of Sert has always been unpopulous. To illustrate this physical-demographic factor it might be mentioned that a traveller along the Gulf's coast would have to travel a distance of 466 n.m., almost 40.6 per cent of the total coastal length of Libya; whereas a direct route across the Gulf's mouth is 304 miles in length, that is, a coastal reduction of 32 per cent. This is a major factor in maintaining the country's national unity and in defending the unpopulous, yet economic, heartland of Libya to the south. At present there is a scheduled passenger service between the ports of Benghazi and Tripoli which takes about twenty-four hours. Without the line of enclosure Libya would be the only non-island State in the world to have its two major population centres separated by international waters.

(iv) The Gulf's 'encroachment' effect

International adjudication on maritime boundary delimitation affirms the position that a State's security interests might be regarded as a relevant factor. Both the *Guinea–Guinea Bissau* and the *Libya–Malta* decisions confirmed this explicitly, though in neither case was it made very clear in what circumstances they might be relevant.[206] However, adjudication seems to identify this factor with the principle of non-encroachment. This was discussed earlier.[207]

The application of this principle in future maritime boundaries between Libya and its neighbours in the central Mediterranean is of some interest. The Italian and Greek coasts facing the Gulf of Sert project a little further southwards than the rest of the northern Mediterranean coast east of Crete and west of Sicily. The effect is to push a median line drawn in this area a little further south; and, if the concavity of the Gulf of Sert coast were not to be regarded as a 'special circumstance',[208] the projection of the median line southwards would be exaggerated, possibly to such an extent that it might be regarded as encroaching on maritime areas that

[206] See *supra*, Ch. 4 s.3(*b*)(iii)(3).

[207] Ibid.

[208] See the statement of the Libyan delegation at the 2nd Caracas Session who argued for the inclusion of, *inter alia*, 'historic conditions' as a special circumstance for the 'subdivision of the Exclusive Economic Zone'. UNCLOS III, *Official Records*, iii (1974), 175.

more naturally belong to Libya.[209] The analogy of this situation to that of the German coast in the *North Sea* cases is of interest.[210]

(v) The Gulf's strategic vulnerability

As noted earlier, the Gulf's deep incision into the land mass of Libya and the oilfields' proximity to its coast has made it the focal point in the economic survival of Libya. Ironically, perhaps, these same intertwined geologico-geographical factors have placed Libya's economy at a strategic risk. The crucial point in Libya's argument seems to centre on the fact that if the Gulf were not closed, not only would the oilfields and other oil-related industries be within striking distance of the high seas but so would the whole central land mass of Libya. To demonstrate this fact, if a 313-km. arc was drawn from a mid-point on the line representing 12 miles from the coast (as shown on Map 15), the area covered would circumscribe the oilfields, pipelines, and all industrial centres in the region.

As the Gulf of Sert protrudes into this critical area, a hostile vessel or aircraft freely entering the Gulf could proceed to within striking distance of Libya's heartland. The risk is increased by the nature of the underwater and subjacent seabed formation,[211] which would enable missile-carrying submarines to lie there undetected.[212] The great extension of the Gulf represents, from this point of view, a considerable and permanent risk for the Libyan nation, which lacks an adequate navy which could control and patrol such an immense area (23,531 sq.n.m.). For this reason Libya should now be expected to see the Gulf of Sert as more intimately connected with her national security than in the past.

To counter this situation Libya considers the enclosure of the Gulf a key safeguard for its national security. This, in Libya's view, would facilitate Libya's surveillance capability in respect of warning time and discovery of intended targets. This, of course, is the principle behind the early warning systems such as those the United States and Canada have deployed across the northern American continent since the early 1950s.[213]

[209] For the relevance of this factor to a future maritime boundary between Italy and Libya, see *infra*, Ch. 9 s.2(*g*)(iii)(2).

[210] *Supra*, Ch. 3 s.3(*a*).

[211] See *supra*, s.4(*a*).

[212] Detection of small submarines lying among the Gulf's seabed ledges and rocks would be extremely difficult. The spaces offered between such features are due to the slow rate of sedimentation in the Gulf, mainly because of the absence of rivers flowing into it. The same argument is considered by Strohl, *The International Law of Bays*, 250, in respect of Canada's claim over Hudson Bay.

[213] The USA and Canada maintain air defence identification zones, for areas over land and water in which the identification and control of aircraft are required for purposes of national security. For the States' respective regulations see the *Gulf of Maine* case, *Documentary Annexes* submitted by the USA, ii (27 Sept. 1982), annexe 37.

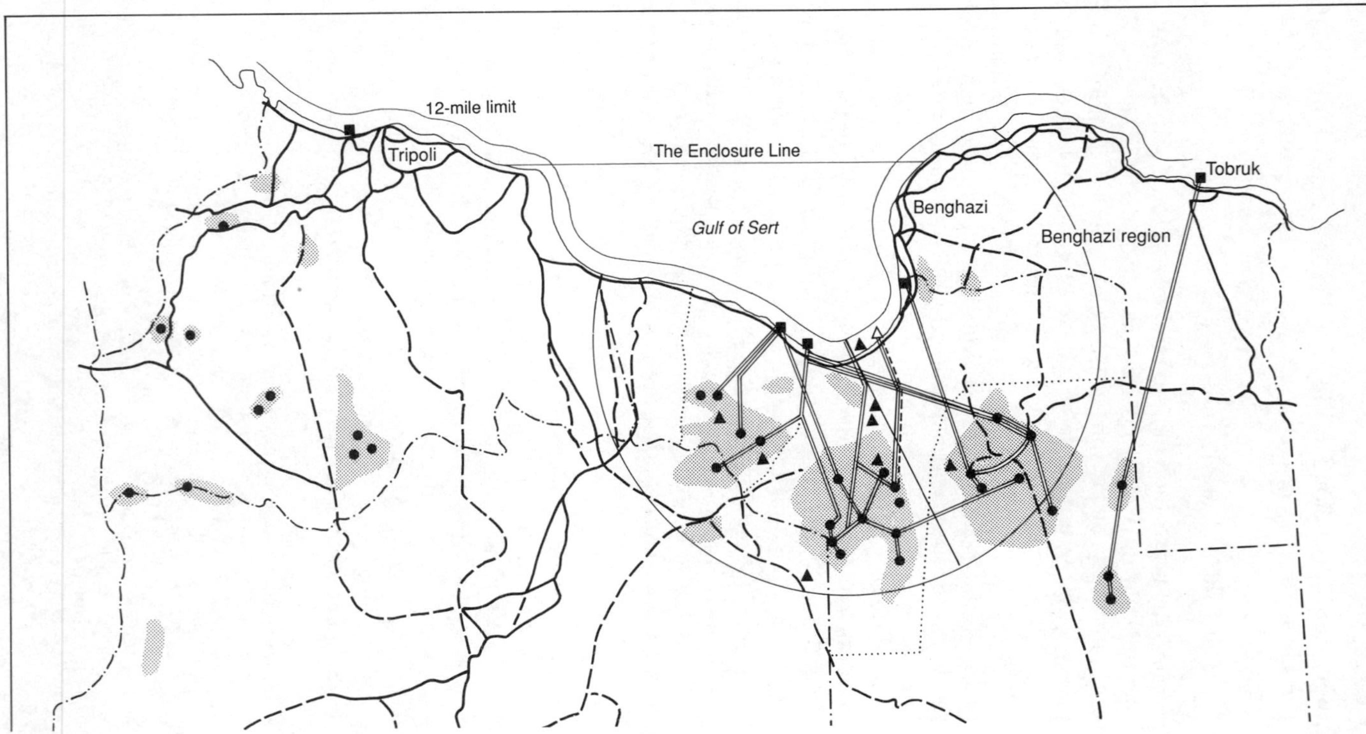
12-mile limit
Tripoli
The Enclosure Line
Gulf of Sert
Benghazi
Benghazi region
Tobruk

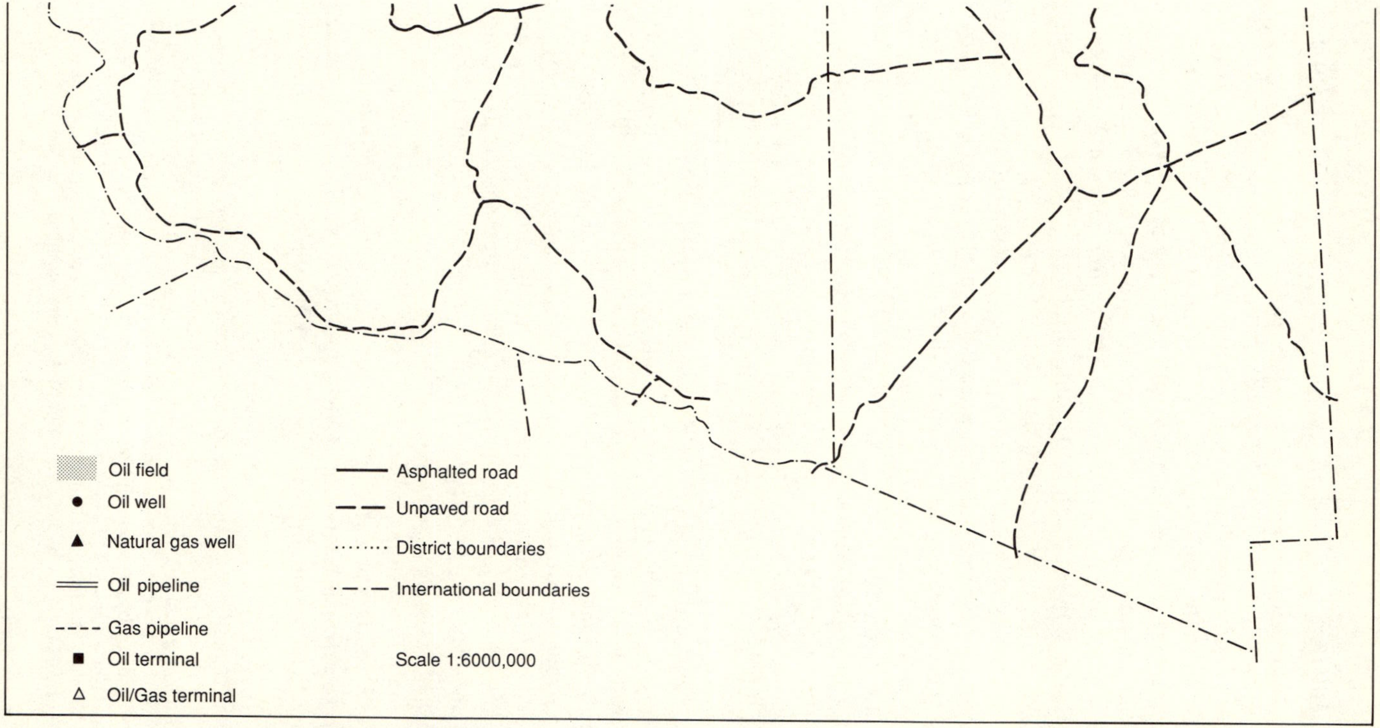

Map 15 The enclosure line of the Gulf of Sert

The argument that today's rapid improvement in the technology and capability of weaponry has made all distances on the ocean somewhat irrelevant[214] is indeed a relative argument. While it might be held valid *vis-à-vis* States possessing advanced technological capability, the same is not true in relation to the majority of nations. On this ground Libya would argue that in view of the increased inequity in military strength between nations, it is *a fortiori* that a State like Libya should be permitted, in situations where so much is at stake, to express its vital need for maximum warning time as to an impending attack.[215]

However, whilst the appreciation of a national security risk is a matter germane to a State's own perception and evaluation of the relevant world, regional, and internal circumstances, it is still pertinent that Libya demonstrates the existence of some factors which would, in normal circumstances, justify a State's argument of security risk. This is submitted to be relevant for the analysis of the general guidelines of 'good faith' and 'reasonableness' suggested above.

The most effective global policy-making apparatus—affecting the world's social process—is the military. Its effective power is laid in its ability to generate instability. It is this very instability 'which feeds a pervasive insecurity and which tends to aggravate the demand for self and grouping mobilization, constantly heightening the expectation of violence'.[216]

The Mediterranean region is perhaps one of the regions of the world most associated with conflict. The classical ingredients for this are all present: intra-regionally, the Mediterranean has been an arena of local conflicts of various ideological implications. Although the Arab–Israeli conflict is the major source of instability in the region, the Aegean dispute, the Cyprus crisis, the Lebanon civil war, the Moroccan–Spanish dispute over the 'Spanish *presidios*', as well as the dramatic impact of oil politics on world economics are all contributing factors.

Inter-regionally, the superpowers and regional organizations such as NATO, the Warsaw Pact, the European Economic Community, the Organization of African Unity, and the League of Arab States all 'engage in a gigantic chess game in the basin'.[217] This outside interest in the

[214] e.g. McDougal and Burke, *The Public Order of the Oceans*, 335.

[215] The dog fight that took place between US and Libyan aircraft in August 1981 and the US air raid on Libyan combatants' fast-attack vessels and missiles and radar sites within the Gulf's closing line and along its shore in Mar. 1986 illustrate that even the relatively short time that may be allowed by the enclosure of the Gulf is vital for anticipatory purposes. The F.14, or Tomcat, which was used in both incidents against Libya, is one of the USA's main tactical fighter aircraft. It can reach a maximum speed of Mach 2.4 (approximately 1,850 miles per hour). 'Le F.14 et le Su 22: Deux avions à flèche variable', *Le Monde* (21 Aug. 1981), at 4.

[216] McDougal and Burke, *The Public Order of the Oceans*, 105.

[217] Burnett, 'Mediterranean Mare Clausum in the Year 2000?', 88.

Mediterranean will continue, because the basic values at stake there will remain substantially unchanged, given the area's rich endowments, culture, oil, and commercial transport, its high population, and its geostrategic significance sitting as it does at the crossroads of three continents between the Indian and Atlantic Oceans. To protect their interests there, the superpowers especially have engaged in a grotesque competition—demon strated in the concentration of the US Sixth Fleet and the USSR Fifth Eskadra[218]—with a view towards shaping events, including small-power conflicts in the area, to their benefit. Since their introduction into the Mediterranean in the 1940s, the fleets have been used in operations related almost exclusively to political developments and conflict in the area.[219]

The inevitable result has been the further escalation of the highly tense economico-political situation in the Mediterranean. What aggravates this situation is that, in the immediate future, there would seem to be few incentives for the superpowers as well as for major Mediterranean powers to agree to substantive naval arms control in the Mediterranean. The dearth of arms control initiatives is equalled by the determination of most European governments and those of the superpowers to exclude considerations of Mediterranean problems from their deliberations on Europe.[220]

In such a situation it becomes inevitable that all instruments of the Mediterranean States' policy—not merely military but also legal, diplo-

[218] The two superpowers' naval capabilities in the Mediterranean are designed to fulfil all military purposes across the entire spectrum of armed forces: from the maintenance of an unobtrusive presence, to the deliberate show of force for political purposes, to limited war, to general conventional war, to a launch of a strategic nuclear attack. For detailed statistics see *Truver*, *The Strait of Gibraltar*, 72–81.

[219] For instance, a study of the USA's use of its Sixth Fleet as a political instrument in the period between 1946 and 1975 concludes that US armed forces 'have been directed primarily at political and military crises in and between Mediterranean States rather than at Soviet threats to Europe security—excluding, of course, those threats directed at Greece, Turkey and Yugoslavia prior to Stalin's death'. B. Blechman and S. Kaplan, 'The Political Use of Military Power in the Mediterranean by the United States and the Soviet Union', 13 *SI* (1978), 29, at 46.

[220] A clear example is the Helsinki Conference on Security and Co-operation in Europe in 1973. Although many Mediterranean nations stressed vigorously the importance of arms reduction in the Mediterranean, the majority of participating States, including the USA and the Soviet Union, decided against including the Mediterranean as a subject for negotiations. See C. Zoppo, *Naval Arms Control in the Mediterranean*, Research Paper No. 57, California Seminar on Arms Control and Foreign Policy (Santa Monica, Calif., 1979). Whilst the Final Act of the Conference (signed in 1975) included a section concerned with 'Peace and Security in the Mediterranean', neither this section nor the Final Act in general constituted a legal commitment upon the signatories. See O. Schachter, 'The Twilight Existence of Nonbinding International Agreements', 71 *AJIL* (1977), 296. The idea of the Mediterranean as a zone of peace was again raised in GA Resolution 36/102 (1981). In voting on this resolution there were 20 abstentions, including 4 Mediterranean States. A consensus resolution on co-operation and security in the Mediterranean adopted two years later (38/189) does not make any reference to the creation of a zone of peace in the Mediterranean.

matic, economic, and ideological—are continuously deployed in an ever more sophisticated process of preparation, attack, and defence. Most vulnerable in this situation are the newly independent States of the Mediterranean. Living in a constant fear of losing their independence once more to politically stronger States, and involved as they are in a process of nation-building which often necessitates their adopting an internal as well as an external policy of nationalism, they are naturally cautious of the situation in the Mediterranean harbouring a severe threat to their security. This sense of insecurity will continue to grow, driven by fear and hostility towards the warships of either or both the superpowers passing near a country's coast and perhaps trying to affect the course of events there.[221] This, with the socio-economic phenomenon of contending ideologies and fundamental differences between States as to the appropriate institutional modalities for effecting decisions and pursuing values, would lead such nations to identify their effective power process largely with 'national territory'. It is against this psychological, social, economic, and political background that we should analyse the legal foundations of Libya's claim over the Gulf of Sert.[222]

5. APPRAISAL

The above account indicates that, upon the strict application of the generally assumed requirements for an historic claim, the striking economic/strategic characteristics of the Gulf of Sert may be of no special legal value in the assessment of Libya's 'historic' claim. In particular, the restricted interpretation of the word 'historic' would seem to reinforce such a conclusion. This conclusion is clearly traceable to a widely pervasive misconception of the specific process of legal sources through which controversies concerning historic waters have been resolved. Highly abstract and ambiguous doctrines, derived from a distorted description of

[221] This concern was stressed by Libya and other Mediterranean States on numerous occasions. For instance, at UNCLOS III Libya suggested that, under the doctrine of enclosed and semi-enclosed seas, the Mediterranean should be 'a sea of peace, free from any foreign fleets that might threaten the security and health of the coastal people'. Statement of the Libyan delegate, 1 UNCLOS III, *Official Records*, Plenary Meeting (33d met.) at 133, para. 81 (1974). See also proposals submitted, *inter alia*, by Algeria and Turkey, A/CONF. 62/C. 2/L.20, L.62, L.56, respectively. The transformation of the Mediterranean into a zone of peace has been listed among the aims of the Non-aligned Movement ever since the Algiers Summit of 1973. See generally A. Abbadi, 'Security and Co-operation in the Mediterranean Basin', 14 *ODIL* (1984), 55.

[222] In the *Anglo-Norwegian* case Judge Alvarez stressed the nature of modern international law as neither static nor 'exclusively juridical' but 'dynamic', discovering in it 'aspects which are political, economic, social, psychological, etc.' (149).

certain past decisions[223] and authoritative pronouncements,[224] are put forward as the most rational policy for future decisions, which must, nevertheless, be reached under vastly changed conditions.

The suggestion which has recently been made to amend this unrealistic approach is that claims to title based upon vital interests should be distinct from 'historically' based claims to title.[225] The problem with this solution is that it does not reflect State practice or authoritative judicial pronouncements on the subject.[226]

A sound, recommended policy, however, may seek a reasonable approach which, while making the necessary link between such requirements and the characteristics of a claim to a historic bay, does not at the same time allow the use of the claim as an escape valve for unwarranted expansionism. What is suggested above is a careful rejuvenation of the traditional 'legal requirements' in the light of the governing principle of community policy objectives.

The law of the sea is basically the art of distributing authority between the inclusive and exclusive uses of the oceans.[227] 'Historic waters' is but one of several means whereby a kind of equilibrium can be maintained between the legitimate interests of individual coastal States in preserving their respective shares of natural resources and the security of their territories and the maximum sea area possible for navigation, flight, communication, and other inclusive interests.

Throughout history, community policy has taken the form of a strong presumption in favour of inclusive authority and use of the oceans, balanced with the recognition that a certain degree of exclusivity is essential if each State 'is to thrive within the community of States and the companion realization that community interests are as well-served by a policy which encourages the existence of independent nation-States, secure in their own borders and economically viable, as by maximum open Seas'.[228]

[223] e.g. the requirement of 'immemorial' or 'well-established' usage, often quoted by recent commentators (see *supra*, s.2(*b*)), is believed to be largely envisaged in the light of well-known *but limited* instances, such as the ancient exercise of sovereignty over the Strait of Mannar and Palk Bay between India and Sri Lanka. Such instances, as Goldie has noticed, pre-date 'contemporary public international law' and 'have little or nothing to do with our ideas of the Roman law rights upon which Hugo Grotius and his contemporaries and followers relied when they acclaimed it as *ratio scripta*', 'Historic Bays in International Law', 225.

[224] e.g. the insistence of prominent writers on explaining a historic waters title by reference to the traditional concept of prescription, which is largely connected with contesting claims between States over land areas (see *supra*, s.2(*d*)). The legal requirements for historic bays were accordingly analysed in a strict fashion which reflected the exceptional nature of a prescriptive title.

[225] *Supra*, s.3(*a*).

[226] *Supra*, s.3(*b*).

[227] See *supra*, s.2(*a*).

[228] Westerman, *The Juridical Bay*, 21.

Within this general framework, however, a specific rule of community policy has been able to crystallize, namely that indentations protruding into the territory of a single State are perceived by coastal States as so vitally intertwined with life on shore as to be valid components of their own territory.[229] Thus whenever such a presumption comes under scrutiny it is the close connection between the sea areas and the land domain in question that must constitute the proper basis for a decision that pre-eminent community inclusive interests may be waived and the waters incorporated within a State's sovereignty. This close connection may be identified in two ways. First, geographical penetrations, even of relatively wide expanse, may be effectively occupied, utilized, and controlled by the adjacent coastal State.[230] Secondly, because these penetrations do not serve as ordinary routes of international navigation and communication but merely as means of access to points within the penetration, inclusive community interests may not be effected.[231]

This in fact is the whole basis upon which international community policy regarding exclusive State jurisdiction has been established. When, for instance, exclusive State jurisdiction in the territorial sea was found to be severely prejudicial to a much stronger inclusive community interest, institutions such as innocent and transit passage were formulated to maintain the equilibrium. The same is true of the rest of a State's jurisdictional zones: the greater the impact on inclusive community use the stricter the latitude granted to individual coastal States and vice versa.[232] In other words, the 'balancing of interests' provides the key to 'good faith' and 'reasonableness' and these are considered essential criteria for identifying in a concrete case a presumed community policy.

A claim to historic waters is expected to be an expression of this policy. The claim is expected to be generated by interrelated factors none of

[229] As put by Hyde, when 'nature has lodged a bay in the very bosom of a maritime State', she has encouraged that State to claim the bay as its own and other States to respect that claim. *International Law, Chiefly as Interpreted and Applied by the United States*, 468.

[230] See e.g. analysis of the Court in the *Gulf of Fonseca* case; the opinion of the ICJ regarding the Norwegian Lapphavet in the *Anglo-Norwegian* case, *supra*, s.3(*a*).

[231] e.g. in the *North Atlantic Coast Fisheries* Arbitration the Permanent Court of Arbitration pointed out the element of 'exclusivity' of a bay 'secluded from highways of nations on the open sea'. J. B. Scott, *Hague Court Reports*, 1 (1916), 187; the decision of the US Supreme Court in *Alabama and Mississippi Boundary* case, *supra*, text at n. 136.

[232] As a concrete example of compliance with this policy, it is of interest to mention the recent decisions of India and Sri Lanka to attribute the status of internal waters to the Palk Strait and Palk Bay and of *territorial sea* to the Gulf of Mannar, though both the Bay and the Gulf have long been recognized as having the status of internal historical waters. Palk Bay, in contrast to the Gulf of Mannar (which is open in its southern sector to the Indian Ocean), is practically land-locked and not easily navigable. For a detailed account see P. Rao, *The New Law of Maritime Zones with Special Reference to India's Maritime Zones* (Millind Publications, New Delhi, 1983), 74.

which could have an independent legal effect. But they should together point out the pre-eminence of the exclusive interest.

The absence of usage of 'long-standing' quality should not arrest the dynamic movement of all other factors towards formulating a successful 'historic claim'. As O'Connell pertinently observed,

> history cannot be ossified at one moment in time. The whole theory behind the doctrine of historic rights is that special situations crystallize as the result of historical evolution. To arrest this process is to produce an artificially stable international situation, which is unlikely to be tolerated by those States which feel discriminated against, as most countries which became independent after 1958 must feel when they reflect upon the broad expanses of water effectively enclosed by some great powers, and compare the restrictions now impassed upon them when their history is just beginning.[233]

It is not this writer's suggestion that the vital interest factor *alone* ought to generate a claim to 'historic title' and that history can be altogether discounted—it is a simple fact that the land–sea social and political relationship can figure prominently only after the lapse of a certain period of time. What is suggested is that the old rigid requirement of 'long-standing usage' must evolve into a more flexible notion whose crucial element is the 'vitality of the usage to the claimant'. Time is necessary in so far as it indicates the *existence* and *continuity* and therefore the *vitality* of the usage. This should suffice to point out (1) the pre-eminence of the exclusive use or interest and (2) the presence of 'good faith' and 'reasonableness' as general guidelines for the proper use of vital interests.[234]

Similarly, the requirement of 'acquiescence' would prove more beneficial for clarification of the law if it were taken as an objective means for identifying community policy. What is suggested in this regard is a concentration on the major concerns of community policy, explaining the rationale for the requirement of 'acquiescence'.[235] The criterion for an 'effective Protest' should, therefore, spring from this background. The concentration on other alternatives, such as the number of protesters or the attitude of 'the directly interested' States, is submitted to be subjective, inconclusive, and misleading.[236]

233 O'Connell, *The International Law of the Sea*, i. 425.

234 *Supra*, s.2(*c*).

235 *Supra*, s.2(*d*).

236 A good example is the extremely different reaction of 'Western powers' towards two claims with remarkable similarity in location, geographical proximity, time of proclamation, and, above all, reasons for enclosure. These are the Libyan claim over the Gulf of Sert and the Italian claim over the Gulf of Taranto. Like the Gulf of Sert, the Gulf of Taranto lies in the central Mediterranean, intrudes into the heartland of Italy, and was closed in 1977 exclusively for security reasons. Its enclosure is not supported by long history (see *supra*, s.3(*b*)). Yet, whilst for reasons shown earlier none of the NATO members has protested against the Italian claim, six of them have strongly protested against or reserved their position regarding the Libyan claim (see *supra*, n. 102).

The above contention for 'reform' may also find support in the obvious premiss that it is essential for the persistence of a customary rule to have an evolutionary character. The customary rules, such as those relating to the law of the sea, are the sources of living law only in so far as, by adhering to the flexibility of their own process, they are capable of providing States with an adequate basis which could be adapted to the diverse facts in question.[237] Far from denoting a disintegration of the law, these adaptations, as long as they are inspired by the realities of international practices and trends, are proof of the process of the law.[238] Exclusive claims of coastal States, including maritime historic claims, must not, therefore, be regarded as exceptions to the principle of the freedom of the high seas. They are 'functional adaptations' inherent in the flexible virtue of the principle itself. This function reveals itself as soon as we abandon the notion of the freedom of the high seas implying an absolute right to absolutely unimpeded navigation, a concept which, 'when rigidly applied, may be as obnoxious as the uncompromising doctrine of absolute sovereignty'.[239]

It is submitted that it is within this framework of community policy, prompting the desirability of the evolution of rules of law,[240] that the legality of contemporary claims such as that of Libya over the Gulf of Sert should be assessed. A proponent of international law of small as well as 'big' nations would be more likely to understand the demands of newly emerging States, arising out of unhappy past experiences, present regional crises, and present and possible appraisal of national needs.

A relevant system, such as 'historic bays', will therefore recognize that there are today rival systems of public order aspiring toward completion on a global scale and will explicitly compare these rival systems in terms of their consequences for the shaping and sharing of reasonable legal concepts. A relevant legal concept, that is, postulated for preferred world public order, cannot be representative only of the exclusive values of some particular segment of the larger community of States, but should admit elements of moderation and reasonableness by which the inevitable diversity in community practices is resolved and secured.

This is not to recommend a flabby obeisance to unjustified and flamboyant assertions of national prerogatives—the solution postulated above, based on an objective analysis of each case in accordance with presumed

[237] C. de Visscher, 'Reflections on Present Prospects of International Adjudication', 50 *AJIL* (1956), 467, at 473.

[238] See *supra*, s.3(*b*).

[239] Lauterpacht, 'Sovereignty over Submarine Areas', 403.

[240] As shown earlier, this desire for 'reform' (though within a domestic context) was perhaps the reason behind the US Supreme Court's mitigation of its rigid standard of proof regarding historic claims. See *supra*, s.3(*a*).

community policy, should exclude this result. Rather, it is to suggest here that rejection or approval of a claim (such as that postulated by Libya) should be a most carefully considered decision, not swayed by overly immediate suspicions and prejudices.[241]

[241] e.g. apparently unaware of the Gulf of Sert's implications for Libya, the UK representative during the 1986 Security Council debates described the Libyan enclosure as 'eccentric' (UN Doc., S/PV.2669, at 32 (1986)). See also the hasty analysis of Spinnato, 'Historic and Vital Bays', esp. 78–82.

8

Delimitation of the Territorial Sea in the Mediterranean

1. INTRODUCTION

All along the Mediterranean, out of twenty potential territorial sea boundaries, only four have so far been delimited. The absence of formal agreements does not necessarily indicate disagreements between the coastal states over the principles of delimitation applicable. In fact, as many as nine Mediterranean littorals adopt, as will be seen shortly, the delimitatory rule of Article 12 of the 1958 Territorial Sea Convention. In many cases there has not been much urgency to delimit the boundary. The growing awareness among the Mediterranean States of the hazards of pollution[1] and, more importantly, their increased interest in offshore hydrocarbons and other non-energy resources[2] would contribute to accelerating the desirability of reaching agreements to settle the territorial sea boundaries. This would perhaps take place in association with shelf/EEZ or single maritime boundary delimitations. Problems associated with these delimitations will be considered in the next chapter.

Of relevance to the territorial sea delimitation is the status of the land boundary terminus between the neighbouring States. This is all the more important in the Mediterranean, for as many as eighteen potential territorial sea boundaries are to be (or have been) delimited between adjacent coasts. As illustrated in Table 2, the land boundary system in the Mediterranean is relatively recent and in a few places remains to be properly delimited. This is due to a number of disputes which are more related to sovereignty over the land than to boundary allocation or demarcation. In this respect, the Arab–Israeli conflict, the Moroccan–Spanish dispute over the 'Spanish *presidios*', and the Anglo-Spanish controversy over Gibraltar are the most prominent sources of boundary instability in the region.[3] Needless to say, the settlement of such disputes is a prerequisite to any maritime boundary agreement.

[1] See *supra*, Ch. 7 s.4(*c*)(ii).

[2] See *infra*, Ch. 9 s.3.

[3] It is also worth mentioning the political conflict in Cyprus which resulted in the declaration of an independent 'Turkish Republic of Northern Cyprus' on 15 Nov. 1983. The State remains recognized only by Turkey. See *Keesing's* (1984), 32638–41.

TABLE 2. *Status of land boundary termini in the Mediterranean Sea*

Boundary terminus	Category	Status	Year of final designation
Turkey–Greece	International	Demarcated	1923
Turkey–Syria	International	Demarcated	1939
Lebanon–Syria	International	Demarcated	1923
Lebanon–Israel	Armistice	Demarcated but sovereignty uncertain	1949
Gaza Strip–Israel	Armistice	Demarcated but sovereignty uncertain	1949
Libya–Egypt	International	Demarcated	1926
Libya–Tunisia	International	Demarcated	1910
Algeria–Tunisia	International	Demarcated	1968
Algeria–Morocco	International	Demarcated	1845
Morocco–Spanish enclaves	International	Sovereignty disputed	1860
Spain–Gibraltar	International	Sovereignty and location disputed	1713
France–Spain	International	Demarcated	1868
France–Italy	International	Demarcated	1947
France–Monaco	International	Demarcated	1963
Yugoslavia–Italy	International	Demarcated	1975
Yugoslavia–Albania	International	Demarcated	1926
Albania–Greece	International	Demarcated	1925
Cyprus–UK (Sovereign Base Area)	International	Demarcated	1960

Source: *IBS*, Special Issue: *International Boundaries and Disclaimers* (Dec. 1967), and other sources.

Another issue giving the question of territorial sea delimitation in the Mediterranean special importance is that of straits. In retrospect, there was at UNCLOS III strong resistance expressed by the Mediterranean straits States against the adoption of a system of passage allowing more freedom than was already established by the long-standing regime of innocent passage.[4] It was obvious, however, that the major maritime

[4] Several straits States which were aligned in the UN Seabed Committee and at UNCLOS III expressed their rejection of the concept of free transit. Initially they opposed any reference to 'free transit' and only 'innocent passage' was in the list of subjects prepared by the Group of 77 to be considered in Subcommittee II. See UN Doc. A/AC. 138/66. For the US representative's complaint about the adoption of such a list without reference to the

powers were prepared to accept the overwhelming tendency in the conference to extend the territorial sea limit to 12 miles only in exchange for the adoption of a free 'transit passage' in the international straits which would lie entirely in territorial seas;[5] and, as a result, the major straits States, including Morocco and Spain, reluctantly agreed to the adoption of the new regime of transit passage.[6] As a semi-enclosed sea, comprising other subsidiary seas, such as the Adriatic and the Aegean, the Mediterranean includes eight international waterways, three of which are bordered by two (or more) littoral States. The delimitation of territorial sea boundaries in these straits is of particular importance, both for the littoral States and for the interests of the international community, resting in the peaceful use of such straits for passage.

The following section examines the principles claimed in Mediterranean States' legislation for delimiting territorial waters. This will be followed by another section discussing the boundary agreements which have so far been concluded. Another section will discuss some outstanding Mediterranean disputes over, or having direct effect on, the question of territorial sea delimitation. The last section will examine the status of territorial sea delimitation in the Mediterranean straits which are important for international navigation.

2. UNILATERAL PRACTICE

Eight Mediterranean States have established in their legislation rules for delimiting the territorial sea.[7] The legislation of Albania,[8] France,[9] Morocco,[10] and Yugoslavia[11] expressly refers to a median line, whereas the legislation of Spain[12] and Turkey[13] provides for the same criterion

question of free transit, see UN Doc. A/AC. 138/SC.II/SR.29. For proposals submitted by coastal and straits States aiming at retaining the system of innocent passage in straits, see e.g. the 'Eight Power Draft', including Cyprus, Greece, Morocco, and Spain. UN Doc. A/AC. 138/SC.II/L.18.

[5] See e.g. comments by the delegates of the UK, USA, Germany, and Italy in UN Docs. A/AC. 138/SC.II/SR.27; 62/C. 2/SR.3; 62/C. 2/SR.12, respectively.

[6] For details see K. Koh, *Straits in International Navigation* (Oceana Publications, London, 1982), 129 ff.

[7] None of the rest of the Mediterranean States provides for an *ad hoc* legislation for the delimitation of the territorial sea.

[8] See *infra*, text at n. 145.

[9] Art. 2 of the Law of 1971, *supra*, Ch. 6 n. 26.

[10] Art. 5 of Law No. 1.73.211 of Mar. 1973, text in *New Directions*, vii. 116. See also *infra*, n. 185 and accompanying text.

[11] Art. 219 of the Yugoslav Customs Law as amended by Art. 229 of the Finance Law of 1924/1925. See *EESDLS* (1979) (Binder I), 353.

[12] Art. 4 of the Act of 1977, *supra*, Ch. 6 n. 18.

[13] Art. 3 of the Law of 1964, *supra*, Ch. 6 n. 49.

only in the event of a failure of agreement with the other State on another boundary line. The legislation of Egypt[14] and Syria,[15] on the other hand, refers to agreement according to the principles of international law.

Of the nineteen littorals (including Gibraltar and the Sovereign Base Area in Cyprus), Spain, Italy, Malta, Israel, Yugoslavia, and the United Kingdom are parties to the 1958 Territorial Sea Convention. None of these is recorded to have made reservations or declarations regarding the application of Article 12 of this Convention.

3. SETTLED BOUNDARIES

Five territorial sea boundaries have so far been delimited. These are, chronologically, the territorial sea boundaries between Albania and Yugoslavia, Cyprus and the British Sovereign Base Area, Italy and Yugoslavia in the Gulf of Trieste, France and Monaco, and France and Italy in the Strait of Bonifacio. For practical purposes, only the first three are discussed in this section. The fifth boundary will be discussed in the final section. The fourth one constitutes only a small segment of the shelf/water-column boundary, which will be discussed in the next chapter.

(*a*) *Albania–Yugoslavia*

The territorial sea boundary between the adjacent coasts of Albania and Yugoslavia was established by a Protocol of Delimitation in 1926, annex III of which provides that the boundary, 'starting from the limit of the territorial waters in the Adriatic Sea, follows first a straight line perpendicular to the general direction of the coast and ends up at the mouth of the principal arm of the Boyana'.[16]

The Protocol's reference to the outer limit of the territorial sea as the starting-point of the boundary line, though unusual, does not seem to raise at present any particular problem, since both States are now claiming identical widths of territorial sea of 12 n.m.

Even before 1990, when Albania was claiming a 15 n.m. territorial sea,[17] it was reported that the 1926 Protocol 'is considered to be in force

[14] Art. 8 of the Royal Decree of 1951, *supra*, Ch. 6 n. 63.

[15] Art. 7 of the Legislative Decree of 1963, *supra*, Ch. 6 n. 59.

[16] *Protocole de délimitation de la frontière serbe–croate–slovène–albanaise (juillet 1926)*, annexe III, quoted in A. G. de Lapradelle, *La Frontière* (Étude de Droit International) (Éditions Internationales, Paris, 1928), 216 n. 1 (trans.).

[17] Decree No. 4,650 of 1970, as amended by Decree No. 5,384 of 1976, *UNLS* (ST/SER. B/19) (1980), 3.

even today', and that 'Albania did not want this extension [of its territorial sea to 15 n.m.] to cause delimitation problems with Yugoslavia.'[18]

This is reinforced by the amendment of the Albanian Decree of 1976 which extended Albania's territorial sea to 15 n.m. The new amendment[19] brings the Albanian territorial sea extension back to 12 n.m. The same amendment retains the 1976 clause which provided that the 'width of the territorial sea from the estuary of the Buna river to the cape of Rondoni extends up to the Albania–Yugoslavia State boundary line'.

(*b*) *Cyprus–the United Kingdom (the Sovereign Base Area)*

The delimitation of the territorial sea between Cyprus and the British military bases in Dhekelia and Akrotiri is of interest, not only because of the eventual impact of areas of such nature (military bases) on the applicable methods of delimitation, but also because it was delimited between two different limits of territorial waters.

TABLE 3. *Bearings of the territorial sea boundary: Cyprus–UK (SBA)*

Lines	Degree	Distance (m.)	Degree	Distance (m.)	Degree	Distance (m.)
Line I	163°	6.85	207°	3.0	204°	Indefinite
Line II	108.5°	7.8	136°	Indefinite		
Line III	170°	3.8	136°	3.1	156°	Indefinite
Line IV	103°	0.9	150°	6.3	176°	Indefinite

When Cyprus became independent under the Treaty of 1960,[20] Britain retained full sovereignty over these bases as a 'Sovereign Base Area' which altogether comprises 3 per cent of the island. The limits between these areas and the Republic of Cyprus were precisely defined as international boundaries, and provisions were made for the demarcation of the lines. Section 3 of the Treaty also provided for territorial sea boundaries between the Republic and these non-ceded areas. Four specific lines have been established, all running from the low-water mark off the international boundary terminals, seaward at angles of different bearings as shown in Table 3.

[18] Informal letter (dated 14 Oct. 1987) to the writer from Professor Budislav Vukas, of the Institute of International Law and International Relations, University of Zagreb, Yugoslavia.

[19] Decree No. 7,366, modifying decree No. 4,650, dated 9 Mar. 1990. See *LSB* 16 (1990), 2.

[20] Treaty concerning the Establishment of the Republic of Cyprus, 16 Aug. 1960, 382 *UNTS* (1960), 10.

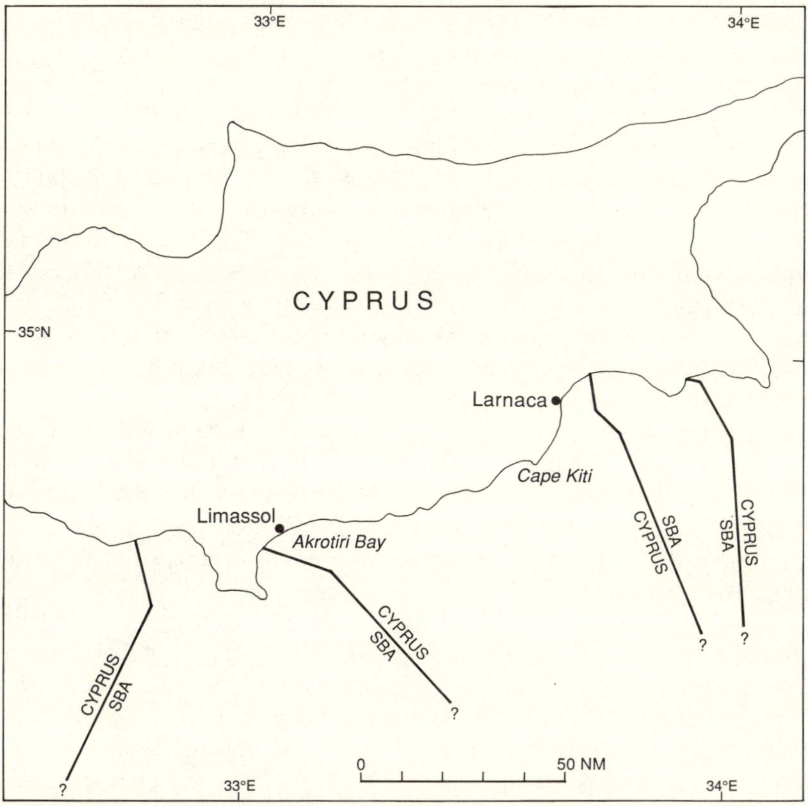

MAP 16 Territorial sea boundary: Cyprus–UK (Sovereign Base Area)

Unfortunately, there exist no published records of the negotiations which led to the delimitation agreement. It is not therefore possible to determine precisely the principles used for the delimitation of these four lines.

However, the first segments of the four lines within which the British claim of 3 territorial sea miles is defined evidently coincide with the equidistance line. Further seawards, the directional alignments of the lines change dramatically.[21] In the eastern area (Dhekelia) the two lines gradually approach each other, and, if continued, would converge at a distance of approximately 32 miles from the low-water line. In contrast, in the western area (Akrotiri) the more the delimitation lines extend seaward the further the distance between them increases. These segments

[21] See Map 16.

do not follow a specific principle. Rather, they appear to have been drawn in such a way as to secure British military requirements at the time of concluding the Agreement.

What is of special contemporary interest is the fact that the seaward terminals of these delimitation lines have not been designated. Since, at the time of signing the Treaty, the 3-mile territorial seas of Britain and Cyprus could have been determined precisely, the non-determination of the terminal points must have been deliberate. The United States Geographer comments that this decision could be based 'on the concern that one party might extend its territorial sea into a distance which would envelop or enclose the sea of the other; or the anticipated need for the limits to serve as a continental shelf or seabed boundary beyond the territorial sea'.[22]

With regard to the first possibility, Cyprus extended its territorial sea to 12 miles in 1964.[23] This development did not affect the status of the area which lies beyond the 3-mile territorial sea claimed for the Base Areas but is enclosed between the boundary lines. Section 3 of the 1960 Treaty provides that 'Cyprus shall not claim, as part of its territorial sea, waters lying between Line I and Line II . . . or between Line III and Line IV.' This seems to imply that the present seaward extension of these lines, although not formally determined, is intended by the United Kingdom to serve as the maximum limit for a prospective territorial sea (or territorial sea plus contiguous zone) jurisdiction attributed to those areas.

With respect to future shelf/EEZ claims, it is obvious that section 3 (mentioned above) is *lex specialis* confined to claims of territorial waters; and the question of a shelf/EEZ delimitation (should the United Kingdom opt to claim such jurisdiction for the Areas) with Cyprus remains an issue which requires a separate agreement. The special characteristics of these Sovereign Areas (as small enclaves, at present used for exclusively military purposes) would perhaps be regarded as a 'special circumstance' influencing the principles applicable to the delimitation of such boundaries.

(*c*) *Italy–Yugoslavia*

The boundary line between Italy and Yugoslavia in the Gulf of Trieste was established by Article 2 of the 'Treaty of Osimo' of 1975.[24] According to annexe III of the Treaty, the line is determined by reference to

[22] *LIS* 49: 2.

[23] Law No. 45 of 6 Aug. 1964, *supra*, Ch. 6 n. 2.

[24] The Treaty (signed on 10 Nov. 1975) together with an agreement concerning economic co-operation and associated documents is repr. in 60 *RDI* (1977), 674 ff.

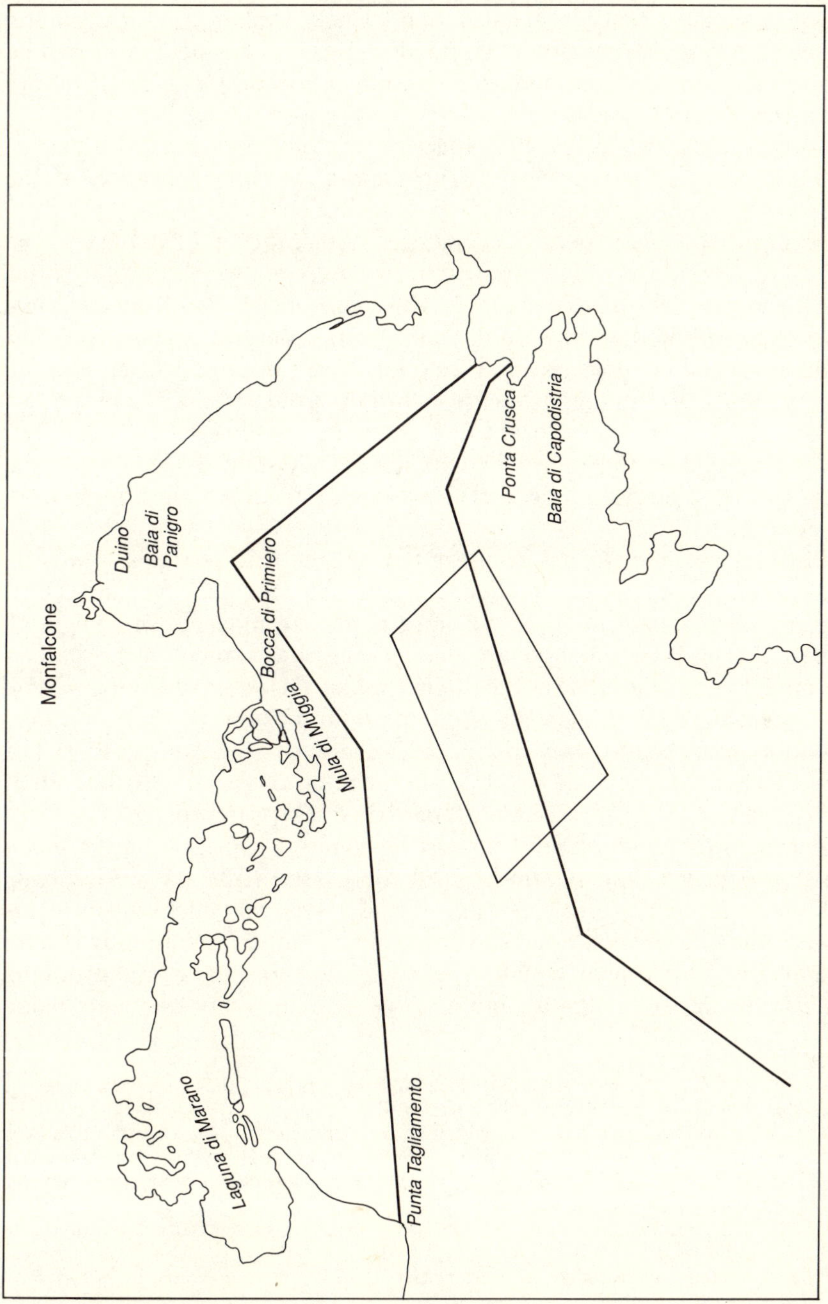

Source: Gazzetta ufficiale della Repubblica Italiana, 188 (13 Aug. 1987).

MAP 17 Territorial sea boundary and common fishery zone in the Gulf of Trieste

geographic co-ordinates,[25] but in the exchange of notes reproduced in annexe v to the Treaty it is stated that: 'En Procédant à la délimitation des eaux territoriales dans le Golfe de Trieste, Chaque Partie a tenu compte des principes découlant de la Convention de Genève sur la mer territoriale . . . du 29 avril 1958.'

As has been seen, Article 12 of this Convention postulates the relevance of 'Historic title' and 'Special circumstances' to the application of the rules of law relating to territorial sea delimitation.

During the debate on the formula of Article 12 at UNCLOS I, the Yugoslav delegate strongly opposed such doctrines; for, according to his view, 'the granting of a right to establish an unspecified boundary line other than the median line would cause confusion and encourage States to claim special circumstances for reasons of self-interest'.[26] Italy, on the other hand, generally supported the inclusion of the 'special circumstance–historic title' formula.[27] These differences in opinion between the two States emerged at the time of negotiating the Osimo Treaty, when the geographical/socio-economic characteristics of the Gulf of Trieste were considered.

The Gulf of Trieste has roughly the shape of a rectangle, with an opening of about 12 miles—between the Istrian peninsula and Porto Grado—and a penetration of no more than 13.5 miles.[28] The Yugoslav side of the Gulf is indented by several small bays, such as the Perano (Peranski) and Capo d'Istria Bays. The Italian side, starting in the middle of S. Bartelomeo Bay, is mostly straight and has extensive shallow waters along the entire shore with one major indentation, the Muggia Bay. The deepest waters in the Gulf are closer to the coast of Istria, with a depth of between 20 and 30 metres, and are particularly rich in fish.

Well before the adoption of the Osimo Treaty Yugoslavia had included in its straight baseline system of 1965 the enclosure of the waters lying within the aforementioned Bays, and had taken this enclosure into account to draw a median line of the 'maritime frontier' unilaterally.[29] This median line was drawn with reference to the normal coastline on the Italian side. Just after the adoption of the Osimo Treaty Italy had 'made

[25] With regard to the land boundary, including the terminus, the parties approved the provisional boundary established in the London Memorandum of 1954. For details see G. Gaggiano, 'Some Reflections on the Treaty of Osimo between Italy and Yugoslavia', 2 *IYIL* (1976), 248; B. Vukas, 'Solution définitive de la "question de Trieste"', 22 *AFDI* (1976), 77; V. Udina, 'Les Accords d'Osimo entre I'Italie et Yugoslavie et le règlement du problème de Trieste', 62 *RGDIP* (1979), 301.

[26] UNCLOS I, *Official Records*, iii. 187, para. 8. See also the Yugoslav proposal in A/CONF. 13/C. 1/L.60.

[27] See statement by the Italian delegate in UNCLOS I, *Official Records*, iii. 191–2.

[28] See Map 17.

[29] For the Yugoslav baselines system, see *supra*, Ch. 6 s.3(*e*).

known its intention to draw straight baselines in the Adriatic Sea'.[30] Accordingly, in 1977 Italy settled its straight baseline system,[31] which established, *inter alia*, straight baselines along the Bays of Muggia, Panzano, and the Marano lagoon.

Before the Osimo Treaty, however, Yugoslavia enforced its unilateral median line, and kept it guarded by armed patrol boats, along a salient which intersected the deep/navigational routes leading to the Italian port of Trieste. Accepting this situation 'temporarily',[32] Italy argued during the negotiation of the Treaty that it was entitled to enclose the Marano lagoon and the waters in front of it up to the line joining Punto Tugliamento to Grado under the regime of internal waters. This would have shifted the median line to the east, particularly in the vicinity of the port of Trieste, thereby including waters which were safer for the passage of oil tankers of great tonnage directly to their destination in the terminal of the Trieste oleoduct.

In addition to access to deep waters, there was also at stake access to the rich 'rectangle area' of fishing which was crossed by the Yugoslav maritime boundary and where incidences of arrest of Italian fishing boats by the Yugoslav authorities had frequently occurred.[33] Italy feared that the acceptance of such a line would ultimately deprive the Italian local fishermen of their traditional fishing area[34] and, consequently, pressed for the continuation[35] of Italian fishing activity within the area which would eventually become Yugoslav waters with the assurance that this 'would not in any way call in question what is already an established fact'.[36]

Balancing all these relevant factors, the two parties agreed on a modified equidistant/median line, which, while following the general trend of the unilateral boundary line established by Yugoslavia, thus preserving Yugoslav interests, also guaranteed to Italy access for shipping

[30] Annexe v of the Osimo Treaty, 60 *RDI* (1977), 680.

[31] See *supra*, Ch. 6 s.3(*c*).

[32] See F. Florio, 'Problemi della frontiera marittima nel Golfo di Trieste', 60 *RDI* (1977), 467, at 469.

[33] Florio, ibid. 476, reports that the number of arrests and seizures between 1955 and 1964 was about seventy.

[34] See statement by the Italian Under-Secretary for Foreign Affairs before the Chamber, in 4 *IYIL* (1978–9), 237.

[35] Fishing by Italians across the unilateral line guarded by Yugoslavia has been regulated by a series of agreements between the two States starting in 1958 when 'a Common Fishing Zone', known as the rectangle, was defined by geographical co-ordinates and this had been confirmed, with slight variations, by the Treaty of 1973. The latter Treaty was renewed several times until it was abrogated by the Fishery Agreement of 1983 (see *infra*, n. 41). For previous arrangements, see Florio, 'Problemi della frontiera marittima', 475–6. See also 'Italian Practice Relating to International Law, Diplomatic and Parliamentary Practice', 5 *IYIL* (1980–1), 310.

[36] See statement of Under-Secretary for Foreign Affairs, 4 *IYIL* (1978–9), at 237.

to ports via waters deep enough for the transit of high tonnage vessels[37] and thus corrected 'the present need to cross only Yugoslav territorial waters'.[38] With regard to fishing, the new line restored one strip of 'the rectangle fishing zone' permanently to Italy.[39] Furthermore, as a result of the 'friendly atmosphere' following the conclusion of the Treaty, the Yugoslav Government apparently gave an assurance that fishing by Italians would be tolerated in that part of the rectangle which became Yugoslav waters.[40] This latter arrangement, however, raised serious problems and it was not until 1983 that the two parties agreed on the creation of 'a fishery zone' overlapping the boundary line, within which fishing by a certain number of local fishing vessels of both sides would be allowed.[41]

4. UNSETTLED TERRITORIAL SEA BOUNDARIES

(*a*) *General appraisal*

Six territorial sea boundaries in the Mediterranean are not expected to give rise to any particular problems. These are the territorial sea boundaries of Syria with Lebanon and Turkey, of Libya with Egypt, Algeria with Morocco and Tunisia, as well as between the continental coasts of France and Italy.[42] The segments of the coast which are relevant to these potential boundaries are generally regular and no questions of 'special circumstance' or 'historic title' appear to be involved; and the coastal States, except Syria,[43] all claim identical belts of 12 territorial sea miles.

[37] According to the Italian Parliamentary Commission of Foreign Affairs, the boundary line crosses waters of 19–24 metres in depth and this, according to the Commission, would ensure access to the Port of Trieste without vessels continuing to have to sail through Yugoslav waters exclusively. *Atti Parlamentari*, VIIth legislation, Chamber of Deputies, No. 440A.

[38] Declaration of the Italian Minister for Foreign Affairs to the Chamber on 1 Oct. 1975, in *Relazioni internazionale* (1975), 988. See also an extract of the Declaration in 2 *IYIL* (1976), 423–4.

[39] Florio, 'Problemi della frontiera marittima', at 476.

[40] See statement by Under-Secretary for Foreign Affairs before the Chamber, 4 *IYIL* (1978–9).

[41] Agreement by exchange of notes, done in Rome on 18 Feb. 1983, entered into force on 16 June 1987. See *Gazzetta ufficialle della repubblica italiana*, 188 (13 Aug. 1987). For location, see Map 17.

[42] The land frontier between Italy and France terminates at a right angle in the Bay of Menton. The coastal States have never fixed a territorial sea boundary, although they signed a 'fishery boundary' Agreement in 1892, providing for a boundary line perpendicular to the coast (UN Doc. A/CN.4/71/Add. 2 (Sept. 1953), 2). The Agreement, however, was never ratified by either party. For recent negotiations between the two States on, *inter alia*, their territorial sea boundary see *infra*, Ch. 9 s.3(*d*)(ii).

[43] In accordance with the Decree of 1981 Syria has extended its territorial waters to 35 miles. See *LSB* 1 (1983), 61.

It should also be noted that these States' respective land boundaries are well established[44] and all end at the coast on bearings almost at right angles.

It is certainly true that a given boundary is often unique and too complex to be amenable to general categorization. But, with boundaries like those identified above, and given their limited extensions seawards, there is little doubt that the drawing of perpendicular lines would be regarded as equitable territorial sea boundaries. In a limited number of cases, however, outstanding disputes do exist.

(*b*) *Outstanding disputes*

(i) The Greek–Turkish territorial sea boundary

Both Greece and Turkey still maintain limited territorial sea claims of 6 miles for the Aegean.[45] The problems generated by these limited claims are to be considered first. The repercussions that may ensue from the decision of either party, Greece or Turkey, to extend its limit to 12 miles are discussed second.

1. *Delimitation within the 6-mile limits*

The delimitation of the territorial sea in the Aegean would be effected seaward of the two States' land boundary terminus at the mouth of the Martisa River, for a few miles and then between the 'opposite coasts' of Turkish Anatolia and the most eastern Greek islands adjacent thereto. Within this limited area the delimitation of the territorial sea *per se* raises no specific problem. First, title to sovereignty over those islands does not at the moment seem to be a disputed issue. In accordance with the Lausanne Peace Treaty of 1923[46] and the 1947 Peace Treaty with Italy,[47] the eastern Aegean islands, with the exception of Imyros, the Rabbit Islands, and Tenedos, were all ceded to Greece,[48] thus, in effect, placing the Turkish–Greek frontier on or near the mainland coast of Turkey.[49]

[44] See Table 2.

[45] Turkey claims 12 miles of territorial sea in the Black and Mediterranean Seas.

[46] See Treaty Ser. No. 16 (1923), Cmnd. 1929.

[47] Treaty of Peace with Italy, 10 Feb. 1947, 49 *UNTS* (1950), 126, at 134.

[48] There appear to be, none the less, Greek fears that Turkey seeks to reopen the issue of sovereignty over these islands. For a Statement by the Greek Press and Information Office in London in 1976 to that effect, see *Keesing's* (1976), 27987. See also Keesing's (1976), statement attributed to the Turkish Deputy Prime Minister in which he claimed that all the islands within 50 km. of the Turkish coast should belong to Turkey.

[49] On the circumstances in which sovereignty over the islands was accorded to Greece, see *History of the Peace Conference*, vol. vi, pt. II; *Lausanne Conference on Near Eastern Affairs, 1922–23, Records of Proceedings and Draft Terms of Peace, Turkey No. 1* (1923) (HMSO, 1923), 91–111. See also R. Beeley, 'The Greek–Turkish Boundary: Conflicts at

note 49 cont. on p. 266.

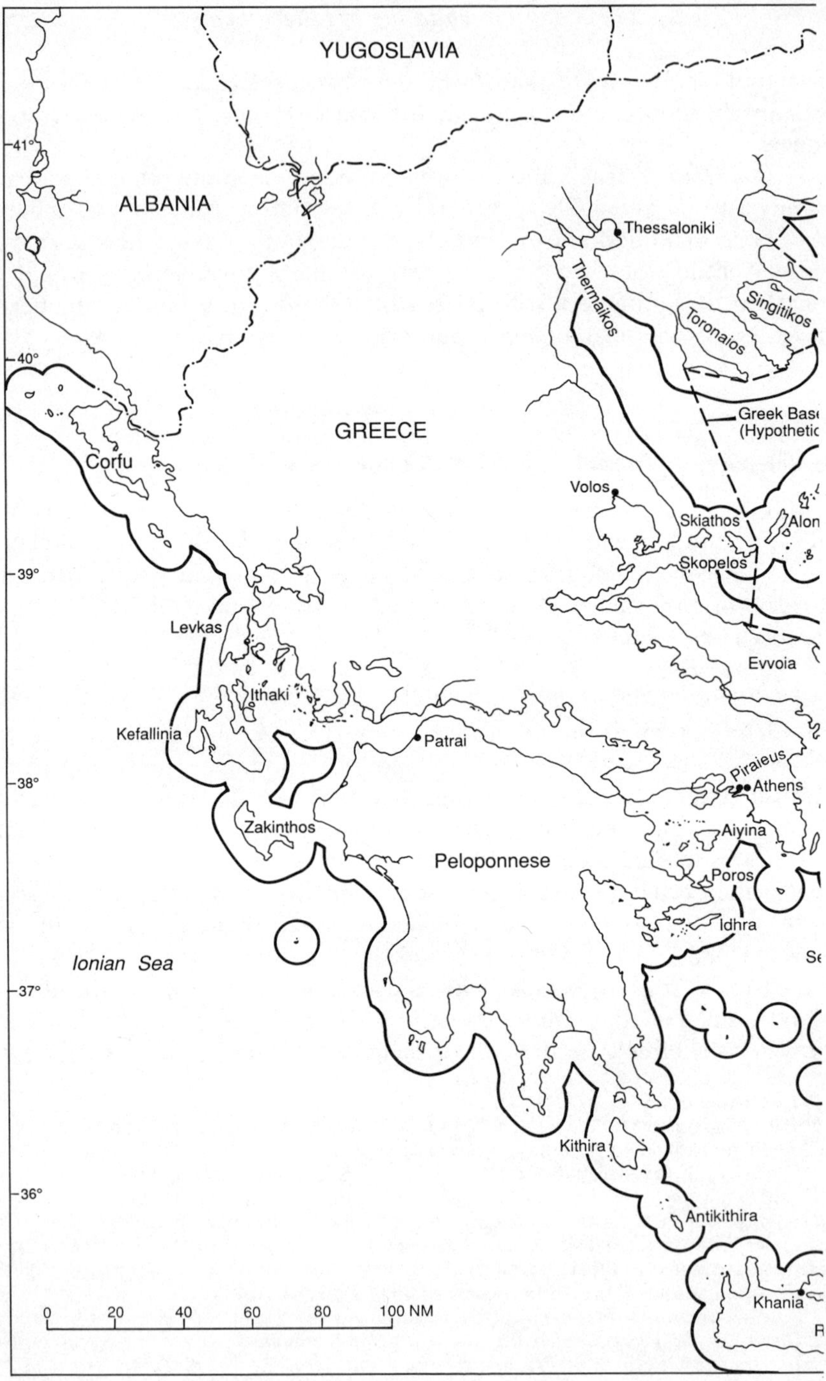

YUGOSLAVIA
ALBANIA
GREECE
41°
40°
39°
38°
37°
36°
Thessaloniki
Thermaikos
Toronaios
Singitikos
Greek Base
(Hypothetic
Corfu
Volos
Skiathos
Skopelos
Alon
Levkas
Ithaki
Kefallinia
Patrai
Evvoia
Piraieus
Athens
Zakinthos
Aiyina
Peloponnese
Poros
Idhra
Ionian Sea
Kithira
Antikithira
Khania
0
20
40
60
80
100 NM

Source: After *Aegean Sea Continental Shelf*, *Pleadings*, ICJ Reports (1980), map attached to p. 20.

MAP 18 The Aegean Sea and territorial sea belts of 6 miles.

Second, as the distance between the two States' mainland opposite coasts exceeds 24 miles, the situation becomes one which does not call for the difficult task of producing a boundary line between mainland coasts, with various island locations in between. Third, the delimitation of this limited territorial sea boundary would not be much affected by any future territorial sea extension to 12 miles; for the full impact of the Greek islands is already established by the existing limits of 6 miles, which, due to the closeness of the islands to the Turkish coast, overlap and, if the 12-mile limits were ever claimed, a median line would have to be only slightly adjusted, especially around Samothrace, Imyros, Limnos, Lesvos, Chios, and Samos (see Map 18).

One difficulty in the way of a simple equidistance/median line, however, might be related to security considerations, given the frequent occurrence of political conflicts between the two States, partly as a result of overlapping claims of sovereignty over the Aegean Sea and its islands.[50] This is especially true for Turkey in the northern Aegean, where a simple equidistant territorial sea/shelf boundary would tend towards the Dardanelles, thus raising the problem of unimpeded access to the straits and the Turkish ports on the coast of Anatolia. This will be futher considered elsewhere.[51]

The precise determination of a future equidistant/median boundary line may, however, be described in relation to the previous arrangements which settled the sovereignty question. The Treaty of Lausanne does not provide for an *ad hoc* maritime boundary delimitation between the ceded islands and the Turkish mainland coast, but a map prepared by the Official International Boundary Commission of 1925–6[52] shows, *inter alia*, an unconnected 'international boundary line' which starts at the mouth of the Martisa River and runs seaward in a south-south-easterly direction, thus ignoring the general direction of the Greek–Turkish adjacent coasts. Thence the line runs southwards between the Greek islands of Samothrace, Limnos, Lesvos, Chios, and Samos and the Turkish mainland coast. The line follows a roughly median line, except east of Chios where it almost touches the mainland of Turkey. This 'international boundary line', if now measured with reference to the Turkish straight baselines system, would certainly attach to Greece more territorial waters than it does to Turkey.

the Interface', 3 *IBGT* (1978), 351; D. Bowett, *The Legal Regime of Islands in International Law* (Oceana Publications, New York, 1979), 249.

[50] See e.g. *infra*, Ch. 10, text at nn. 23–9.

[51] See *infra*, Ch. 10 s.4(*d*)(viii).

[52] *Carte de la frontière gréco-turque*, prepared by the Boundary Commission of 1925–6 as provided in the Treaty of Lausanne. See also Treaty Series No. 16 (1923), Cmnd. 1929, map facing p. 243.

It is obvious that the Commission's line was established to show the 'Greek–Turkish frontier' as determined by the Lausanne Treaty rather than to determine a territorial sea boundary as such. An *ad hoc* maritime boundary agreement would perhaps adjust the present 'frontier line' in many localities so as to produce an accurate median/equidistant territorial sea boundary.

The territorial sea boundary between the coast of Anatolia and Kastellorizo Island was established in Article 5 of the 1932 Treaty between Italy and Turkey.[53] This boundary was also more concerned with settling the ownership of the islands, islets, and rocks on either side of it than with determining an equidistant territorial sea boundary precisely; for although described as a 'territorial sea boundary', it diverges in many places from the course of a precise median line between the Turkish coast and Kastellorizo Island. Sovereignty over this group, together with the rest of the Dodecanese archipelago, was turned over to Greece in accordance with the 1947 Peace Treaty with Italy.[54]

The present situation in the Aegean shows, already, how a relatively narrow belt of 6 miles turns the Aegean into a 'Greek lake'. Statistically, the 6-mile limit would turn 35 per cent of the Aegean basin into Greek territorial waters, in contrast to only 8.8 per cent for Turkey. The situation will be more acute if either party, Greece or Turkey, extends the territorial waters to 12 miles, a possibility that we shall now consider.

2. *The decision to extend the territorial sea to 12 miles*

It will be recalled that, when islands and islets are situated no more than 24 miles off their own mainland coast and from each other, the normal outer limit of the territorial waters may be extended so as to include those waters lying between the islands and the mainland coast.[55]

Because of the geographical characteristics of the Aegean (a narrow, semi-enclosed basin dotted with numerous islands throughout its length),[56] the application of this rule (combined with a claim to 12 miles of territorial sea) would reduce the area of the Aegean high seas dramatically. The allocation of 12-mile belts to the Greek islands within 24 miles of the

[53] Convention for the Delimitation of the Territorial Waters between the Coasts of Anatolia and the Island of Kastellorizo, 4 Jan. 1932, 138 *League of Nations Treaty Series* (1933), 243.

[54] The relinquishment of such sovereignty does not affect the boundary's status as established by the 1932 Agreement. Under both positive and customary law rules of succession now valid, the Agreement is still in force and continues to govern the territorial sea boundary between the coast of Anatolia and Kastellorizo Island. See Art. 12 of the 1978 Vienna Convention on Succession of States in Respect of Treaties; I. Brownlie, *Principles of Public International Law* (Clarendon Press, Oxford, 1979), 667; D. O'Connell, *State Succession* (Cambridge University Press, Cambridge, 1967), ii. 273.

[55] See *supra*, Ch. 1 s.2(*a*).

[56] See *infra*, Ch. 10 s.1.

mainland coast or from each other would produce territorial seas which of necessity coalesce and would consequently extend the sovereignty of Greece over approximately 64 per cent of the total area of the Aegean, with only 8.3 per cent left to Turkey. This would, in turn, reduce the proportion of remaining high seas from 56 per cent to 26 per cent.

The entitlement of Greece to bring about such results is, however, opposed by Turkey on the basis that, according to the concept of the 'semi-enclosed sea', a sea area may be subject to special geographical circumstances which exempt it from certain general rules, in this case the rule of a 12-mile territorial sea.[57] Turkey also argues the matter on the basis of the 1923 Lausanne Treaty, which, according to Turkey, recognizes economic, commercial, navigational, and security rights of both Greece and Turkey in the Aegean. Turkey argues that the 'enclosure' of its western Anatolian region by extended Greek territorial waters would upset the equilibrium which was established between the two States by the Lausanne Treaty.[58]

Although the 1982 Convention recognizes the concept of an 'enclosed or semi-enclosed sea', it does not in fact provide for any special rights or limitations on States bordering such seas, other than to require them to 'co-ordinante' their activities in the area.[59] It is clear, therefore, that UNCLOS III rejected the notion that enclosed or semi-enclosed seas should be permitted *codified* exemptions from universally applicable rules. Nor can it be claimed that the maintenance by both parties, Greece and Turkey, of the old Mediterranean limit of 6 miles[60] for several decades reflects their conviction of a binding *local custom* between them. The legislation of Greece and Turkey clearly rules out this reasoning. For Greece, the Legislative Decree of 1973 provides that 'The territorial sea comprises a sea zone, the breadth of which is fixed at six nautical miles, whereas it may be fixed differently by Presidential Decrees, issued upon proposition of the Council of Ministers.'[61] More explicitly, the 1964 Law of Turkey states that: 'Against States claiming wider territorial waters (i.e. wider than 6 miles), the width of Turkish territorial waters is deter-

[57] See e.g. Turkish proposal at UNCLOS III: 'draft Article on the breadth of the territorial sea; global or regional criteria . . . semi-enclosed and enclosed seas', A/CONF. 62/C. 2/L.8; 'draft Article on enclosed and semi-enclosed seas', A/CONF. 62/C. 2/L.56.

[58] See Background Paper, 'Current Issues in Turkish–Greek Relations' (Turkish Embassy, Washington, DC, 1982), 2; statement by Turkish delegate to UNCLOS III, quoted *infra*, text at n. 67.

[59] Art. 123 of the 1982 Convention.

[60] Until May 1970, out of nineteen territorial sea claimants in the Mediterranean eight were claiming 6 miles.

[61] Art. 139 of Legislative Decree No. 187 of 29 Sept./3 Oct. 1973, promulgating the 'Code of Public Maritime Law'. Text in *Aegean Sea Continental Shelf*, *Pleadings* (1980), 151.

mined on the basis of reciprocity.'[62] Article 1 of the 1982 Act, which superseded the 1964 Act,[63] states that

> The width of the Turkish territorial waters is six nautical miles. The Council of Ministers has the authority to fix limits in excess of six miles for certain waters after taking into account all the relevant features and the circumstances of those waters, provided that this is compatible with the principles of equity.

The reason for the two States' observance of a 6-mile territorial sea limit is, therefore, reciprocity, necessitated by the area's geographical limitation. As was stated by the Turkish delegate to UNCLOS III:

> There were . . . regions where several States bordered a narrow sea and where a very delicate balance had been established in regard to the territorial sea on the basis of the old norms. That could be seen from the example of the actual six-mile territorial waters of Turkey and Greece in the Aegean Sea . . . The extension of the territorial sea by one of the coastal States in such a sea was liable to upset the existing balance and harm the legitimate interests of other States, interests which were not limited to navigation alone.[64]

There seems, therefore, no legal reason which prevents either Greece or Turkey from extending their territorial sea limits to 12 miles. This action is not, however, likely to take place because of the dramatic effect it would have on the relations between the two States. This would appear in at least three different respects.

There is, first, the effect on the right of innocent passage to and from Turkish Aegean ports. Access by sea to İzmir, Istanbul, and the Straits would have to be effected through Greek waters. Although the right of innocent passage would be preserved, Turkey considers that in certain circumstances she would be vulnerable to total enclosure.

A second effect would be the change in the present *de facto* territorial sea boundary between the Turkish coast and the eastern Aegean islands. However minor this change would be,[65] it would certainly upset the status quo respected by both parties for several decades.

A third, but most important, effect would be the reduction of the area of the Aegean continental shelf. By definition, such areas lie beyond the territorial waters. As the sovereign State of effectively all the Aegean islands, Greece argues that all these islands and islets are entitled to their own continental shelf jurisdiction. Maintaining the same argument for areas beyond territorial sea belts of 12 miles would effectively deny

[62] Art. 2 of the Law No. 476, *supra*, Ch. 6 n. 49.

[63] See *supra*, Ch. 6 n. 50.

[64] UNCLOS III, *Official Records*, i. 169.

[65] As noted earlier (*supra*, s. 4(*b*)(i)(1)) the adjustment of the median line would be limited to certain areas.

Turkey any continental shelf (or EEZ) jurisdiction along its Anatolian coast. Aware of these effects, the two States seem to exclude the possibility of such an extension. Greece, while maintaining the position that this is a 'question [involving] the sovereign will of the nation', declares that this 'does not mean that Greece will extend its territorial sea to 12 miles'.[66] Turkey for its part attaches 'fundamental importance . . . to the maintenance of the *status quo* with respect to the territorial sea limits'.[67] This *de facto* abstention would seem to serve as 'an essential requirement' for arriving at a 'package deal' between Greece and Turkey to solve their differences in the Aegean, including the continental shelf and air traffic.

A situation somewhat analogous to the Aegean Islands exists in the Torres Strait between Australia and Papua New Guinea. Here a considerable number of Australian islands lie close to the Papua New Guinean coast. A claim to a 12-mile territorial sea for these islands would effectively turn the entire Torres Strait into an area of Australian jurisdiction. In 1978, Australia and Papua New Guinea signed a treaty in which they agreed, *inter alia*, to reduce the territorial sea jurisdiction attributed to these islands to 3 miles, which may not be extended should Australia opt at a future date for a 12-mile territorial sea.[68]

Notwithstanding the fact that the application of the Torres Strait solution in the Aegean would require accommodation of qualitatively different issues, the similarity between the two cases is best illustrated by the following comment of Burmester on the Torres Strait Treaty:

> just as with the Aegean, the consequence here of a general claim to a 12-mile territorial sea [instead of 3] around the islands needs to be kept in mind. Such a claim would effectively have turned the Torres Strait area into a sea of Australian jurisdiction extending up to within a mile or so of the Papua New Guinean mainland. . . . Such a result was not seen as satisfactory or equitable by either side.[69]

This agreement, it must be noted, however, was not limited to the delimitation of the territorial sea, but was part of an overall settlement of several maritime jurisdictions between the parties, including, in addition to the islands, continental shelf, fisheries, and navigation.

[66] Statement of Mr Tzounis (head of the Greek delegation) in Minutes of the Meetings of the Greek and Turkish Continental Shelf Experts (Berne, 31 Jan.–2 Feb. 1976). Text in *Aegean Sea Continental Shelf*, *Pleadings*, 152–3.

[67] Statement of Mr Bilge (head of the Turkish delegation), ibid.

[68] Treaty Concerning Sovereignty and Maritime Boundaries in the Area between the Two States, 18 *ILM* (1979), 291.

[69] H. Burmester, 'The Torres Strait Treaty: Ocean Boundary Delimitation by Agreement', 76 *AJIL* (1982), 321, at 343.

(ii) The Libyan–Tunisian territorial sea boundary

The Libya–Tunisia international boundary meets the Mediterranean at Ras Ajdir, almost in a north–south direction. The coastline immediately east and west thereof runs in a north-west–south-east direction. Within this relevant segment, the territorial seas of both States are measured from the low-water line.[70] There is as yet no agreement between Libya and Tunisia regarding the delimitation of the territorial sea. In the late 1960s, following the granting by both States of oil concessions in overlapping areas of their respective territorial seas and continental shelves, the two States engaged in diplomatic activities to reach agreement regarding these overlapping areas. Unable to do so, they agreed in 1977 to submit their dispute over the continental shelf to the ICJ. In February 1982 the Court delivered its Judgment. The purpose of this section is to examine the possible impact of the two States' diplomatic practice as well as the Court's Judgment upon a future territorial sea delimitation.

1. *Bilateral Practice*

Diplomatic activities regarding boundaries between the two States probably started with the conclusion of the Tripolitania–Tunisia Boundary Agreement of 1910. Article 1 of the Agreement provides that the boundary between the two States 'Shall start at Ras Ajdir, on the Mediterranean, in a generally north–south direction'.[71] As will be seen, Libya relied on this provision in support of its view on the maritime boundary with Tunisia, namely, that the maritime boundary was envisaged as continuing in a northerly direction.

Since 1913, France and Italy, acting as colonial powers in Tunisia and Libya respectively, engaged in a dispute regarding their respective jurisdictions over the fisheries and the sponge banks situated seaward of the land frontier of their colonies. Following a protest presented by France to the Italian authorities arising out of the latter's arrest of fishing boats possessing Tunisian licences westward of the Tunisian-claimed line of '*zone de surveillance*' (extending north-east of Ras Ajdir),[72] Italy proposed in a *Note verbale* a line extending 'normally to the general direction of the coastline', being approximately north-north-east.[73] The Italian proposal

[70] See *supra*, Ch. 6 ss. 2, 3(*i*).

[71] 'Convention relative à la frontière entre la régence de Tunis et la vilayet de Tripoli, signée à Tripoli de Barbarie le 19 mai 1910.' English text in *IBS* 121.

[72] 'Instruction sur le service de la navigation et des pêches maritimes du 13 décembre 1904.' Text in *Memorial* submitted by the Government of Tunisia, vol. ii: *Annexes* (1980), annexe 77. The arrest occurred 11.7 miles offshore at longitude 11° 42′ 14″. See *Reply* submitted by the Libyan Arab Jamahiriya (1981), i. 17–18.

[73] *Note verbale* of 2 Oct. 1913. Text in Libyan *Reply* (1981), annexe 1-18. The word 'normally' is generally used to mean 'perpendicular to'. See the *Tunisia–Libya* case, ICJ Reports (1982), para. 93.

was not expressly rejected by Tunisian authorities, and, consequently, the Italian Government of Tripolitania stated in its Fishery Regulations of 1919 that 'as far as the sea border between Tripolitania and Tunisia is concerned, *it was agreed* to adopt as a line of delimitation the line perpendicular to the coast at the border-point, which is . . . the approximate bearing north-north-east from Ras Ajdir'.[74] This was reiterated in similar instructions in 1931[75] and, several years after Libya's independence, in 1961.[76]

Contrary to these regulations, however, Tunisia claimed in 1951 and again in 1963 a reserved *fishing zone* off the frontier with Libya and bounded by a line of 45° running north-east from Ras Ajdir to the point of intersection with the 50-metre isobath.[77] The important point about these regulations of both States is that they only referred to a fishing zone—contiguous to the territorial sea—and never purported to determine, by a unilateral act, the territorial sea boundary between the two States.[78]

Therefore, when representatives of both States met in July 1968 to discuss their 'maritime boundaries', the Tunisian delegate, unwilling to consider the question of demarcating the territorial waters of Tunisia, took the position that the ZV (*Zénith vertical*) 45° line constituted the eastern boundary of the historical fishery zone of Tunisia. Libya, on the other hand, contended that the maritime frontier should extend northwards from Ras Ajdir.[79]

These meetings were obviously inspired by the granting by both States of oil concessions to explore for and exploit petroleum resources within sea areas off Ras Ajdir; Tunisia was the first to do so in October 1964. The south-eastern boundary of the Tunisian concession was determined by a stepped line running north-north-east from Ras Ajdir. In 1968 Libya followed suit by granting several concessions. On the basis of the information available to the ICJ in the 1982 Judgment, the Court ruled that the States' respective concessions were aligned with each other at an angle of 26° to the meridian.[80]

[74] Italian Instructions of 16 Apr. 1919, text in *Documentary Annexes* submitted by the Libyan Arab Jamahiriya (1981), annexe 43 (emphasis added).

[75] Italian Instructions of 25 June 1931, ibid. annexe 44.

[76] Libyan Decision No. 1 of 1961, ibid. annexe 47.

[77] Decree of 26 July 1951, *Official Journal of Tunisia* (July 1951), 950; Tunisian Law No. 63-49 of 30 Dec. 1963, ibid. (31 Dec. 1963), 1870.

[78] Neither the Libyan Law of 1959 (*supra*, Ch. 6 s.2) nor the Tunisian Law and its executive Decree of 1973 (*supra*, Ch. 6 s.3(*i*)) includes provisions regarding the lateral boundaries of the territorial sea.

[79] See *Memorial* submitted by the Libyan Arab Jamahiriya (1980), 17.

[80] See the *Tunisia–Libya* case, para. 117. For a request by Tunisia to revise the Court's finding, see *infra*, text at nn. 90–3.

From 1968 it was evident to both States that a dispute was arising with regard to overlapping areas of continental shelf and consequently discussions were held between the two States between 1973 and 1976. Although it was obvious that both sides held different views with regard to the delimitation of the continental shelf,[81] the problem of the delimitation of their respective territorial waters was never discussed independently.

It was only in 1976 (in a memorandum circulated to diplomatic missions) that Tunisia reiterated, with a slight adjustment, that its 45° fishery line was also valid for the purposes of sea boundary delimitation with Libya.[82]

2. *The relevance of the conduct of the parties as reflected in the Court's Judgment*

The first step taken by the Court towards ruling on a method of delimitation was to divide the area of delimitation into two sectors, in accordance with their proximity to the coast. The Court found that the proper appreciation for the relevant circumstances characterizing the area 'call[ed] for the area close to the coasts of the Parties to be treated differently from the areas further offshore'.[83]

In this first sector, the Court found that a line drawn from the outer limit of the territorial sea which formed an angle of 26° was appropriate. Much of the Court's reasoning with regard to this line was founded on the conduct of the parties, for, as the Court put it, account must be taken of 'whatever indicia are available of the line or lines which the Parties themselves may have considered equitable or acted upon as such'.[84] By reviewing the respective legislative and diplomatic conduct of the two parties, the Court concluded that a *de facto-modus vivendi* line corresponding with the 26° line was equitable. The 1913 line established by the Government of Tripolitania which, as has been shown, was not formally contested by Tunisia for a long period was regarded as a 'historical justification' for the choice of a line to the extent that Tunisia could not claim historic rights opposable to Libya to the east of the *modus vivendi* line.[85] The 26° line was also found to correspond with a *de facto* line created by the alignment of petroleum concessions granted by the two States.[86]

The findings of the Court with regard to the precise direction of this *de*

[81] See *Memorial* submitted by the Libyan Arab Jamahiriya (1980), 17–18.
[82] In this latter memorandum Tunisia claimed a line of 43° 21′ rather than 45°. For text see *Memorial* submitted by Libyan Arab Jamahiriya (1980), 18.
[83] Para. 114.
[84] Para. 118.
[85] Para. 95.
[86] Para. 117.

facto line were disputed by Tunisia in the 1985 *Revision* case.[87] Tunisia requested the Court to revise its 1982 Judgment on the ground, *inter alia*, that according to newly emerging facts the real course of the concessions granted by Libya was different from that decided by the Court, and therefore there was reason to replace the co-ordinantes of the first sector as given by that Judgment with others reflecting the concessions' real course.[88] However, the Court in the *Revision* case maintained its view on the equitableness of the 26° line. It stated that the line resulting from the grant of the concessions was 'by no means the sole consideration taken into account by the Court'.[89] The method chosen by the Court 'for ensuring an equitable delimitation [was] derived in fact from a balance struck between a number of considerations, a process which has always been regarded as inherent in the application of equity in this domain'.[90]

3. *The relevance of the 26° line to the territorial boundary*

The 26° line required, in effect, selection of a termination point for an imaginary line of territorial sea delimitation. Recognizing such a correlation, the Court elaborated that the starting-point for this line must be 'from the boundary of the territorial sea off Ras Ajdir, the exact point (and thus the relationship of the delimitation line to the unsettled lateral boundary of the territorial sea) depending upon the direction of the line with respect to Ras Ajdir'.[91] Constructing the initial segment of the 26° line, the Court explained that:

> There should first be determined what point on the outer limit of the territorial sea corresponds to the intersection of that limit with a line drawn from the terminal point of the land frontier through the point 33° 55′ N, 12° E, thus at an angle [corresponding to 26°].[92]

The Court had no hesitation in pointing out the relevance of the 26° line to a possible territorial sea delimitation. Emphasizing the non-arbitrary nature of this line, the Court stated: '[i]t should be recalled that in the context of delimitation of the territorial sea the methods of delimitation . . . examined by the Committee of Experts . . . in 1953 were the continuation in the seaward direction of the land frontier . . . and the drawing of a line perpendicular to the line of general direction of the coast.'[93] The 26° line was found to be roughly perpendicular.

[87] *Application for Revision and Interpretation of the Judgment of 1982 in the Case Concerning the Continental Shelf* (*Tunisia–Libyan Arab Jamahiriya*) (*Tunisia* v. *Libyan Arab Jamahiriya*), Judgment, ICJ Reports (1985), 192.

[88] Ibid. 6. See also paras. 15, 16.

[89] Para. 35.

[90] Ibid.

[91] The *Tunisia–Libya* case, para. 116.

[92] Ibid. para. 121.

[93] Para. 119.

While the 26° line is not a territorial sea boundary line, and indeed neither the Court nor the parties purported to claim that to be so,[94] it is nevertheless appropriate to ask whether this determination by the Court of a line with a precise angle from the land frontier would still render necessary any future negotiation or discussion relating to territorial sea delimitation between the two States. A possible answer would certainly depend on the position of the two States with regard to the equity of this proposed line.

The parties' awareness of the confinement of the case to the dispute over the continental shelf did not prevent them from giving their views on a possible territorial sea delimitation. Their views, being inevitably influenced by the rather important issue at stake, i.e. the continental shelf delimitation, took consequently different approaches.

In its *Memorial* Libya initially reiterated its position that, in light of the boundary direction established by the 1910 Convention, it might be assumed that the maritime boundary between Libya and Tunisia would continue seaward from Ras Ajdir in 'a northerly direction'.[95]

However, during the pleadings Libya declared its acceptance of the 26° line or any line which is roughly 'perpendicular to the coast' at the land frontier.[96] It is therefore evident that a shift from a strict south–north line to an angulation of 26° would now be acceptable to Libya as a possible territorial sea boundary.

The Tunisian position is far more complicated. Tunisia introduced a 'sheaf of lines'[97] which, although based on two different criteria, were, according to Tunisia, capable of producing similar results. The first of these criteria was the natural prolongation of the two States based on geological, geophysical, and bathymetric data introduced by Tunisia. The second was the configuration of the coasts measured on the basis of geometry. The relevance of these criteria to territorial sea delimitations is doubtful. First, these criteria are based on lines of crests and the location of the abyssal plain and thus take into account geophysical formations of areas of seabed well beyond that enclosed within a limited territorial sea stretch of 12 miles. Secondly, they take into account, according to the explanation submitted by Tunisia, coastal areas of length greater than that required for the same narrow belt of territorial waters.[98]

However, another factor brought by Tunisia in support of those lines was the existence in the inner-shore area of an 'established' fishing line of

[94] Para. 116.

[95] *Memorial* submitted by the Libyan Arab Jamahiriya (1980), 14.

[96] The *Tunisia–Libya* case, para. 119.

[97] See *Memorial* submitted by the Government of Tunisia (1980), ch. 9. For a map showing these lines, see the *Tunisia–Libya* case, 81.

[98] The lines proposed by Tunisia utilize extensive segments of the Tunisian–Libyan coast, i.e. between Ras Mustapha (south of Cape Bon) in Tunisia and Ras Zarrouq in Libya. See *Memorial* submitted by the Government of Tunisia (1980), 244–7.

45° 'running in a northeast direction from Ras Ajdir to the point of intersection with the 50 metre isobath'.[99] As has been seen, the concept of 'historic title' is part of the territorial sea delimitation formula produced in Article 12 of the Territorial Sea Convention. Although neither Tunisia nor Libya is a party to this Convention, it is a generally accepted rule of customary law that 'historic titles must enjoy respect and be preserved as they have always been by long usage'.[100] Within this context the ZV 45° line must now be examined.

It has first to be noted that the main thrust of the argument of Tunisia was directed towards the existence of established fishing rights in the Gulf of Gabes, around the Kerkennah Islands and the shoals surrounding them.[101] These areas are certainly irrelevant, for they are situated well beyond any area subject to the delimitation of the territorial sea between the two States. In fact data submitted by Tunisia regarding the location of its fisheries and sponge banks indicated that none of these zones existed within Tunisia's overlapping territorial sea waters with Libya. The most eastern 'fixed fishery' of Tunisia within the belt of 12 n.m. is found between the Tunisian coast west of Ras Ajdir and Sadok Belhiba buoy, approximately at longitude 8° 50′ 14″ E.[102]

Also relevant in this respect is the Court's conclusion regarding Tunisian submissions on the ZV 45° line. The Court in the 1982 Judgment concluded that 'taking all the stages of Tunisian–Libyan relations into account, up to the time when the Special Agreement was concluded, the ZV 45° north-east line is not opposable to Libya, even as a mere inchoate maritime boundary between the two countries'.[103] It is pertinent to recall that the Libyan fishery regulations which had been issued since 1913 had always claimed the northward line from Ras Ajdir as the only recognized line. In this regard Judge Aréchaga goes as far as saying that 'Libyan proven historic rights are as worthy of respect as those invoked by Tunisia.'[104]

It seems obvious, therefore, that the Tunisian argument for a line other than the 26° line is not well established. The reasonableness of the latter line is also reinforced by the fact that it roughly corresponds, as noted by the Court, to a perpendicular/equidistant line.[105] If drawn precisely, this equidistant line would break into, as Judge Oda suggests, 'a plurality of

[99] See *supra*, n. 82.
[100] *Tunisia–Libya* case, para. 100.
[101] See *Memorial* submitted by the Government of Tunisia (1980), para. 4.14–4.18. See also the *Tunisia–Libya* case, para. 99.
[102] See Tunisian *Memorial* (1980), fig. 4.04, facing p. 89.
[103] *Tunisia–Libya* case, para. 90.
[104] His Sep. Op., ibid. 132.
[105] Para. 126.

TABLE 4. *Islands and enclaves in the Mediterranean Sea*

	Area in sq. km.	Population	Political status
Gibraltar	6.50	300,000	Dependant of Britain, sovereignty long claimed by Spain
Alboran	1.00	n.a.	Spanish islet
Ceuta	18.50	8,000	Remnant of Spanish colonies on the North African coast of Morocco
Melilla	4·75	65,000	Remnant of Spanish colonies on the North African coast of Morocco
Chafarinas	2.50	n.a.	Remnant of Spanish colonies on the North African coast of Morocco
Velez de la Gomera	0.20	n.a.	Remnant of Spanish colonies on the North African coast of Morocco
Al Hocaima	0.20	n.a.	Remnant of Spanish colonies on the North African coast of Morocco
Balearic Islands	4,992.00	685,088	Form Balearics province of Spain
Corsica	8,721.00	240,000	Dependant of France
Sardinia	24,092.00	1,628,690	Island of Italy
Sicily	25,708.00	5,051,413	Island of Italy
Maltese Islands	315·59	345,636	Form the Republic of Malta
Cyprus	9,251.00	642,731	Republic of Cyprus[a]
Crete	8,332.00	502,165	Island of Greece
Rhodes	1,404.00	87,831	Island of Greece
Dhekelia and Akrotiri	265.00	n.a.	UK Sovereign Base Areas

n.a. = Not available.

[a] Turkish invasion in 1947, followed by partition and declaration in 1983 of the independent Turkish Republic of Northern Cyprus.

Source: *Annual Abstract of Statistics* (Malta, 1989); *Anuario estadístico de España* (Instituto Nacional de Estadística, Madrid, 1988); *Annuaire statistique de la France (1987)* (Institut National de la statistique et des Etudes Économiques, Paris, 1987); *Annuario statistico italiano* (Istituto Centrale di Statistica, Rome, 1988); *Statistical Abstract 1985 and 1986* (Republic of Cyprus, Department of Statistics and Research, Ministry of Finance, 1987); *Statistical Yearbook of Greece (1987)* (National Statistical Service of Greece, Athens, 1989); *The Longman Encyclopedia* (London, 1989).

equidistant lines', drawn between the coast of Jerba in Tunisia and Ras at-Tarqa on the Libyan coast.[106]

(iii) The Spanish enclaves question

The prospect of a maritime boundary delimitation between Spain and Morocco faces difficulties of a political nature stemming from the dispute of sovereignty over the 'Spanish *presidios*', existing on the north coast of Morocco and consisting of four insular formations and two enclaved territories.[107] These enclaves are a remainder of the former Spanish colonial territories along the coast of North Africa.[108] Morocco, however, takes the position that such enclaves still hold a colonial status, and it formally requested the relinquishment of Spanish sovereignty over them,[109] because they form, according to Morocco, part of 'its national territory'.[110]

Spain on the other hand maintains the position that these enclaves are part of Spanish territory and has condemned the Moroccan position as 'a threat to Spain's national unity and integrity'.[111] Tension between the two States has mounted several times, particularly off the Moroccan coast on the Gibraltar Strait,[112] where Morocco expressed fears of 'a Spanish plan' to establish 'barriers', 'lines of interception', and 'naval defence' in the region surrounding the Strait of Gibraltar and for the 'control of sea traffic' in that region, in which the 'enclaves of Ceuta and Melilla and Moroccan islands were assigned a most important and a most dangerous role'.[113]

While the settlement of the dispute is of course a prerequisite to the consideration of the delimitation question, it is still relevant to consider certain factors of potential legal impact in determining whether a balance of equity in a given delimitation situation involving isolated insular formations and territories (clustering against or situated on the mainland of a foreign State) can be attained.

A preliminary question is whether these features are entitled, in accord-

[106] Judge Oda, his Dissent, ibid. 272.

[107] See Table 4; Map 19.

[108] Ifni, for instance, a Spanish enclave on the Moroccan Atlantic coast, was transferred to Morocco on 4 Jan. 1969. See *Keesing's* (1969–70), 23146.

[109] See a letter dated 26 Jan. 1975 addressed from Morocco to the Chairman of the UN Special Committee on the Situation with Regard to the Implementation of the Declaration on Granting of Independence to Colonial Countries, A/AC. 109/475. See also *Keesing's* (1973), 27414.

[110] See statement by the Moroccan representative to the Second Committee, UNCLOS III, *Official Records*, iv. 75–6.

[111] A statement by the Spanish Minister of Information and Tourism on 7 Feb. 1975. *Keesing's* (1975), 27415.

[112] For incidents which occurred in 1975, see the statement of the Moroccan representative, UNCLOS III, *Official Records*, iv. 75–6, and *Keesing's* (1975), 27415, 27418.

[113] Statement of the Moroccan representative in UNCLOS III, *Official Records*, iv. 75–6.

ance with the law of the sea, to fully fledged maritime jurisdictions. With regard to shelf and EEZ jurisdictions, there seem to be at least three guiding principles emerging. First, State practice indicates that the solution of the problem of disputed islands is often considered as part of the boundary settlement involving other areas beyond the disputed features. If no solution is reached, however, States may agree on disregarding those features for boundary purposes, thus leaving the issue of entitlement undecided.[114] Second, insular formations which are detached from the sovereign State may be ignored in delimiting the principal boundary and only given a limited belt of maritime jurisdiction, often equivalent to the territorial sea; thus determining, in effect, their maritime jurisdiction.[115] Third, in terms of entitlement, 'rocks which cannot sustain human habitation or economic life of their own shall have no exclusive economic zone or continental shelf'.[116]

Whilst the latter principle is of course irrelevant to territorial sea zones, the other two are obviously associated with situations where the feature is incidental to a delimitation between the *mainlands* of the parent State and the foreign State. In the case under consideration, the enclaves themselves comprise (except along a small segment of the Gibraltar Strait) the only foreign territories, situated off or on the coast of another State. The above principles are, therefore, of little help unless the effect of these enclaves is considered within an overall shelf/EEZ delimitation between the mainland coasts of Spain and Morocco.

However, dealing with the questions of territorial sea entitlement and delimitation alone, the relevant body of law seems to be less certain. Considering the question of territorial sea entitlement, legal pronouncements, since the nineteenth century, have contributed to the norm affirming the entitlement of islands to their own belt of territorial sea as distinct from that of a mainland territory.[117]

[114] e.g. Kachchativu Island appears to have been disregarded by the parties (India and Sri Lanka) as a point of measurement primarily on account of its disputed status (see *LIS* 66). The disputed island of Machias and the North Rock have been disregarded in the delimitation of the single boundary between Canada and the USA in the Gulf of Maine (see the *Gulf of Maine* case, ICJ Reports (1984), para. 20). Hans Island was disregarded in the delimitation of the continental shelf between Canada and Denmark (Greenland) in the Kennedy Channel. See *infra*, Ch. 10 s.4(*b*)(i)(3). The islands of Greater and Smaller Lubainah were also disregarded in the shelf boundary between Bahrain and Saudi Arabia. See *infra*, Ch. 10 s.4(*b*)(i)(3).

[115] The classic example is the enclave solution for the Channel Islands. The Court was careful to point out that the Channel Islands were 'not only "on the wrong side" of the mid-channel median line but *wholly detached geographically from the United Kingdom*'. 18 RIAA, para. 199 (emphasis added). For further examples, see *infra*, Ch. 10 s.4(*b*)(iii)(2).

[116] Art. 121(3) of the 1982 Convention. See *supra*, Ch. 2 s.3(*d*).

[117] See e.g. Art. 10(2) of the 1958 Territorial Sea Convention. For relevant judicial decisions and other relevant legal materials, see H. Jayewardene, 'The Regime of Islands in International Law', Ph.D. thesis (Cambridge, 1981), 11–19.

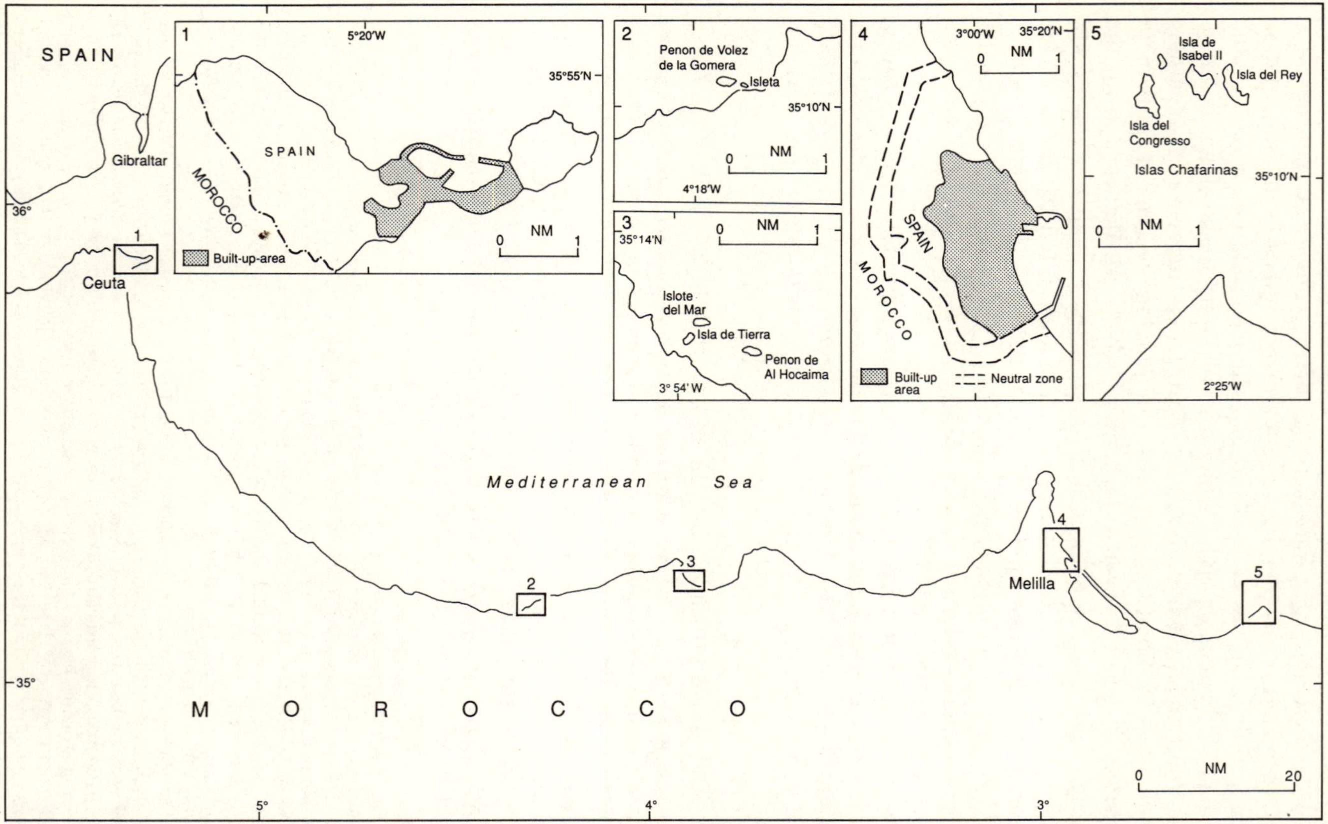

Source: G. Blake and A. Drysdale, *The Middle East and North Africa: A Political Geography* (Oxford University Press, New York, 1985), 53.

MAP 19 Spanish enclaves on Morocco's Mediterranean coast

A careful examination of these materials, however, does not confirm that all *insular formations*, irrespective of their political status, location, and habitation, can generate their own belt of territorial waters. For instance, in the dispute between England and Spain in 1852 on the ownership of 'unoccupied keys' adjacent to Cuba, Britain admitted Spain's title to these keys, but asserted that only the inhabited islands possessed territorial sea zones.[118] Furthermore, in all the 'precedents' which are often cited in this regard, there was hardly any case which dealt with such peculiar circumstances as those encountered in the Spanish enclaves. The small size of these enclaves, their close proximity to the mainland coast of Morocco (as will be seen below), and the inability of the islets to sustain human habitation make their analogy to these 'precedents' unreasonable.[119]

At the Hague Conference and also in the ILC, although the requirements of 'habitation' and 'capable of effective use and control'[120] were rejected, these proposals were never discussed in association with the peculiar situation where the feature clusters against or literally forms part of the foreign mainland coast. In fact, the discussions in the 1958 Conference (and its preparatory reports) always focused on the more common situation wherein an island is situated near, or at some distance from, the sovereign State, or where it is 'further away from the State to which it belonged than from the other State'.[121] The question whether the rights of isolated features (comprising the only foreign territory) should be counterbalanced by the rights of a foreign State in a situation such as that under consideration was never raised.[122] As shown earlier, the only relevant discussions at UNCLOS III affirm the trend, expressed in several proposals, which aimed at curtailing the traditional view that every island, properly so called, is entitled to territorial sea jurisdiction.[123]

[118] See H. Smith, *Great Britain and the Law of Nations* (King & Son, London, 1935), ii. 226–7, for the British reply to a Spanish dispatch; 228–31, for the opinion of the Queen's Advocate (Harding).

[119] See e.g. the decision of the High Court of Admiralty in the *Anna* case of 1805. Although the islands involved were uninhabited, they were so close to the mainland coast of the sovereign State that they were regarded by the Court as 'the natural appendages of the coast on which they border, and from which indeed the area formed'. *C. Robinson's Admiralty Reports* (1805), 373, at 385. Similarly, in the case of the *US* v. *Middleton*, the Court's decision in 1929 that it did not make any difference (for territorial sea entitlement) that the islands in question were not inhabited was primarily based on the fact that they were keys just off the coast of California. For relevant discussions in the Hague and Geneva Conferences, see text at next note.

[120] See the proposal of Great Britain in respect of Basis of Discussion No. 14; the suggestion of Lauterpacht in 1954 in *Yearbook of the ILC* (1954), i. 92; Bowett, *The Legal Regime of Islands in International Law*, 7–9.

[121] See *supra*, Ch. 2 s.3(*c*)(i).

[122] See e.g. the report of the Sub-Committee of the Hague Preparatory Committee, prepared by Schucking, LN Doc. C.196. M. 70, 1927, V.; *Yearbook of the ILC* (1954), i. 91 and (1956), ii. 253, 270.

[123] See *supra*, Ch. 2 s.3(*d*).

It cannot therefore always be assumed that all islands (constituted by 'naturally formed areas of land, surrounded by water, which is above high tide'),[124] irrespective of location, size, and other circumstances, are entitled to their belts of territorial sea. A more plausible view seems to suggest that the peculiar geography of, as well as the sensitivity of, the politico-legal considerations associated with the 'Spanish enclaves' may not lead to the application, *mutatis mutandis*, of the general rule mentioned above.

It is unreasonable to argue, for instance, that the rock of Perejil (situated west of Ceuta) should be entitled to a full belt of territorial sea. While Perejil is situated at the entrance of the Atlantic side of the Strait of Gibraltar (and could thus be argued to have a strategic importance), it is a small, uninhabited rock and situated only 200 metres from the coast of Morocco. The same is also true in relation to the rocks of Velez de la Gomera and Al Hocaima. They can indeed be regarded as 'natural appendages' of the coast of Morocco. They are uninhabited, although Spain keeps on them a reduced number of military personnel, renewed periodically.[125] The above considerations may make it equitable to attribute to these rocks reduced areas of security zones, delimited from the south by a line half-way from the Moroccan coast and curving northward to circumvent them.[126]

Although the Chafarinas group[127] is slightly bigger than the previously mentioned ones, the islets are in some places only 1.94 n.m. from the Moroccan coast. A similar security zone solution may also be equitable.

Ceuta and Melilla have different characteristics. As they are non-insular features, they bear little relevance to the law of the sea 'policies' regarding 'distant' and 'detached' islands. In fact, there appear to be no positive rules of delimitation that may be related specifically to this situation. The only relevant discussion was that advanced by several newly independent and developing States at UNCLOS III regarding islands and territories pending decolonization. The main thrust of the various proposals submitted in this respect was to curtail the question of extended maritime zone jurisdictions for islands and territories subject to colonial or foreign dependence.[128] Morocco was one of the sponsors of the proposal submitted by the Group of 77 on 13 August 1981, which effectively became part of the Draft Convention (ICNT/Rev.3). It provided that with regard

[124] Art. 10(1) of the 1958 Territorial Sea Convention; Art. 121(2) of the 1982 Convention.

[125] E. Orihuela, 'La delimitación de espacios maritimos', ii. doctoral thesis (University of Murcia, 1986), 611.

[126] This solution is also suggested by Orihuela, ibid. 600, 612.

[127] The group comprises three islets: Congresso, Isabel II, and del Rey.

[128] The issue was included in the original list of subjects for discussion in accordance with General Assembly Resolution 2,752 (xxv). For details see Jayewardene, 'The Regime of Islands in International Law', 25–30.

to disputed territories, such jurisdictions should not be exercised 'except with the prior consent of the parties to the dispute until such dispute is settled in accordance with the purposes of the Charter of the United Nations'.[129]

The discussion of this item was later dropped and the Convention was eventually adopted without any particular rule on the subject being included. The short life-span of this item in the Conference indicates not only the impracticality of the notion it advances but also its failure to reflect any emerging principle of international law.

The questions of territorial sea entitlement and delimitation for Ceuta and Melilla can still, however, be addressed in the light of their socio-economic characteristics, which distinguish them from the other enclaves. Ceuta and Melilla have their own established international boundaries,[130] and they have a mixed population of a few thousand of Moroccan and Spanish nationality. Ceuta occupies a strategic position on the eastern entrance of the Strait of Gibraltar. Their entitlement to territorial waters can be overshadowed only by Morocco's increasing pressure to convince the Madrid Government to negotiate on the political future of these enclaves.[131] As to the delimitation of this zone with Morocco, one solution (by analogy to distant islands) might be to draw an equidistant line from the terminal points of the land boundaries. The screening effect of the equidistance line, however, might be mitigated either by granting decreasing volume as the lateral extensions of the line are developed on either side or enclaving the features within a limited space of territorial sea. These solutions were adopted in some cases of distant or detached islands, including the Dutch Antilles,[132] the two Australian island groups off the coast of Papua New Guinea,[133] and the three Chilean islands at the eastern entrance of the Beagle Channel.[134]

In conclusion, the delimitation of the territorial sea between Spain and Morocco remains speculative until the dispute over sovereignty between

[129] Doc. A/CONF.62/WP.10/Rev.3, p. 126.

[130] See G. Blake and A. Drysdale, *The Middle East and North Africa: A Political Geography* (Oxford University Press, New York, 1985), 53, 100.

[131] See e.g. a statement by King Hassan of Morocco on Spanish television on 8 Feb. 1985, implying the possible support of the Soviet Union for Morocco's claims over Ceuta and Melilla, in C. Rousseau, 'Chronique des faits internationaux', 89 *RGDIP* (1985), 765.

[132] Territorial sea/continental shelf Boundary Agreement of 21 Mar. 1978, text in *LIS* 105. The islands are situated within 16–35 miles of the Venezuelan mainland. The delimitation line runs in an east–west direction and, at the two extremities of this segment, curves northward to encompass the maritime area around the Antilles from three sides.

[133] According to the Agreement of 1978, the Abusi and Saibi groups are given a non-extendable territorial sea belt of 3 miles. See *supra*, text at n. 68.

[134] According to the 1985 Agreement between Argentina and Chile, the islands of Picton, Lennox, and Nueva were confirmed as Chilean, but given only a territorial sea belt of three miles. See *supra*, Ch. 2 n. 21.

the two States over the enclaves is solved. The hypothetical solutions suggested here are only meant to reflect the kind of factors deemed relevant in any future legal discussion aiming at solving the questions of territorial sea entitlement and delimitation. The underlying rationale for these suggestions stems from the settled principle that, objectively, the location and size of islands are important factors in the determination of the equity of a given solution. An equally important rationale is that any solution must also be sought in the light of the *specific, subjective* characteristics of the enclaves, especially their contested political status and, above all, their unique encroachment upon the mainland coast of Morocco.

5. DELIMITATION OF THE TERRITORIAL SEA IN THE MEDITERRANEAN STRAITS

In a study submitted to the United Nations General Assembly in 1957 Commander Kennedy included seven Mediterranean straits among others which were regarded as international waterways.[135] Another Mediterranean passage, the Northern Corfu Channel, has also been 'considered as belonging to the class of international highways.'[136] The extension of most Mediterranean territorial sea claims to 12 miles, combined with the provisions of the 1982 Convention regarding straits, has inevitably manifested important peculiarities in the Mediterranean, wherein the legal status of these straits can be divided into three legal categories.

(*a*) *Straits through which passage is guaranteed by an* ad hoc *agreement*

Passage through the Straits of Dardanelles and Bosporus which join the Mediterranean via the Sea of Marmara to the Black Sea is subject to the regulation of the Montreux Convention of 1936.[137] The Straits, including

[135] J. E. Kennedy, *A Brief Geographical and Hydrographical Study of Straits which Constitute Routes for International Traffic* (UN Doc. A/CONF. 13/6 and Add. 1, 1958), 114–64. See also L. Alexander, 'Regionalism and the Law of the Sea: The Case of Semi-enclosed Seas', 2 *ODIL* (1974), 151.

[136] See the *Corfu Channel* case (Merits) (*UK* v. *Albania*), ICJ Reports (1949), 29.

[137] Convention Regarding the Regime of the Straits, 20 July 1936. Text in 31 *AJIL* (1973) (supplement), 1. According to C. Colombos (*The International Law of the Sea* (Longman, London, 1967), 200), the freedom of navigation through the Strait of Gibraltar is secured by a special agreement, i.e. the Anglo-French Declaration of 8 Apr. 1904. The prevailing view, however, is that the Declaration does not amount to an international agreement organizing for a special regime of passage. The Declaration merely affirms a right to a free passage which is already guaranteed in accordance with the general rules of law. For details, see S. Truver, *The Strait of Gibraltar and the Mediterranean* (Sijhoff & Noordhoff, The Hague, 1980), 178–9.

the Sea of Marmara, are about 160 miles in length and have various widths, from 750 yards to 40 miles. The entire Straits are situated within Turkish territory. They (including the 40-mile-wide Sea of Marmara) are enclosed as Turkish internal waters.[138]

(*b*) *Straits for which the regime of innocent passage is retained*

These are straits between islands and mainlands, if a safe alternative route through a high sea or an exclusive economic zone exists in the seaward direction.[139] Of the eight Mediterranean straits only the Northern Corfu Channel qualifies for this category. Although the Channel is regarded as 'a useful route of international maritime traffic',[140] there exists west of Corfu Island a more direct and broader sailing route into and out of the Adriatic and Ionian Seas.[141]

The delimitation of the territorial sea in the Northern Corfu Channel might well be considered as part of an overall agreement to delimit the Albanian–Greek shelf/EEZ in the whole area between the Albanian coast and the Greek island of Corfu.[142] With regard to territorial sea delimitation, however, there are two fundamental facts furnishing a basis for a legal consideration of a boundary line in the Northern Channel.[143]

First, there is the special geographical relation between the coasts of the two States. The most striking geographical characteristic of the coastlines fronting on the area relevant to the territorial sea delimitation is the existence of two different 'coastal relations'. At the Bay of Phtelias, where the land boundary between the two States meets the sea, the delimitation of the territorial sea has to be effected between two *adjacent coasts*. Following the conclusion of the Paris Peace Conference of 1921, an International Boundary Commission finalized in 1925 the details of the Albanian–Greek boundary. It was stipulated that the land boundary ended in Phtelias Bay and from there 'the boundary follows a line perpendicular to the general direction of the coast up to the limits of territorial waters.'[144] The determination of the terminus of this latter line, and indeed its overall equitableness for both States in a future boundary

[138] See *supra*, Ch. 6 s.3(*g*).

[139] Arts. 38(1) and 45(1)(a) of the 1982 Convention.

[140] *Corfu Channel* case, 28.

[141] Of this opinion: S. Ibler, 'The Changing Law of the Sea Affecting the Adriatic', 20 *GYIL* (1977), 174, at 180; Truver, *The Strait of Gibraltar*, 202–3.

[142] After the death of the Albanian leader Enver Hoxha, Albania and Greece sought to improve their mutual relations. Delegates of the two countries agreed in July 1985 to restore their markers along their common land border and also agreed to plan for negotiations respecting the delimitation of their continental shelf. *The Times* (10 July 1985).

[143] Other special circumstances arising out of socio-economic factors, such as established fishing grounds or a navigable channel, do not seem to exist in the Northern Corfu Channel.

[144] For the text of the statement of the Commission see *IBS* 113.

settlement, should properly be assessed in the light of the delimitation of the boundary line between the *opposite coasts* of the mainland of Albania and the island of Corfu, which requires the application of a different method (or methods).

The second fundamental fact is the general configuration of the relevant coastlines, as well as the presence of small islands in the Channel. In this respect, the adjacent Greek–Albanian coasts are generally regular, although small indentations exist and alternate with headlands which may be utilized as 'basepoints' for determining the general direction of the coast. The western coast of the Channel (i.e. the eastern coast of the island of Corfu) is marked by a major curvature, forming the Bays of Ipsou and Krevatsoula. The enclosure of this major indentation, which would conform to the requirements of law for enclosing juridical bays if enclosed by a line of 6.7 miles, would reduce the maximum distance between the Albanian coast and a median line measured from this closing line to no more than 1.6 miles. However, if the two States agree on a line equidistant from the low-water line, it is likely that the two biggest Greek islets of Vidho and Gouvinon, situated off the coast of Corfu, would be used as basepoints for determining the boundary line.

However, the delimitation of the territorial sea from the low-water line rather than from any closing or straight baselines seems to be implicit from the practice of the two States. Although claiming a belt of 12-mile territorial sea measured, as has been seen, from straight baselines, the Albanian Decree of 1990 stipulates that the territorial sea of Albania extends 'between the Albanian shore and the Greek islands up to the middle of the Corfu Channel'.[145] The low-water line is still the normal applicable rule in Greece.[146]

(*c*) *Straits subject to the new regime of transit passage*

This category includes all straits serving international navigation 'between one part of the high seas or an exclusive economic zone and another part of the high seas or an exclusive economic zone'.[147] The Straits of Karpathos, Kithira, Messina, Minorca, Bonifacio, and Gibraltar are the most important Mediterranean straits which come within this category. As a result of the coastal States' right to a territorial sea belt of 12 miles, passage through these straits is already, or expected to be, effected through the territorial waters of the bordering State(s).

[145] Decree No. 7,366, modifying decree No. 4,650, dated 9 Mar. 1990. See *LSB* 16 (1990), 2. See also *Keesing's* (1976), 27745. Albania has firmly maintained this position. For an incident which occurred as a result of this position see *Keesing's* (1985), 33558.

[146] See *supra*, Ch. 6 s.2.

[147] See Arts. 37 and 45(1)(b) of the 1982 Convention.

At present, the Strait of Karpathos, lying between the west coast of the island of Rhodes and the south and west coast of Khalki and the east coasts of Karpathos and Saria islands, still contains a stretch of high seas, as a result of Greece's maintenance of a 6-mile territorial sea claim. Similar to this is the passage through the Strait of Kithira which is situated between the Greek islands of Kithira and Antikithira, which are about 16 miles apart. The Strait of Messina lies between the Italian island of Sicily and the south-western coast of the Italian peninsula. Passage through the Strait is wholly effected in the internal/territorial waters of Italy (with a width which varies from 1.75 to 9 miles). The *Freu de Minorca* is situated between the Spanish islands of Minorca and Majorca. The breadth of the Strait, measured from the straight baseline system of Spain, varies from 19.75 to 25 miles. The whole Strait, with the exception of a narrow stretch of about 2 miles at its northern entrance, lies within the territorial waters of Spain.

The first four straits mentioned in this category are, as has been seen, bounded only by one State and, consequently, there exists no question of territorial sea delimitation. The Straits of Bonifacio and Gibraltar are different.

(i) Strait of Bonifacio

Joining the high seas of the Tyrrhenian basin with those of the Balearic Sea eastwards and westwards of the islands of Corsica (France) and Sardinia (Italy), the Strait extends for about 10.5 miles. The width varies between 3.4 and 9 miles. The Strait contains numerous islands, islets, and rocks which are divided into two groups by the Bocca Grande.

In 1908 France and Italy concluded a fishery agreement providing for 'a fishing line' crossing approximately the middle of the Bocca Grande.[148] However, it was not until 28 November 1986 that the two States were able to sign an agreement (not yet in force) which provided for the delimitation of the territorial sea in the Channel.[149] The Agreement refers to no specific principle, but the line, determined by geographical co-ordinantes, closely follows the 1908 fishing line. It is a modified equidistant line which seems to give effect to the most southern French islets of Lavezzi and Cavallo on the one hand and the opposite Italian islets of Budelli and S. Maria on the other. The latter islands were enclosed, amongst others, within Sardinia by the Italian baselines system of 1977.[150] Points 1 and 6 of the line represent the outer limit of the two States'

[148] Fishing Agreement of 1908. Text in 101 *BFSP* (1908), 1059.

[149] Agreement signed on 28 Nov. 1986. Copy obtained from the Office of the Legal Adviser, Ministry of Foreign Affairs, Rome, May 1988 (see Map 20).

[150] See *supra*, Ch. 6 s.3(*c*).

respective 12-mile territorial waters. East and west of these terminal points the parties are still unable to agree on a mutually acceptable continental shelf boundary for the rest of their overlapping jurisdictions.[151]

However, the same Agreement abrogates the 1908 fishing line and provides for the creation of a common fishing zone which overlaps the boundary line. Within this zone the 'traditional' fishermen of both Sardinia and Corsica would be allowed to continue their fishing activity irrespective of the boundary line.[152]

(i) Strait of Gibraltar

There are over 100 straits in the world less than 25 miles wide. If one ranked them on the basis of the number of passing ships, volume of oil cargoes, alternative routes, and intensity of military use, the result might highlight the extraordinary significance of the Strait of Gibraltar (the only natural gateway between the Mediterranean and the world oceans) as one of the top six waterways, with the Turkish Straits, the Suez Canal, Bab el-Mandab, and the Straits of Dover and Malacca its closest rivals.[153]

It is therefore not surprising that rival political claims over the narrow shores of the Strait of Gibraltar have been operative from time immemorial. At present, the disputes over sovereignty of the strategic points of Ceuta and the Rock of Gibraltar, two key points on this vital Strait, are among the most outstanding disputes in the Mediterranean and, consequently, have brought the possibility of delimiting the waters of the Strait to a stalemate.

The origin of the dispute between Britain and Spain over Gibraltar lies in the interpretation of the ambiguous terms of Article X of the 1713 Treaty of Utrecht, whereby Spain yielded to Britain 'the full and entire propriety of the town and castle of Gibraltar, together with the port, fortifications, and forts thereunto belonging . . . without any territorial jurisdiction'.[154] Two controversial issues are particularly relevant to the question of delimitation.

There is, first, the problem of defining the land frontier between Gibraltar and Spain. Britain has been of the view that it has jurisdiction over certain 'neutral ground' north of the Rock proper, and argued that this was a necessary arrangement of the act of cession itself.[155] During the

[151] See *infra*, Ch. 9 s.2(*d*)(ii).

[152] Arts. 2, 4.

[153] It is estimated that about 150 ships a day pass through the Strait. See Blake and Drysdale, *The Middle East and North Africa*, 131. For the economic and strategic importance of the Strait of Gibraltar see also Truver, *The Strait of Gibraltar*; A. J. Cottrell and J. Theberge (eds.), *The Western Mediterranean: Its Political, Economic and Strategic Importance* (Praeger, New York, 1974).

[154] Art. X of the Treaty, reproduced in *Red Book on Gibraltar* (Madrid, 1965), 157.

[155] Throughout the 18th, 19th, and early 20th centuries, British diplomacy has advocated the position that the sovereign rights of Britain in Gibraltar extend north of the Rock up to

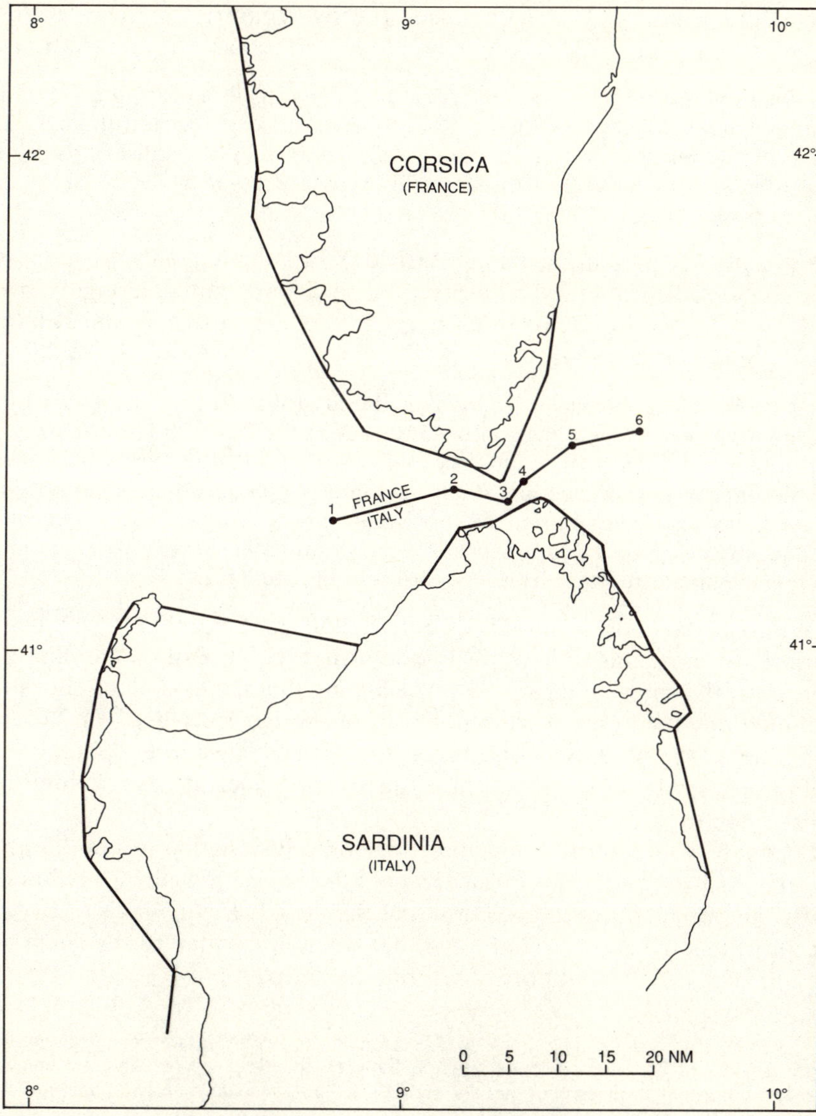

Source: Leanza, U. (ed), *Mediterranean Continental Shelf: Delimitations and Regimes*, Oceania Publications, New York 1988.

MAP 20 Territorial sea boundary in the Bocche de Bonifacio

the adjacent space lying within the range of its guns, 'for it is unquestionable that when a town is yielded, there is tacitly yielded at the same time, all the ground commanded by its artillery, since otherwise the cession would be of little use'. Letter from the Minister in Madrid, Stanhope, to the Spanish Secretary of State, 19 Aug. 1723, text in *Red Book on Gibraltar*, 30. See also *infra*, text at n. 170.

discussions which took place between the two countries in 1966, Britain argued that it had a title to the whole area on the basis of prescription:

Her Majesty's Government do not accept that the ground between the Gibraltar frontier fence and the foot of the Rock is Spanish sovereign territory . . . The whole of the territory has in any case been under exclusive British jurisdiction since at least 1838, by which British sentries were established along the line of the present frontier fence.[156]

Spain on its part argued that Article X was meant only to grant to Britain 'a British military base installed in Spain', not sovereignty over the territory of Gibraltar.[157] With regard to the northern limit of this 'base', Spain argued that

The Spanish Government has continually declared that its acts of tolerance did not imply any extension of the concessions made in the Treaty of Utrecht, and the Government of Great Britain has repeatedly assured it that in their actions there was no intention of altering the 'status quo ante'. Consequently it is not possible to speak to any legitimization of the British presence in part of the neutral ground, by reason of the long series of Spanish protests and refusals which have been made uninterruptedly from 1713 down to the present day.[158]

Whilst Britain still maintains essentially the same position,[159] the negotiation history seems to leave little doubt that 'Great Britain had and has not [now] the slightest legal basis for claiming that the cession of Gibraltar included any land on the isthmus north of the Rock itself.'[160] In fact, as early as 1803 the Law Officers of the Crown gave the opinion that Gibraltar's civil courts did not have jurisdiction over offences committed on the neutral ground.[161]

Secondly, there is the problem of maritime jurisdiction in Gibraltarian waters. Article X does not explicitly yield the adjacent waters of Gibraltar to Britain. Spain takes the position that the act of cession yields to Britain 'no jurisdictional zone . . . nor any other boundaries than those which are comprised by the actual port of Gibraltar'.[162] Spain only ceded to Britain

156 Cmnd. 13131, p. 62. The 'fence' was erected by the British authorities in 1909 almost in the middle of the 'neutral' zone situated between the Rock proper and the Spanish-guarded points northwards. For details see H. Levie, *The Status of Gibraltar* (Bowker Publishing Co., Epping, 1983), 76–8.

157 Cmnd. 3131, p. 8. Britain denied this limited reading of Art. X, ibid. 53.

158 Ibid. 27–8.

159 See e.g. formal British proposal to Spain (to submit the dispute to the ICJ) made on 12 July 1966, in which it was stated that ']t}his proposal is made without prejudice (a) to the view of Her Majesty's Government that Great Britain has sovereignty over the whole territory of Gibraltar, which includes all the ground up to the frontier fence'. Cmnd. 3131, p. 42.

160 Levie, *The Status of Gibraltar*, 82.

161 Public Records Office, Colonial Office Papers, 91/42/21, as quoted ibid. 80.

162 *Red Book on Gibraltar*, 238.

'the port of Gibraltar without any territorial waters'.[163] Britain, on the other hand, is just as adamant in asserting that the 'cession of a territory automatically carries the cession of the appurtenant territorial waters'.[164]

Britain's argument on this second issue appears to be more in line with the rules of law than that advanced by Spain. Reference to the many treaties of cession concluded in the past as well as the present disclose that rarely, if ever, is there any explicit mention of territorial waters to accompany a land cession which includes a coastline—the concurrent cession of the territorial waters has always been considered as implicit. As was held by the Permanent Court in the *Grisbadarna* case, it is a settled principle of internatioal law that 'the maritime territory is an essential appurtenance of land territory', so it follows that the cession of land territory would include the accessory maritime territory.[165] Particularily relevant in this regard is the Treaty of Utrecht itself, which included other land cessions. In no case was there any mention of territorial waters, 'but never has the fact that the relevant territorial waters were implicitly included in the cession been challenged'.[166] In this respect, Article 1(1) of the 1958 Territorial Sea Convention is to the point. It provides that: 'The sovereignty of a State extends, beyond its land territory and its internal waters, to a belt of sea adjacent to its coast, described as the territorial sea.' There is nothing in the Treaty of Utrecht which would make this general provision of international law inapplicable to Gibraltar.

What is of interest in this regard, however, is the opinion of the Law Officers, which seems to be that Britain's territorial sea right should not be extended under any circumstance beyond the limits of the port of Gibraltar:

> Any action which might be taken by the Governor, the Secretary of State, or Parliament in extending the limits of the port at all, and specially, to bring those limits within the three miles of any Spanish ground, would therefore be, at least, a doubtful exercise of authority, and would certainly be offensive to the Spanish Government.[167]

More stringent was the opinion of the King's Advocate for the Colonial Office, in which he said, *inter alia*,

[163] See *Negotiations on Gibraltar (A New Spanish Red Book)* (Madrid, 1968), 493; declaration made by Spain in its instrument of accession to the 1958 Convention on the Territorial Sea, 766 *UNTS* (1971), 335. Spain made the same declaration when it signed the 1982 Convention.

[164] Cmnd. 3131 10, pp. 84, 117 (1966–7).

[165] *Grisbadarna* case, 4 AJIL (1910), 226, at 231.

[166] Levie, *The Status of Gibraltar*, 82. He gives the cession of Minorca—Art. XI of the Treaty—as an example. See also the cession treaties affecting the eastern Aegean islands, *supra*, s.4(*b*)(i)(1).

[167] Report of 1878 of the Law Officers and Dr J. Parker Deane, Cmnd. 1313 10, p. 84.

This Treaty [of Utrecht] . . . expressly withholds the right of exercising territorial jurisdiction . . . His Majesty has, thus, supreme and exclusive authority over all persons . . . within the Town and Fortress, and the Port of Gibraltar, but . . . beyond those Limits He does not possess the right of sovereignty; and that foreign ships lying in the Bay or in the Anchorage Ground, beyond the limits of the Port, are not amenable to British jurisdiction.[168]

The British Government, however, did not take account of these opinions; they opted for claiming sovereignty over the maximum area of the waters of the Bay of Gibraltar. In addition to the 'range of guns' thesis mentioned earlier,[169] Britain also claimed that all of the waters of the Bay to the east of a line drawn from Punta Mala to the old (northern) mole in Gibraltar were encompassed within the terms of Article X, 'Gibraltar, together with the Port thereunto belonging'.[170]

However, as the above opinions clearly indicate, the controversy over territorial sea jurisdiction has been more acute with regard to the waters of the Bay of Gibraltar[171] than to the Mediterranean side of the Rock, though Spain maintains the same contention, as will be seen shortly, with regard to this area.

The above clearly shows that the settlement of these two issues is a prerequisite to the pending question of the territorial sea delimitation in Gibraltar. Indeed, because of these two disputed issues, the two States were unable even to agree on a *modus vivendi* line proposed by Spain in 1878 in order to prevent collision and contraband trade in the Bay of Gibraltar.[172] As they proceeded to negotiate on this line, the two Governments advanced proposals which reflected their differences regarding the land frontier and maritime jurisdiction.

While agreeing in principle on the *modus vivendi* line solution, Britain stressed that it had no intention of

abandoning [its] right to a 3-mile jurisdiction around the Rock . . . But, as the limits of the Bay of Gibraltar, which in places is considerably less than 6 miles across, precludes the possibility of each Power enjoying an exclusive jurisdiction of 3 miles, Her Majesty's Government are not averse to the consideration of a provisional scheme for arriving at a *modus vivendi* . . . by a division of the bay between the two Powers, to be effected by a conventional line based on the

[168] Opinion of the King's Advocate, Jenner, 16 Mar. 1831, as quoted in Levie, *The Status of Gibraltar*, 87.

[169] See *supra*, n. 158.

[170] Canning, Secretary of State, to Alcudia, 30 Nov. 1826, *Red Book on Gibraltar*, 176. See also Palmerston to Howden, 16 Dec. 1865, ibid. 193–4.

[171] See Levie, *The Status of Gibraltar*, 83–92.

[172] See 'Correspondence between Great Britain and Spain, Respecting Maritime Jurisdiction in Gibraltar Waters 1878–1883', 74 *BFSP* (1883), 1319–52.

medium filum aquae, and marked out . . . by a line of buoys, as originally proposed by the Spanish Government.[173]

The Spanish Government accepted this proposition, with the reservation:

[it] being understood that any arrangement which may be accepted will be only of a temporary character, and as a *modus vivendi* between Spain and Great Britain in times of peace, and that the arrangements come to, however long their duration, are not to cause the prescription of the rights of Spain, which she has not renounced.[174]

Aware of the implications of this reservation, the British Government instructed their Mission in Madrid

to propose to the Spanish Government that the arrangement in contemplation should provide that the 3-mile jurisdiction claimed by Great Britain over the coastal waters of Gibraltar outside the bay should not be questioned, and that the line dividing the Spanish and British waters on the Mediterranean side of the Rock should be settled and defined.[175]

The Spanish Government's answer to this proposal was not difficult to predict. They 'repudiated all claims to the 3-mile jurisdiction on the east of the Rock', i.e. on the Mediterranean side; and they 'would be forced to take up their position upon the Treaty of Utrecht, which . . . conferred no territorial or maritime jurisdiction whatsoever upon Great Britain when put in possession of the fortress by that Act'.[176] Even when the two Governments agreed to discuss the *modus vivendi* 'without prejudice to each others' position on the question of jurisdiction', they differed (again due to their different interpretation of Article X) on the principle of delimitation to be followed. Britain's *medium filum aquae* raised Spanish fears that it would encroach upon the waters of the Bay bordering their coast.[177] Spain insisted on negotiating another formula, 'whereby the waters [are] to be divided in a convenient manner and in such a way that no part of the coast should remain without jurisdictional waters'.[178] For Britain, this Spanish proposal, aiming at replacing the *medium filum aquae*, 'would appear to imply that Great Britain must give up that portion of the waters comprised in the limits of the part which wash the portion of the Spanish coast'.[179] At this point it was clear that no prospect

173 Mr West to Señor Silvela, 12 Feb. 1879, ibid. 1322.
174 The Marquis de Molins to Mr West, 13 May 1879, ibid. 1322–3.
175 The Marquis of Salisbury to Mr West, 5 July 1879, ibid. 1323–4.
176 Mr West to the Marquis of Salisbury, 16 Oct. 1879, ibid. 1325–6.
177 See the Marquis de la Vega de Amijo to Morier, 1 June 1882, ibid. 1348–9.
178 Ibid. 1349.
179 That is the portion of the waters which wash the Spanish coast from the neutral ground to Punta Mala. Earle Granville to Mr Langley, 29 Aug. 1881, ibid. 1343.

of any agreement between the two States could be contemplated; consequently, it was decided to 'drop negotiation upon the subject'.[180]

The dispute over the maritime jurisdiction in Gibraltar has never been formally resolved. However, while recent events have led both States to renew their conflicting claims,[181] it appears that new developments in the relations between the two States may help to modify their position, especially with respect to Britain's extended claims over the waters of the Bay of Gibraltar, and to follow the delimitatory rule laid down in Article 12 of the 1958 Territorial Sea Convention.[182]

Pertinent in this respect is the two Governments' Lisbon Declaration of 1980 which aimed, as declared in the Brussels Joint Communiqué of 1984, *inter alia*, at the 'establishment of a negotiating process aimed at overcoming all the differences between [the two Governments] over Gibraltar . . . Both sides accept that the issue of sovereignty will be discussed in that process.'[183]

However, in so far as the geographic characteristics of the Strait are concerned, the question of an equitable criterion for the delimitation of the overlapping territorial waters of the Strait does not seem to raise a special problem. An initial median line would run along the Strait's Mediterranean coasts between the eastern entrance, about 14 miles wide, and the narrowest point of 7.5 miles, between Point Marrouque (Spain) and Point Cires (Morocco). Two further equidistant lines, defining the territorial waters of Ceuta and Gibraltar, would terminate as they reach this initial median line.[184]

Morocco unilaterally specifies the outer limit of her territorial waters in the Strait by a line of geographical co-ordinantes which is determined 'selon la ligne médiane entre les côtes marocaines et les côtes étrangères

[180] See Earl Granville to R. Morier, 15 Feb. 1883; R. Morier to Earl Granville, 18 Mar. 1883, ibid. 1352.

[181] Following the incursion by a Spanish Navy flagship into the disputed waters of Gibraltar on 20 Mar. 1986, Britain presented a formal protest which was rejected by Spain in a formal reply affirming the Spanish view that the waters adjacent to the Rock are under Spanish jurisdiction (*The Times* (3 Apr. 1986); *The Guardian* (4 Apr. 1986)). When questioned in the House of Commons about the 'United Kingdom territorial waters off the Dependent Territory of Gibraltar', the Secretary for Foreign and Commonwealth Affairs spoke, albeit cautiously, of 'Gibraltarian Waters', 95 *Parliamentary Debates* (House of Commons, Official Reports), Written Answers, 10 Apr. 1986, col. 163.

[182] Levie, *The Status of Gibraltar*, 91, mentions that Britain's new moderate position was already inferred from an official map published by the DIS in 1968. It showed a boundary line which 'continues the line of the Fence out into the waters of the Bay (and the Mediterranean) and then substantially follows the *medium filum aquae* in a southeasterly direction, thus dividing the waters of the Bay roughly in half'.

[183] Joint Communiqué Agreed by the Foreign Ministers of the United Kingdom and Spain in Brussels on 27 Nov. 1984. Text in *Revindicación espanola de Gibraltar* (Ministerio de Asuntos Exteriores, Nov. 1987), 63.

[184] Assuming of course that Ceuta is given full effect. See however, argument against this solution *supra*, s.4(*b*)(iii).

qui leur font face'.[185] Two remarks are to be made about this line. First, it is drawn from the Moroccan straight baselines system and this pushes the line further north than a true median line (measured from the low-water line) to a marked degree, if one takes into account the narrowness of the area involved. Second, but more significantly, the line gives no regard to the Spanish enclaves of Perejil and Ceuta. The political status of these enclaves is a matter of controversy between Spain and Morocco. Their possible effect on a boundary line in the area was considered earlier.[186]

In conclusion, the problem of delimiting the territorial waters in the Strait of Gibraltar is inextricably linked to questions of sovereignty. As a result of Spain's restitution of democracy and her recent accession to the European Economic Community, there have been British–Spanish negotiations towards settlement of the question of Gibraltar. The same atmosphere is not as yet envisaged between Spain and Morocco; and until the 'Spanish *presidios*' problem is resolved, the delimitation of the waters of the Strait of Gibraltar remains a distant prospect.

[185] See Art. 5 of the *Décret* of 1975, *supra*, Ch. 6 n. 89.
[186] See *supra*, s.4(*b*)(iii).

9

Continental Shelf, EEZ, or Single Boundaries in the Mediterranean

1. INTRODUCTION

In a sea like the Mediterranean, where there exist numerous islands, and where the coastal States are in such close proximity,[1] the resolution of extended zone delimitations raises difficult problems. There is, firstly, the problem of islands, which has proved to be one of the most difficult factors in maritime delimitation. This will be illustrated in the Aegean dispute, which will be discussed as a case-study in a separate chapter. Secondly, the resolution of any bilateral boundary in the Mediterranean inevitably gives rise to the problem of third States' interests.

Whilst the completion of the Mediterranean boundaries on a bilateral basis will understandably be a long process, the possibility of multilateral settlements could indeed be very complicated because of the rudimentary and underdeveloped nature of the 'rules' applicable to these kinds of boundaries. The law of maritime boundaries, especially that developed by adjudication, is mainly designed for bilateral but not tripartite or quadripartite settlement.[2]

The following analysis of existing shelf boundary settlements in the Mediterranean shows clearly the problems arising in relation to third States' interests. The difficulties discussed, though not exhaustive geographically, are submitted to exemplify the kinds of problems that may be encountered in almost all parts of the Mediterranean. It is therefore thought appropriate to examine first the 'solutions' adopted in these boundary settlements and the impact they might have on delimitations with third States.

The limited number of settled boundaries, however, indicates that the prospect of settling the remaining twenty-six or so maritime boundaries will depend upon several factors. Most relevant is the extent of politico-economic incentives and the type of boundaries that might appropriately meet such incentives. In this respect, the almost unanimous ab-

[1] See Map 1.
[2] See *infra*, s.2(*g*)(i).

stention of the Mediterranean littorals, including the European Economic Community,[3] from exercising their EEZ or EFZ rights is significant. Reasons for this *de facto* policy should be discussed. The final section of this chapter will therefore consider the factors relevant for future maritime claims and boundary delimitation in the Mediterranean.

2. SETTLED SHELF BOUNDARIES AND THEIR IMPACT ON POTENTIAL BOUNDARIES

The pattern of State practice in the Mediterranean basin has so far been evident almost exclusively in the four Shelf Boundary Agreements concluded by Italy and its neighbours, as well as the provisional demarcation between Italy and Malta in the Sicily Strait. The important feature of these Agreements is the modification of the median line so as to reflect the special circumstance of, and indeed the inequity created by, islands located near to or 'on the wrong side' of a land-to-land median line. Indeed, these Agreements can be seen to disclose the broad outlines of a policy of full utilization of the 'equidistance–special circumstances' rule. To these delimitations must be added, however, the two decisions of the ICJ on the delimitation of the continental shelf between Tunisia and Libya and Libya and Malta. Both decisions were recently implemented by the parties concerned.

(*a*) *The Italo-Yugoslav Agreement*

The shelf boundary between Italy and Yugoslavia[4] extends for a distance of 363 n.m. and comprises forty-three turning or terminal points which are separated by an average distance of 8.2 n.m. The northern terminal point of the boundary is now joined with the last segment of the territorial sea boundary drawn in accordance with the Osimo Treaty of 1975 between the two parties.[5] The southern terminal point corresponds roughly with the Albanian–Yugoslav border and was agreed to be provisional, obviously awaiting a final designation of a possible trijunction terminal point between these two States and Albania. The prospect of reaching agreement on this trijunction seems to pose no special problem. There exist no

[3] According to the resolution of the EEC Council of 3 Nov. 1976, from Jan. 1977 'the member states shall, by means of a concerted action, extend the limits of their fishery zones to 200 miles off the North Sea and North Atlantic coasts, without prejudice to similar action being taken for other fishing zones within their jurisdiction such as the Mediterranean'. Text in 15 *ILM* (1976), 1425.

[4] Agreement signed on 8 Jan. 1968, came into force on 21 Jan. 1970. Text in *LIS* 9. See Map 21.

[5] See *supra*, Ch. 8 s.3(*c*).

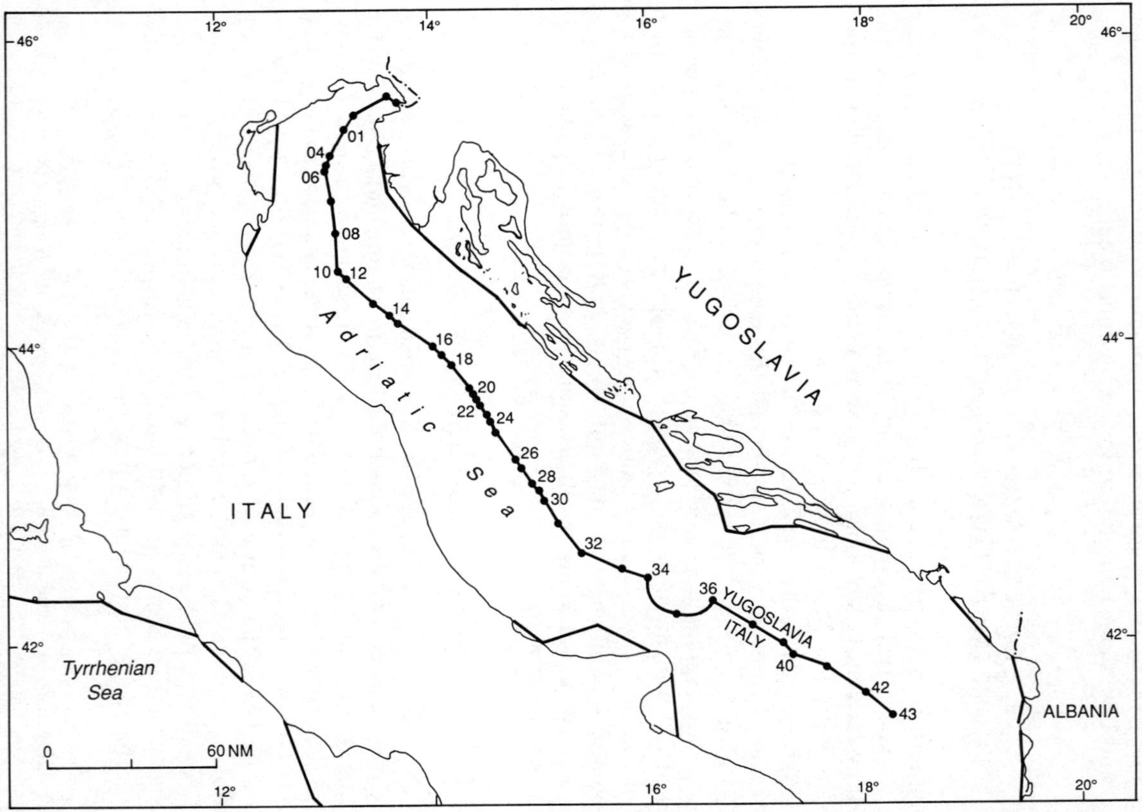

Source: Leanza, U. (ed), *Mediterranean Continental Shelf: Delimitations and Regimes*, Oceana Publications, New York 1988.

MAP 21 Continental shelf boundary: Italy–Yugoslavia

particular geographical or geomorphological features in this limited region of the Adriatic calling for a modification of an equidistant/median line.

The Italian–Yugoslav boundary employs an equidistant line between the Italian mainland and the larger islands located close to the Yugoslav coast, except between Points 26 and 38. The deviation from this simple line was made to compensate Italy for the geographical influence exerted by many Yugoslav islands, including Jabuka, Kajula, and Pelagruz, which are located well out in the Adriatic Sea. Jabuka was used as a basepoint but its full impact was offset by shifting the notional median line eastwards, thus conceding to Italy an area of 1,680 sq.km. The islet of Busi (adjacent to Jabuka) appears to have been disregarded, but the larger islet Lissa, which is adjacent, has been granted full effect. In the area influenced by Lissa, between Points 32 and 34 on the boundary, an exchange of 48 sq.km. has permitted simplification of the boundary immediately north of Point 34.

Further south, Kajula and Pelagruz almost straddled a notional median line between Italy and the outermost fringing islands on the Yugoslav coast. They were ignored in determining the boundary line. Each was given a semi-enclave of 12 n.m. delimited by areas of circles and joined with the median line by tangents.[6] As a result of this arrangement a further area of 1,400 sq.km. was conceded to Italy. On the Italian side, a concession was made by giving no effect to the island of Pianosa—located within 12 miles of the notional median line referred to above. The area conceded in this section was 416 sq.km. The Tremiti islands were also ignored. The areal exchange was of the ratio 7.4:1 with Yugoslavia contributing the larger amount.

A special provision was made, as in the other three Italian agreements, for natural deposits straddling the boundary. As in the other cases, the parties agreed to proceed, with regard to such deposits, by mutual agreement.[7]

(*b*) *The Greco-Italian Agreement*

The Greece–Italy shelf boundary[8] comprises sixteen turning and terminal points and runs in a general north–south direction for 268 n.m. The Agreement provides for the eventual extension of the line northwards

[6] This distance of 12 miles was apparently made in anticipation of the present extension of the two States' territorial sea to 12 miles. At the time of signing the Agreement Italy was claiming 6 miles and Yugoslavia 10 miles.

[7] Art. 2.

[8] Agreement was signed on 4 May 1977 and entered into force on 12 Nov. 1980. Text in *LIS* 96 (see Map 22).

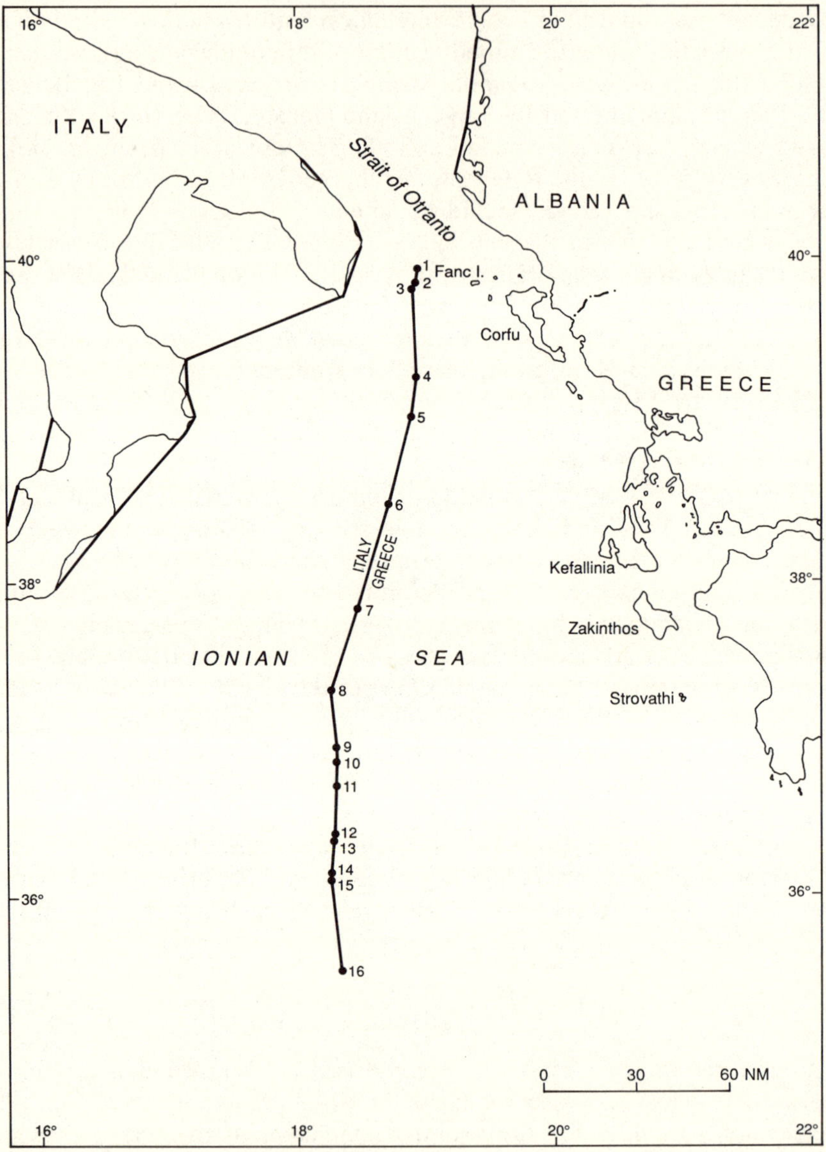

Source: Leanza, U. (ed), *Mediterranean Continental Shelf: Delimitations and Regimes*, Oceana Publications, New York 1988.

MAP 22 Continental shelf boundary: Greece–Italy.

and southwards 'to the points of intersection with the zones of the continental shelf belonging to the respective neighbouring countries'.[9]

[9] Art. I(3).

According to Article I of the Agreement, a median line was used 'taking into consideration any mutually approved minor adjustments'. Significant adjustments appear to have been negotiated at Points 1–3 and 15–16, where the boundary line is closer to the Greek coast than to the coast of Italy. This appears to have been done in order to compensate for the utilization of the fringing Greek islands, in contrast to the corresponding coast of Italy, which is almost devoid of islands. A further concession by Greece appears to be manifested in their acceptance of the Italian enclosure of the Gulf of Taranto. Turning Points 4 and 5 are equidistant from the Italian baselines enclosing, *inter alia*, the Gulf's entrance points at Cape S. Maria di Leuca and Cape Colona. However, at one point (Point 8) the boundary line is 6.1 n.m. closer to the Italian coast than to the Greek island of Kefallinia.

The northern terminus of the boundary is situated 22 and 20.1 n.m. from the Greek and Italian coasts, respectively. It is situated 28 n.m. from the coast of Albania. The location of this terminal point seems to raise no particular problem for a future Albanian–Italian maritime boundary agreement along the Strait of Otranto.[10] The southern terminus of the boundary lies 168.9 n.m. from the coast of Sicily; 163.4 n.m. from the Greek island of Stamfàni; 185 miles from the island of Malta; and 212.6 n.m. from the coast of Libya. The relevance of this point to prospective boundary settlements between the States concerned will be discussed below. It is of interest, however, to mention here that this terminal point is located in the middle of the deepest trough in the Mediterranean, known as the Ionian Abyssal Plain. But no effect was given to this geomorphological factor.

Apart from the usual provision concerning natural deposits straddling the boundary line, a notable feature of this Agreement is a compromissory clause (also present in the Italian–Spanish Agreement) providing for the compulsory jurisdiction of the ICJ over any interpretation or implementation dispute not settled by negotiation.[11]

(*c*) *The Italo-Tunisian Agreement*

The Italy–Tunisia shelf boundary[12] extends for a total distance of 443.36 n.m. and comprises thirty-four turning and terminal points, thus

[10] Italian and Albanian negotiators held a meeting in 1985 and expressed their agreement in principle to implementing an equidistant boundary line, the details of which were left to a formal agreement in due time. An interview with Dr Francalanci, a member of the Italian delegation to bilateral negotiations on the delimitation of the continental shelf, Milan, 28 May 1988.

[11] Art. 4. See also Art. 3 of the Italian–Spanish Treaty, *infra*, n. 25.

[12] Agreement was signed on 20 Aug. 1971 and entered into force on 6 Dec. 1978. Text in *LIS* 89 (see Map 23).

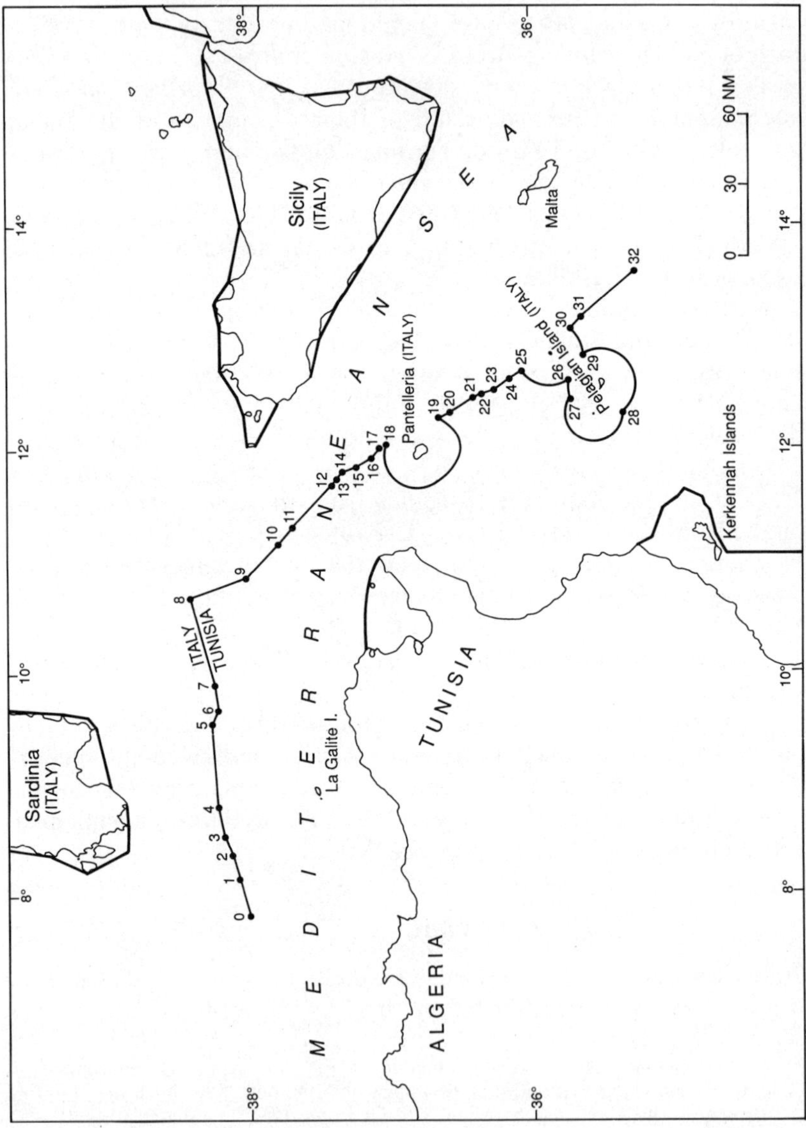

MAP 23 Continental shelf boundary: Italy–Tunisia

delimiting the two States' respective shelves between the Tunisian coast and the Italian islands of Sardinia and Sicily. Article 1 of the Agreement states that the boundary line 'shall be the median line, every point of which is equidistant from the nearest points of the baselines from which the breadths of the Italian and Tunisian territorial seas are measured, taking into account islands, islets and 'low-tide elevations with exception of Lampione, Lampedusa, Linosa and Pantelleria'.

Pantelleria and Linosa lie close to the notional median line drawn disregarding them. Lampione and Lampedusa lie on the 'wrong side' of this line, approximately 60 and 69 n.m. from the Tunisian coast respectively, but 115 n.m. from Sicily. A true median line, established between these islands and the Tunisian coast, would deflect the boundary to within distances of 20 and 33 miles from the Tunisian mainland. To avoid this 'disproportionately distorting effect', the two sides have agreed to disregard these four Italian islands. In the area of the islands, the boundary line deviates from equidistance as developed from basepoints on Sicily and the Tunisian mainland, and follows the outer limits of the maritime zones drawn around each of the four islands.

Pantelleria, Linosa, and Lampedusa have been granted a 13-mile envelope of arcs (thus placing them on the Italian side of the line); this presumably consists of a 12-mile territorial sea and 1-mile continental shelf. Lampione—smaller than the other three—has received only a 12-mile zone, effectively depriving it of a true continental shelf.

However, unlike the other agreements negotiated by Italy, the Italian–Tunisian Agreement does not expressly recognize the provisional nature of the boundary line's terminal Points 0 in the west and 32 in the south-east. This may be interpreted as reflecting the view of Italy and Tunisia that, within these terminal points, the other neighbouring States, i.e. Malta in the east and Algeria in the west, are not entitled to any claim. This attitude is not likely to be tolerated by the States concerned. Malta is reported to have protested against the south-eastern extension of the boundary up to Point 32.[13] The relevance of these two terminal points to prospective boundary delimitations with the other neighbouring States is considered below.

[13] As shown on Map 24, there is an overlap between the area claimed by Tunisia under the Italy–Tunisia Agreement and the area claimed by Malta under the combination of equidistance and the semi-enclave line around the Italian islands. In this area of overlap Tunisia granted concessions in 1974 which prompted Maltese protest. Some preliminary discussions followed, which reached a deadlock when Tunisia rejected the Maltese proposal to apply equidistance, treating the Italian islands as *res inter alios*. See *Notes verbales* between Malta and Tunisia dated respectively 8 Apr. and 9 May 1974; Maltese proposals for a boundary line with Tunisia and Tunisian reply dated 25 Aug. and 18 Nov. 1978, respectively. Copies of these documents were given by courtesy of the Law Commission Office, Valletta, June 1988.

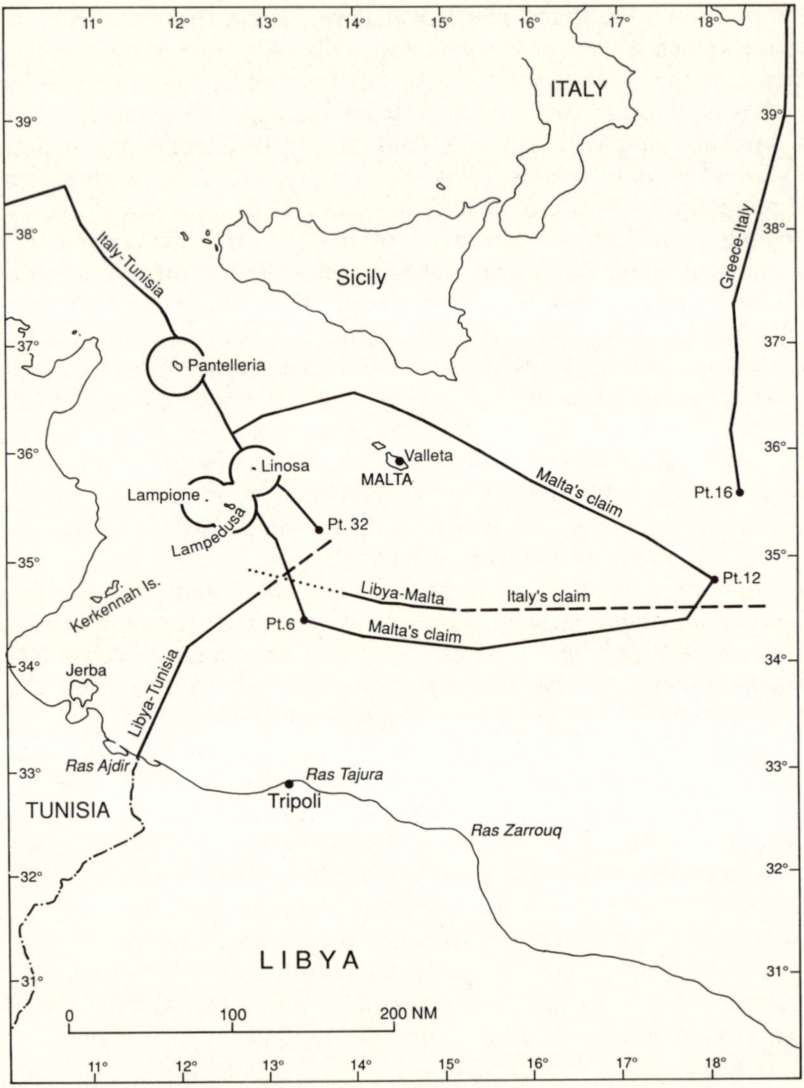

Source: After Libyan Arab Jamahiriya, *Memorial*, *Libya–Malta* case, i (1983).
MAP 24 Overlapping claims in the central Mediterranean

(i) Relevance to neighbouring Malta

Point 32 is approximately 114 miles east of Tunisia's nearest basepoint, 101 miles south of Sicily's, and 135 miles north of Libya's. It is also situated almost half-way between the island of Lampedusa and Malta.

This point is situated in a region where potential boundaries with or between two other States are expected to meet. These are the Libya–Tunisia boundary terminus which should follow from the Court's Judgment of 1982, and the potential boundaries with Malta, which, according to its unilateral equidistance line polygon of 1981,[14] entertain claims *vis-à-vis* Tunisia, Libya, and Italy. As illustrated in Map 24, the result of this situation is a 'grey' area of overlapping claims between four States.

The most difficult problem which may complicate the prospect of agreement on an equidistant tripartite or quadripartite point between the States concerned is the unsettled dispute between Italy and Malta on, *inter alia*, the treatment of the small islands of Pantelleria and the Pelagian Islands.

According to Malta, the equidistant line by which it proposes to delimit its shelf with Italy should be set aside in the area opposite these Italian islands and, instead, the islands should be circumscribed with enclaves corresponding to the breadth of the territorial sea plus 1 mile of continental shelf.[15] This obviously echoes the solution acceded to by Italy with Tunisia. Italy on the other hand rejects the Maltese proposal and contends that the enclave solution in the agreement with Tunisia was appropriate as it reflected the fact that the islands 'were sitting on the extension seaward of the Tunisian landmass';[16] it claims that, as between Malta and the islands, equidistance should govern. This point of disagreement seems to be one of the main factors interrupting the negotiations between Italy and Malta on their boundary problems following their agreement in 1970 on a provisional equidistance line in the channel between the Maltese islands and Sicily.[17] The Italian argument seems to suggest, however, that the islands are entitled to full weight *vis-à-vis* Malta, because (unlike the case with Tunisia) they are geomorphologically separate from the Maltese islands. This selective approach by Italy to geomorphology faces at least three major difficulties.

[14] The Maltese 'polygon' system was claimed for the first time on 19 Mar. 1981 during oral pleadings in *Application by Malta to Intervene*. See pleading of Counsel for Malta, Mr Lauterpacht, CR 81/1, ICJ Reports (1981), 50–4.

[15] See a draft agreement proposed by Malta in the course of negotiations with Italy in 1975, quoted ibid. 69. See also oral pleading of Counsel for Italy, Mr Arangio-Ruiz, in *Application by Italy to Intervene*, CR 84/1, ICJ Reports (1984), 33.

[16] Response of the Italian Government on the Maltese proposal, quoted in Pleading of Counsel for Malta, *supra* n. 14, 70.

[17] For a concise account of the exchange of views and negotiations between the two States, see the statement of the Italian Agent before the Court in *Application by Italy to Intervene*, 10–11. The 1970 provisional Agreement was reached by exchange of *notes verbales* between the two States in the years 1965–70. It provided for provisional exploitation of the continental shelf in a short section of the channel between Sicily and Malta on each side of the median line, subject to any adjustments that might be made in subsequent negotiations. For the texts of the *notes verbales* see *Memorial* submitted by the Republic of Malta, vol. ii (1983), annexes 65–7.

First, the geomorphological characteristics of the area concerned do not seem to allow the Italian islands a better treatment than that established in the Italy–Tunisia Agreement. Although Malta's continental shelf is separated from that of the Italian islands by a major trough in the vicinity, known as the Malta Trough, the same is also true for the Italian islands, as they are bordered by two major troughs extending south-east and east of Pantelleria and the Pelagian Islands, known as the Pantelleria and Linosa Troughs respectively.[18] The latter Troughs might equally be argued to constitute the natural boundary of the shelf of the islands facing Malta.

Second, there is little evidence in the Decisions that such 'incidental' interruption of the morphology of the seabed may be taken into account as a relevant factor.[19] More specifically, in *Libya–Malta* the characteristics of the area under consideration were extensively discussed by the parties to the case, and Libya in particular argued that the aforementioned troughs formed the north-western natural boundary of 'the Rift Zone', which was claimed to constitute 'a fundamental discontinuity' between the shelf area attributed to Libya and that attributed to Malta.[20] Although the extensive data provided by the parties were mainly concerned with the area between Libya and Malta proper, the Court's statement on this point is relevant. It referred to the development of the law which accorded to the coastal State a continental shelf up to 200 miles from its coast: the 'distance' principle. The application of this principle, the Court noted, to claims less than 200 miles apart has created 'juridical difficulties' that stood conclusively against the 'Rift Zone' argument.[21]

Third, in the light of the semi-enclave solution of the Italy–Tunisia Agreement it will be difficult for Italy to argue that Malta should not be entitled to equal treatment, especially since this solution was obviously envisaged, as shown earlier, within the context of the treatment of small and distant islands.

Italy's major reason for rejecting another 'semi-enclave' solution to the islands seems to lie, however, in the concern, already voiced by Italian

[18] The Pantelleria Trough starts at the island of Pantelleria and runs south-east. It reaches a maximum depth of 1,214 metres and a width of 15 miles (at the 1,000-metre isobath). It is about 52 miles long. The Malta Trough starts to the north-east of the Pantelleria Trough and runs in a south-easterly direction. It extends to the south of Gozo and gradually shoals south of the island of Malta. It has a maximum depth of 1,714 metres; width (at the 1,000 isobath) 11 miles; length 87 miles. To the south-east of the Malta Trough and running parallel to it is the Linosa Trough extending in a north-east–south-east direction. Its maximum depth is 1,615 metres; width 8 miles; length 41 miles. The three Troughs are bounded by shallow areas.

[19] See e.g. *Libya–Tunisia*, *infra*, s.2(*e*); *Gulf of Maine*, *supra*, Ch. 4 s.3(*b*)(ii)(1).

[20] *Libya–Malta*, ICJ Reports (1985), para. 36.

[21] Ibid. paras. 40–1. See *infra*, s.2(*f*).

writers,[22] that such a solution would virtually enclose the Pelagian Islands within full enclaves, thus cutting them off from the rest of the Italian territory. Any communication with the islands would have to be effected via the shelf/EEZ zones placed under Tunisian or Maltese sovereignty, furthermore excluding Italy from a substantial portion of the Straits of Sicily and thus raising obvious national security and strategic concerns. This Italian position, however, is undoubtedly weakened by the fact that this is precisely the result produced by the 1977 Award in relation to the Channel Islands, though surely the latter islands have a different (more autonomous) status with respect to Britain.[23]

(ii) Relevance to neighbouring Algeria

The last segment in the Italian–Tunisian boundary's extension westwards is also likely to be challenged by Algeria. It extends between Points 0 and 1 (see Map 23) for about 16.83 miles. Although this segment basically runs half-way between the Algerian–Tunisian border and Isola di S. Pietro (the most western island fringing the coast of Sardinia), the reason for starting the boundary from Point 0 rather than Point 1 is difficult to understand. The potential intersection of this line with an equidistant Algerian–Tunisian boundary would occur at Point 1 rather than any point on the line westwards. Although Algeria does not seem to have expressed any reservation on this Italian–Tunisian terminal point, there appears to be no major difficulty in reaching agreement between the three States on a trijunction point somewhere between Points 0 and 1.[24]

(*d*) *The Italo-Spanish Agreement*

The Spain–Italy shelf boundary between the island of Sardinia and the Balearic island of Minorca[25] runs for a total distance of 137.19 n.m. It comprises ten turning and terminal points. According to Article 1 of the Agreement, the boundary has been established 'following the criterion of equidistance from respective baselines'. Although both Italy and Spain have claimed straight baselines around the islands of Sardinia and Minorca,

[22] See e.g. U. Leanza, *Nuovi saggi di diritto del mare* (Giappichelli Editone, Turin, 1988), 67.

[23] See *supra*, Ch. 3 s.3(*b*).

[24] There is no *ad hoc* Algerian legislation establishing a unilateral principle of delimitation. However, at UNCLOS III Algeria was one of the proponents of the 'equitable principle' formula (see *supra*, Ch. 4 n. 4), and also advocated with other States the proposal that dependent islands should not be attributed shelf or EEZ entitlements. See Doc. A/CONF. 62/C. 2/L.96.

[25] Agreement was signed on 19 Feb. 1974 and entered into force on 16 Nov. 1978. Text in *LIS* 8 (see Map 25).

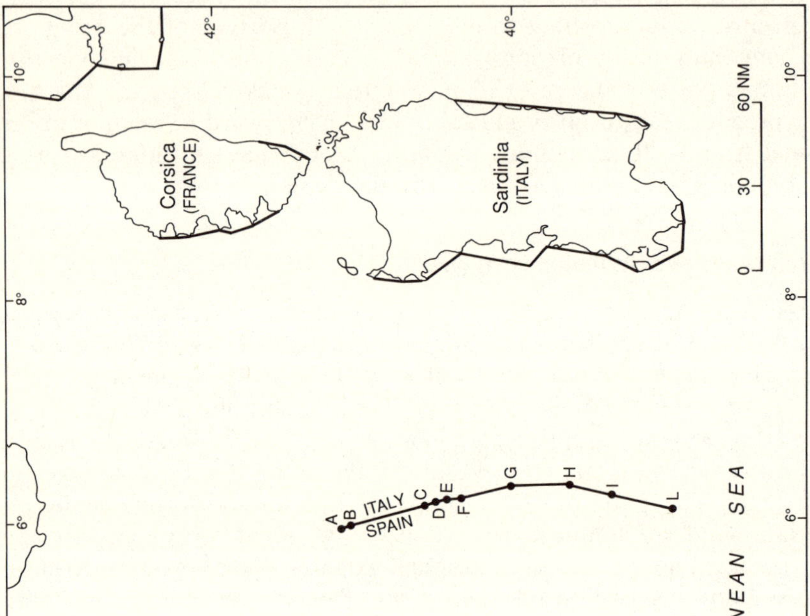

MAP 25 Continental shelf boundary: Italy–Spain

it appears that these straight baselines have not been used in determining the course of the equidistant line. Rather, the parties appear to have sought to develop a concave, 'equidistance' line that takes account of the concave shape of the west coast of Sardinia, being the larger island. As in the Greek–Italian case, the parties have delimited the whole area situated between their territories regardless of abyssal depths.

The two States have agreed that terminal Points A and L are provisional, presumably awaiting final agreements with third States. According to the United States Geographer,[26] assuming that the equidistance line is utilized, the Italy–Spain boundary will extend 5 miles northwards, where a trijunction will be encountered with France. A potential Algerian maritime zone will be encountered approximately 8 miles south of Point L. Whilst an agreement with Algeria on a trijunction point does not seem to raise any serious problem,[27] the prospect of boundary settlement with France is problematic. While Italy and Spain seem generally to favour the application of the equidistance principle, France takes an almost opposite position.

France expresses the view that, owing to the confinement of its coast by neighbouring Italy and Spain from the east, south (Sardinia and the Balearics), and west, the delimitation of the continental shelf in this area should not be decided on a strict application of equidistance, but by following 'equitable principles'. In this respect, France suggests that the problem of delimitation in this area ought to be solved within a regional framework. It specifically proposed to Italy and Spain the creation of 'a zone of economic interest' overlapping the respective continental shelves and in which the three States would co-operate in the exploration and exploitation of mineral resources.[28]

Thus, while at present talks with Spain concerning shelf delimitation are at a standstill,[29] some progress has been achieved in negotiations with Italy. This will be considered below, together with the recently negotiated boundary agreement between France and Monaco.

(i) A prospect of a French–Spanish boundary

Both France and Spain are parties to the 1958 Shelf Convention, with France entering, *inter alia*, a general reservation to Article 6 which states

[26] Ibid.

[27] See however, *supra*, n. 24.

[28] See reply of the Minister for External Relations, *Journal officiel de la République Française, Débats parlementaires Assemblée Nationale*, No. 44A. N.(Q) (1981) (14 Dec. 1981), 3653. See also J. Charpentier, 'Pratique française du droit international, année 1982', 28 *AFDI* (1982), 1069.

[29] Preliminary discussion between the two States took place in 1978, but no progress appeared to have resulted. Information given by Professor Pastor (of Madrid State University and Legal Adviser to the Spanish Ministry of Foreign Affairs), Madrid, 10 June 1988.

that, in the absence of a specific agreement, the French Republic will not accept that any continental shelf boundary determined on the equidistance principle 'shall be invoked against it if it extends beyond the 200-metre isobath'.[30] Most of the shelf area situated off the French–Spanish coasts extends beyond the 200-metre isobath.[31] Although the reservation, as has been correctly described by the Arbitral Court, is a true reservation which 'relates in substance as in form to the regime of delimitation prescribed in Article 6', and is not particularly 'limited to the Atlantic Ocean' or to the English Channel,[32] the present unilateral practice of France seems to renounce this reservation altogether. According to the Law of 1976, the French Government has determined the extension of its continental shelf by the distance criterion of 200 miles, irrespective of any criterion of depth.[33] Thus, knowing that the French reservation was motivated by its opposition in 1958 to the exploitability criterion included in Article 1,[34] one realizes that such a reservation holds no more significance for France.

The difficulty in applying Article 6 of the Shelf Convention is, therefore, confined to the identification of the special circumstances which are derived, firstly, from the particular shape of the two States' adjacent coasts, and, secondly, from the position of the Balearic Islands which are opposite to the continental coast of France. With regard to the first circumstance, the configuration of the coast near the land frontier recalls that of the Bay of Biscay. Due to the protrusion of the Spanish coast at the Cape of Creus (see Map 26), the French coast alternates in a concave shape which is further pronounced by its gradual orientation in a north-easterly direction.

Both France and Spain in their legislation adopt 'equidistance' for the delimitation of their territorial waters.[35] But the application of strict equidistance would put France at a disadvantage, owing to the concavity of the coast referred to above. A modified equidistant boundary could be achieved by, for instance, attributing to the Spanish promontory only half-effect. It is pertinent to recall that in the Bay of Biscay the two States reached a compromise, whereby Spain modified its initial demand for a strict equidistant line.[36]

[30] Text in *Multilateral Treaties in Respect of which the Secretary General Performs Depository Functions* (ST/LEG/SER.D/7), ch. XXI s.4.

[31] See J. P. Flanagan, 'Physiography of the Western Mediterranean: The Alboran Sea', in D. Stanley (ed.), *The Mediterranean Sea: A Natural Laboratory* (Hutchinson & Ross, Stroudsburg, Pa., 1971), 6–7.

[32] *Anglo-French* Arbitration, 18 RIAA, paras. 53, 73.

[33] Art. 2 of Law No. 76-655 of 1976. See *infra*, n. 203.

[34] See the *Anglo-French* Arbitration, paras. 52, 73.

[35] See *supra*, Ch. 8 s.2.

[36] See *supra*, Ch. 4 s.4(*b*)(i).

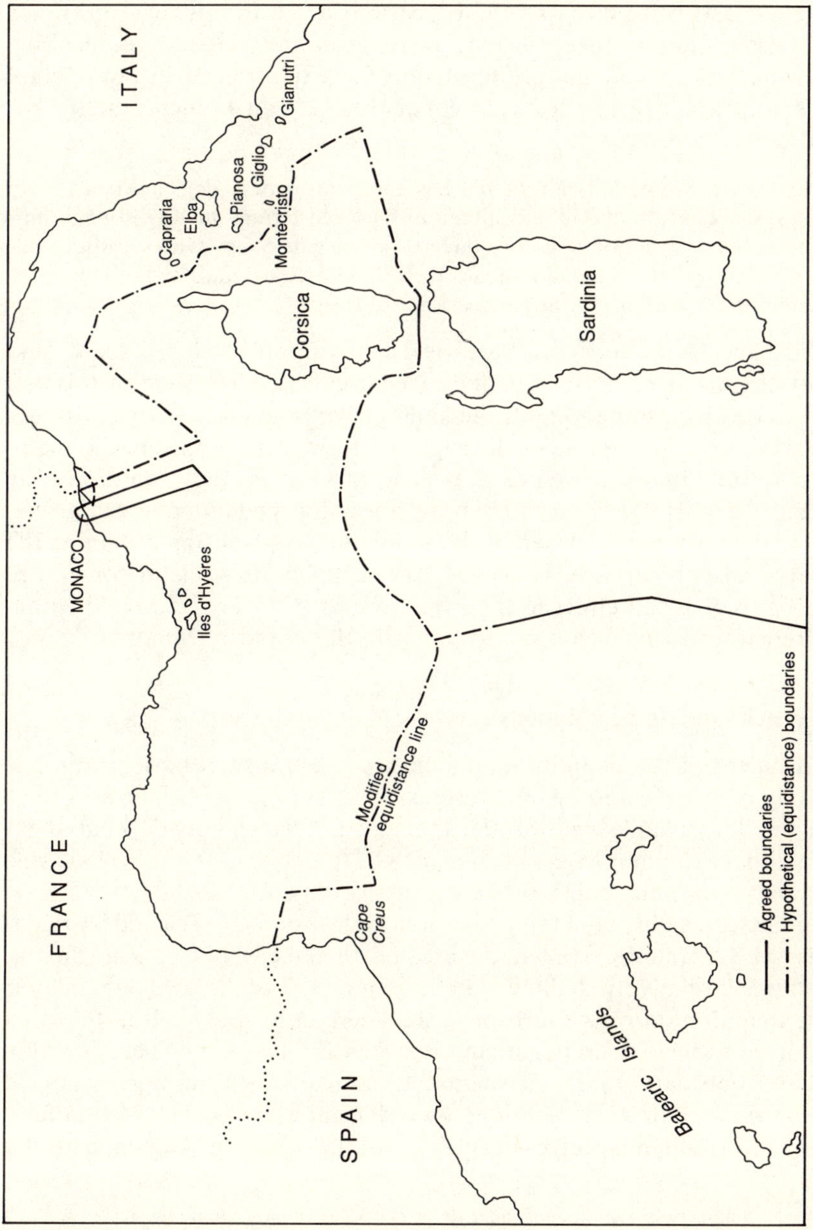

Source: Leanza, U. (ed), *Mediterranean Continental Shelf: Delimitations and Regimes*, Oceana Publications, New York 1988.

MAP 26 French–Italian–Spanish overlapping jurisdictions

The extension of a modified equidistant boundary beyond the 12-mile territorial sea limit depends largely on the boundary envisaged for the opposite coasts of France and the Balearic Islands. In this regard, France might argue that if those islands were given full effect the resulting equidistant line would sharply terminate the extension of its continental shelf southwards. In fact a recent commentator[37] on French practice has stated that, for France,

> La situation est comparable à celle des îles anglo-normandes dans la Manche. On est en présence, selon elle, d'une situation d''espace fermé' qui se distingue juridiquement d'une situation d''espace ouvert' (ces îles de Saint-Pierre et Miquelon, de la Guyane ou de l'océan Indien). Cette distinction aurait été, à son avis, consacrée par la sentence arbitrale de la mer d'Iroise.

Whilst the location of the Balearic Islands in an 'enclosed space' and far off the coast of Spain might be regarded as a relevant factor that ought to be given some weight, the analogy with the Channel Islands does not seem to be appropriate. Although the former are opposite the mainland coast of France as well as the island of Corsica, they can in no way be regarded as situated on the 'wrong side' of a median/equidistant line. In fact, they are much closer to the mainland coast of Spain than to the relevant coasts of France. However, one compensatory factor for France could be to give full effect to the small islands of Hyères, situated within the French territorial sea but also opposite the island of Minorca.

(ii) French–Italian negotiations

The problem of the delimitation of the continental shelf between France and Italy is complicated by the presence of Corsica and Sardinia as well as other smaller Italian islands off the Tuscan coast which impede what might be an otherwise simple application of the equidistance line. However, from the economic point of view, the continental shelf, which is at present inaccessible in the Gulf of Genoa and west of Corsica and Sardinia, seems to be more interesting in the area north and east of Corsica, where it constitutes a relatively shallow plateau joining the island to the mainland.

The delimitation of the territorial sea and continental shelf in this area is reported to have been negotiated between the two States between May 1972 and February 1974. The negotiations resulted in an agreement *ad referendum* on a draft convention,[38] which established an equidistant line, adjusted for administrative purposes, in the Gulf of Genoa and the

[37] A. Reynaud, *Le Plateau continental de la France* (Librairie Générale de Droit et de Jurisprudence, Paris, 1984), 233.

[38] 'Compendium of Negotiations between Italy and France on the Delimitation of the Continental Shelf', meeting of 25–6 Feb. 1974 (unpublished document).

Tyrrhenian Sea (east of Corsica) and in the Bocche de Bonifacio. The line generally gave effect to all islands, except some small islands, such as Giraglia (France), Gorgona, Capraria, and Sc. d'Africa (Italy)[39] (see Map 26).

Whilst the two States were able in 1986 finally to define the median line in the Bocche de Bonifacio proper,[40] the exact delineation of a boundary line for the rest of the area gave rise to several points of disagreement, and the parties were consequently unable to conclude any agreement. The Italian delegation reserved its position as to the exact placing by France of the turning-points in Côte d'Azur/Riviera, Bocche de Bonifacio, and as to the exchange of total areas in the Gulf of Genoa. The French delegation on its part reserved its position on the Italian plan at the time to establish straight baselines around, *inter alia*, the Tuscan archipelago. It recalled the French Government's reservation in 1965 over Article 6 that, in the absence of a specific agreement, France will not accept a continental shelf boundary delimited on the basis of equidistance and by reference to straight baselines established after 29 April 1958.[41] It was agreed at this point, however, that the two parties would consider the establishment of specific 'baselines' on land for delimitation purposes.[42]

There was, however, total disagreement between the parties as regards the method appropriate for delimiting the second zone, namely the area west of Corsica and Sardinia. Such a disagreement was not surprising. The area appears to be one of potential oil interest[43] and France is aware of the fact that Sardinia (extending further westward than Corsica) is situated not only adjacent to Corsica but also opposite the coast of continental France, thus impeding the full effect of this coast's continental shelf southwards. The situation is further complicated by the lack of any room for compensation, because of the immediate effect of the coastlines of the Spanish mainland and of the Balearic Islands. Therefore, when Italy recalled its position that the dividing line in this zone should be equidistant, taking into account the existing Italian–Spanish boundary line, France considered, on the other hand, that, because of the shape of the western Mediterranean, equidistance could not be considered appropriate.[44] It proposed, instead, that the two States consider the establish-

[39] Ibid. negotiations held in Rome on 26 Feb. 1974. See also an earlier report by experts of both States annexed to the 'Compendium' of conclusions of the two States' negotiations on 27–8 Mar. 1973 (unpublished document).

[40] See *supra*, Ch. 8 s.5(*c*)(i).

[41] See *supra*, Ch. 3 s.3(*b*)(i).

[42] *Supra*, n. 38.

[43] See *infra*, s.3(*a*).

[44] *Supra*, n. 38.

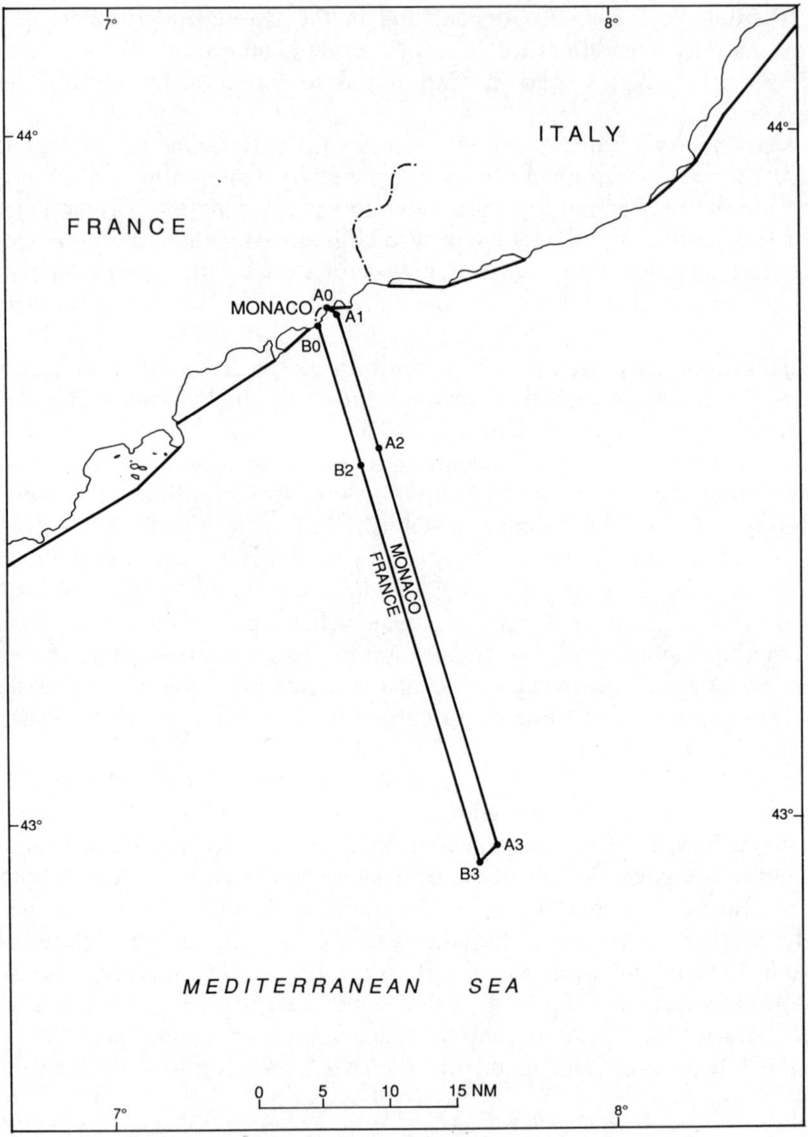

Source: Leanza, U. (ed), *Mediterranean Continental Shelf: Delimitations and Regimes*, Oceana Publications, New York 1988.

MAP 27 Maritime boundary: France–Monaco

ment of a 'zone of economic interests' for the joint exploitation of natural resources in the said zone.[45]

[45] Note of the French delegation of 6 Sept. 1974 on the provisions applicable to the 'Zone of Economic Interest' (unpublished document).

(iii) The French–Monacan Agreement

The 1984 delimitation Agreement[46] between France and Monaco is of interest, as it provides for what appears to be the first single maritime boundary in the Mediterranean. The Agreement establishes the lateral territorial sea boundary between France and Monaco which is determined by five geographical co-ordinates (three on the east and two on the west) and terminates at 'l'arc de loxodromie joignant le point B2 et un point B3', 12 miles from the coast.[47] Further seaward, the boundary proceeds in the same direction for about 35.51 miles, representing 'the limits of the maritime spaces' on which Monaco 'exercises or will exercise its sovereign rights in accordance with international law'.[48] The outer limit of this zone is determined by a line joining the terminal points of the boundary lines. The joining line is 'equidistance from the coasts of France (Corsica) and Monaco'.[49]

At present, Monaco claims only a breadth of 12 miles for its territorial sea.[50] Thus the Agreement could now be said to govern the territorial sea/shelf[51] boundaries between the two States, as well as any prospective EEZ/EFZ jurisdictions.

The 'maritime jurisdiction' of Monaco resulting from this Agreement (which applies no specific principles as far as the adjacent boundaries are concerned) is a small corridor—measuring about 1.69 miles in width and extending for about 47.51 miles. The total area is about 280 sq.km. However, according to Article 4, the present Agreement does not prejudice the fishing activity of the fishermen of both States, who can 'continue to exercise their activity in the traditional fishing grounds situated in the territorial waters of Monaco and the neighbouring French territorial waters'.

(*e*) *The Tunisia–Libya Judgment*

The Court was requested by the parties not only to determine what principles and rules of international law might be applied to the delimitation of the continental shelf, but also 'to clarify the practical method for the application of these principles and rules . . . so as to enable the experts of the two States to delimit these areas without difficulties'.[52]

46 Agreement of 16 Feb. 1984 which entered into force on 22 Aug. 1985. Text in 90 *RGDIP* (1986), 308–11 (see Map 27).

47 Ibid. Art. 1(3).

48 Art. 2.

49 Art. 2(3).

50 See *supra*, Ch. 6 n. 3.

51 Although Monaco does not formally claim a continental shelf jurisdiction, it is entitled to this right in accordance with the *ipso facto-ab initio* principle. See *supra*, Ch. 5 s.2(*c*).

52 Art. 2 of the *compromis*, text in the *Tunisia–Libya* case, ICJ Reports (1982), para. 4.

With regard to the principles of law applicable the Court was adamant that neither equidistance nor any other single method could be obligatory and that several methods might be applied to the same delimitation.[53] Envisaging equidistance as a strict, geometrical method, however, the Court expressed a somewhat elliptical view. It recognized that the strength (as well as the weakness) of the equidistance method lies in taking account of almost all variations in the relevant circumstances, thereby giving it particular value in the outer sector, where the change in direction of the Tunisian coastline and the Kerkennah Islands[54] were prime geographical factors. The Court added that this major change in the direction of the Tunisian coastline altered the relationship of the coasts of Libya and Tunisia to a marked degree 'from that of adjacent States to that of opposite States, and thus to produce a situation in which the position of an equidistance line becomes a factor to be given more weight in the balancing of equitable considerations than would otherwise be the case'.[55] However, nowhere else in the Judgment did the Court consider this 'factor' in its process of the balancing of the equities of the case.

In respect of Article 83 of the 1981 draft Convention (now the 1982 Convention), the Court noticed that although it was bound, according to the *compromis*, to apply the 'new trends' admitted at UNCLOS III,[56] that article lacked 'any indication of a specific criterion which would give guidance to the interested States in their effort to achieve an equitable solution'.[57] Instead, 'emphasis is placed on the equitable solution which has to be achieved'.[58] In this respect, both the Court and the parties were in agreement that the delimitation must be carried out in accordance with 'equitable principles', in the light of the relevant circumstances which characterized the area.[59] However, the Court observed that the parties dealt with its meaning and significance 'in close relationship with the principle of natural prolongation, and have devoted less attention to the

In the course of the proceedings, the two parties disagreed as to the proper interpretation of this clause. According to Tunisia, 'the Court was required to specify the legal rules and principles as well as the appropriate method of delimitation, leaving no room for any difficulty to arise between the two States' experts in point of meaning or method'. The Libyan view was that the Court was not authorized to carry the matter 'right up to the ultimate point before the purely technical work'. The Court, however, did not consider that there existed any substantial distinction between the two positions: 'In any event a careful analysis of the pleadings and arguments of the Parties on the point leads the Court to conclude that there is here no fundamental difference of opinion between them. There is only, in the final analysis, a difference of emphasis as to the respective rules of the Court and of the experts of the two countries.' See paras. 26–9.

[53] Paras. 109–11.
[54] See *infra*, text at nn. 90–2.
[55] Para. 126.
[56] Para. 46.
[57] Para. 50.
[58] Ibid.
[59] Para. 70 and Art. 1 of the *compromis*.

question of what are the equitable principles to be taken into account'.[60] The Court thus had to discuss the relevance of the principle of natural prolongation as considered by the parties.

The Court noticed that the continental shelf area subject to delimitation between the parties was part of the broader submarine region known as the Pelagian Block. This region is inclined at a gentle slope from the west to the east; it is bounded in the north by a series of large depressions (the Troughs of Pantelleria, Linosa, and Malta)[61] and in the east by a change in the slope of the seabed area identified in the parties' arguments as the 'Malta–Misratah Escarpment' or the 'Ionian Flexure'. Within this region, the parties gave different interpretations of the geological and geomorphological formation of certain features in order to support their proposed boundary lines. Libya principally contended that the area in front of its coast, constituting the Pelagian Block, was the natural prolongation northward of the North African land mass to the south, inasmuch as it constituted a typical continental margin produced by plate movement and rifting, explained by the theory of 'plate tectonics'.[62] Therefore, according to the Libyan view, 'the appropriate method of delimitation is to reflect the direction of this prolongation northward of the terminal point of the land boundary'.[63] Tunisia for its part argued for the geological and geomorphological continuity of the Pelagian Block with the land territory of Tunisia, as they were both aligned generally in an east–west direction.[64] The purpose of the Tunisian argument was to establish that 'the natural prolongation of Tunisia . . . extends eastward as far as the area between the 250-metre and 300-metre isobath, and south-eastwards as far as the zone constituted by the Zira and Zuwarah Ridges'.[65]

[60] Para. 69.

[61] See *supra*, n. 19.

[62] According to this theory, 'the outermost structural shell of the earth, the lithosphere, is segmented by a network of faults into a number of "plates", which rest upon the next shell of the earth, the asthenosphere, the boundary between lithosphere and asthenosphere being defined in terms of temperature (by the isotherm of 1,350°C). Each of the plates making up the lithosphere can and does move in relation to the asthenosphere beneath and in relation to the other plates; thus at the points or lines of contact between the plates, various kinds of relative motion may occur.' *Tunisia–Libya* case, para. 52. Libya argued that the effect of this movement in the western Mediterranean has produced, *inter alia*, a continental margin lying to the north of the coastline of Libya and Tunisia, running generally east–west. The subsequent geological events, whereby what is now the greater part of Tunisia was lifted up out of the sea and the Pelagian Block tilted slightly, did not disturb the essential relationship of the Block as a projection to the north of the land mass. See ibid. paras. 56–7 and *Memorial* submitted by Libyan Arab Jamahiriya (1980), paras. 60–5.

[63] Libyan submissions, submission No. 5 in *Memorial* (1980), 70.

[64] *Memorial* submitted by the Government of Tunisia (1980), paras. 5.51 ff.

[65] Tunisian submissions, submission No. I(3), ibid. 251. Tunisia proposed several alternative lines of delimitation based on what it construed as relevant physical features of the 'natural prolongation'. See *supra*, Ch. 8 s.4(*b*)(ii).

Reviewing the development of the law on the continental shelf, the Court expressed the view that

> [w]hile the idea of the natural prolongation of the land territory defined, in general terms, the physical object or location of the rights of the coastal State, it would not necessarily be sufficient, or even appropriate, in itself to determine the precise extent of the rights of one State in relation to those of a neighbouring State.[66]

Thus the Court observed that, in principle, the satisfaction of equitable principles is,

> in the delimitation process, of cardinal importance . . . and identification of natural prolongation may, where the geographical circumstances are appropriate, have an important role to play in defining an equitable delimitation, in view of its significance as the justification of continental shelf rights in some cases; but the two considerations—the satisfying of equitable principles and the identification of the natural prolongation—are not to be placed on a plane of equality.[67]

In the case between the parties, however, the Court observed that in their determination of the 'new accepted trends' which have emerged at UNCLOS III, the 'Parties relied on the principle of natural prolongation: they have not advanced any argument based on the "trend" towards the distance principle [enshrined in Article 76(1) of the 1981 draft Convention]'.[68] The Court then drew the somewhat surprising conclusion that the definition of that paragraph 'affords [therefore] no criterion for delimitation in the present case'.[69] By implication, had the parties considered the distance principle the Court would probably have taken it into account, at least as a relevant factor for the delimitation. Such an argument is submitted to run counter even to the Court's own submission that it would in any event have to apply the available sources or at least 'accepted trends' of international law *proprio motu*, and irrespective of the parties' elaborations thereon.[70]

However, having decided that 'natural prolongation' might be relevant to the law of delimitation, the Court distinguished between its two facets

[66] Para. 43.

[67] Para. 44.

[68] Para. 48.

[69] Ibid.

[70] Para. 24. For a comment on an argument similar to that made by the Court see the Decision of the European Commission on Human Rights (3147/67, *Collection of Decisions* 27, p. 119, at 126–7): 'it is a generally recognised principle of law that it is for the court to know the law (Jura novit curia) . . . the practice of the legal systems of the High Contracting Parties varies in this respect . . . in certain systems the concept of "fair-hearing" is interpreted in the sense that courts are required to invite the Parties to make . . . submissions on those points of law which appear to the courts to be significant . . . However, other legal systems do not make any such requirements . . . consequently, there is no generally accepted practice in this respect.'

as argued by the parties. With regard to the area's geological evolution, the Court observed that it was not possible, for legal purposes, 'to define the areas of continental shelf appertaining to Tunisia and Libya by reference solely or mainly to geological considerations . . . What must be taken into account in the delimitation of the shelf areas are the physical circumstances as they are today . . . not the evolution in the long-distance past.'[71] And, having reviewed the physical evidence and experts' opinions regarding the geomorphology and bathymetry of the area, the Court thought itself 'unable to find that any of [the features examined by the Parties] involves such a marked disruption or discontinuance of the seabed as to constitute an indisputable indication of the limits of two separate continental shelves, or two separate natural prolongations'.[72]

Having said so, the Court, nevertheless, expressed the view that this did not 'necessarily exclude the possibility that certain [of such features], *which do not amount to such an interruption of the natural prolongation of one Party with regard to that of the other, may be taken into account for the delimitation, as relevant circumstances characterizing the area*'.[73] Whilst the *dispositif* did not include any such features as a relevant circumstance,[74] the italicized statement seems to identify 'geomorphological discontinuity' as a possible relevant circumstance which might be taken into account without having to be identified with 'natural prolongation' as an allegedly independent criterion for delimitation. This is considered a significant clarification which does not seem to be reiterated in any of the cases which preceded or followed this Judgment.

However, within this single shelf of the Pelagian Block the Court proceeded to identify other factors which it considered relevant to the delimitation. With regard to the 'relevant coastline', the Court found that it was only the seabed area off the coasts of the two parties between the points of Ras Kaboudia on the Tunisian coast and of Ras Tajura on the Libyan coast that were relevant for the present case.[75] The Court noticed that the seaward limit of this area could be delimited provisionally and only in relation to the criterion of proportionality, to the extent that the rights of other States bordering on the Pelagian Sea would not 'be prejudged by the decision in the present case'.[76] However, in this respect the Court noticed that it was not possible to apply the factor of proportionality, 'by way of touchstone of equitableness, to the method or methods it may indicate, unless it can arrive at a reasonably clear conception of the extent of the areas on each side of the relevant line'.[77] This the Court left to the final stage of its Judgment.

The most marked characteristic of the Tunisian coast was its change of

[71] Para. 61. [72] Para. 66. [73] Para. 68 (emphasis added).
[74] See Para. 133 (*b*(1–5)). [75] Para. 75. [76] Ibid. [77] Para. 108.

direction from roughly east–west at Ras Ajdir into south-west–north-east as it extended further westward. This aspect of geography was considered by the Court as 'largely significant . . . as one of the relevant circumstances which characterize the area'. It modified the situation of lateral adjacency of the two parties, even though it did not 'go so far as to place them in a position of legally opposite States'.

The presence of the Tunisian islands of Jerba and Kerkennah and the surrounding low-tide elevations were also regarded as a special circumstance that could not be excluded in principle.[78] However, the Court established that the practical method applied by it was such that, in the part of the area where the island of Jerba would be relevant, 'there are other considerations which prevail over the effect of its presence; the existence and position of the Kerkennah Islands and surrounding low-tide elevations, on the other hand, are material'.[79]

The other factors found by the Court to be relevant were the settled land boundary terminus at Ras Ajdir, and the parties' past unilateral and bilateral endeavours to establish partial maritime or adjoining concession boundary claims in the area off Ras Ajdir.[80] In this area the Court referred to a *de facto-modus vivendi* line, reflecting (1) the line perpendicular to the coastline proposed by Libyan authority in 1913 and never formally contested by either side throughout a long period of time, and (2) the line of 26° north-east from Ras Ajdir, which had been followed by the two parties in the granting of concessions for the exploration of hydrocarbons during the period 1964–72.

In respect of proportionality the Court considered that 'that element is indeed required by the fundamental principle of ensuring an equitable delimitation between the States concerned'.[81] And in its reflection upon this factor the Court considered in principle the relevance of the 'historic claims' of Tunisia not as an independent circumstance,[82] but in relation to the principle of non-encroachment: 'it is only if the method of delimitation which the Court finds to be appropriate is such that it will or may encroach upon the historic rights area that the Court will have to determine the validity and scope of those rights, and their opposability to

[78] Para. 79.

[79] Ibid. The Court, however, never elaborated on these 'considerations' which justified in its view ignoring the island of Jerba. On this point, see the dissenting opinion of Judge *ad hoc* Evensen, 300, para. 17.

[80] Paras. 85–6.

[81] Para. 103.

[82] Tunisia argued vigorously for the existence of established historic rights justifying the drawing of a line ZV 45° from Ras Ajdir (see *supra*, Ch. 8 s.4(*b*)(ii)). The Court, however, dismissed the relevance of the Tunisia 'historic rights' argument in the delimitation of the continental shelf and abstained, accordingly, from considering the merits of these alleged rights (see *supra*, Ch. 1 s.3(*b*)(ii)).

Libya, in the context of a delimitation of the continental shelf.'[83] However, in the actual circumstances of the case, the Court concluded that the line which it had proposed in the area where such historic rights were claimed to exist did 'not raise any issue which would make it necessary for the Court to decide on the validity or opposability to Libya of the historic rights claimed [by Tunisia]'.[84]

Turning to the practical method for the application of the 'principles and rules of international law' and 'relevant circumstances' identified above, the Court noticed that '[a]ny examination of methods, like the examination of applicable rules and principles, must take as a starting-point the particular geographical situation, and especially the extent and features of the area found to be relevant to the delimitation'.[85] In the view of the Court, the geographical circumstances of the area, including the changes in coastal configurations and the overall effect of any methods applied to the area, called for the 'area close to the coasts of the Parties to be treated differently from the areas further offshore'.[86]

In the area close to the coasts of the parties, that is, off the coasts' immediate extension from Ras Ajdir, eastwards and westwards up to the point of the veering of the Tunisian coast to the north, the Court applied a line drawn from the land frontier at approximately 26° to the meridian. This line was proposed 'on the basis that it was perpendicular to the coast'[87] and was also 'identified both by the Parties and by the States of which they are the territorial successors as an appropriate line'.[88] The practice of the parties relating to this *de facto-modus vivendi* line as well as its proper delineation by the Court was considered in some detail earlier.[89]

The terminal point of this first sector was found to be properly identified with the point reflecting the radical change in the Tunisian coast from a north-west–south-east to north orientation. Thus the Court stated that the first sector of the dividing line should 'veer at the point of intersection with the parallel passing through the most westerly point of the Tunisian coastline between Ras Kaboudia and Ras Ajdir, that is to say, the most westerly point on the shoreline (low water mark) of the Gulf of Gabes'.[90]

[83] Para. 105.
[84] Para. 121.
[85] Para. 114.
[86] Ibid.
[87] Para. 119. In the next paragraph the Court re-emphasized that 'the factor of perpendicularity to the coast and the concept of prolongation of the general direction of the land boundary are . . . relevant criteria to be taken into account in selecting a line of delimitation calculated to ensure an equitable solution'.
[88] Para. 120.
[89] See *supra*, Ch. 8 s.4(*b*)(ii).
[90] Para. 133 (*c*(2)).

The bearing of this parallel was found to be approximately 42° to the meridian.[91]

But east of this parallel lay the Kerkennah Islands, 'surrounded by islets and low-tide elevations, and constituting by their size and position a circumstance relevant for delimitation'.[92] The final location of the corresponding parallel was to be adjusted to reflect this relevant circumstance. The Court found that a line drawn parallel with the most westerly point of the Gulf of Gabes along the seaward coast of the actual islands would clearly run at a bearing of approximately 62° to the meridian. This full effect 'would, in the circumstances of the case, amount to giving excessive weight to the Kerkennahs'.[93] The Court thought it was appropriate to grant the islands 'half effect'.[94]

Therefore, the Court suggested that the second sector was to run parallel to a line drawn from the most westerly point of the Gulf of Gabes 'bisecting the angle formed by a line from that point to Ras Kaboudia and a line drawn from that same point along the seaward coast of the Kerkennah Islands, the bearing of the delimitation line parallel to such bisector being 52° to the meridian'.[95]

As regards the final extension of this line north-eastwards, the Court observed that that was a matter falling outside its jurisdiction in the present case, as it would depend on the delimitation to be agreed with third States on the other side of the Pelagian Sea.[96] However, for the purpose of applying proportionality, the Court found it 'necessary to determine the seaward limits of the area to be taken into account, which is bounded by the coasts of Tunisia as far as Ras Kaboudia and Libya as far as Ras Tajoura'.[97] The Court found that the intersection of the latitude and meridian passing through Ras Kaboudia and Ras Tajura, respectively, would 'afford appropriate seaward limits of the areas to be compared'.[98] Having conducted the factor of proportionality[99] on the 'hypothesis of the whole of that area being divided by the delimitation line',[100] the Court produced on the map attached to its Judgment a line which was not confined within the limits imposed by the claims of the third States, especially those which Malta described precisely in its Application to intervene before the Court.[101] On the contrary, 'the rights of

[91] Para. 128. [92] Ibid. [93] Ibid. [94] Para. 129.
[95] Para. 133 (*c*(3)). [96] Para. 130. [97] Ibid. [98] Ibid.
[99] Having thus defined the seaward area, the Court measured the lengths of both parties' coastlines relevant to the area of delimitation (see *supra*, text at n. 75), and then compared these with the respective shelf areas that would accrue to each by virtue of the boundary line determined by it. It found that the comparisons held up. 'This result', it concluded, 'taking into account all the relevant circumstances, seems to the Court to meet the requirements of the test of proportionality as an aspect of equity.' Para. 131.
[100] Para. 130. [101] See *supra*, s.2(*c*)(i).

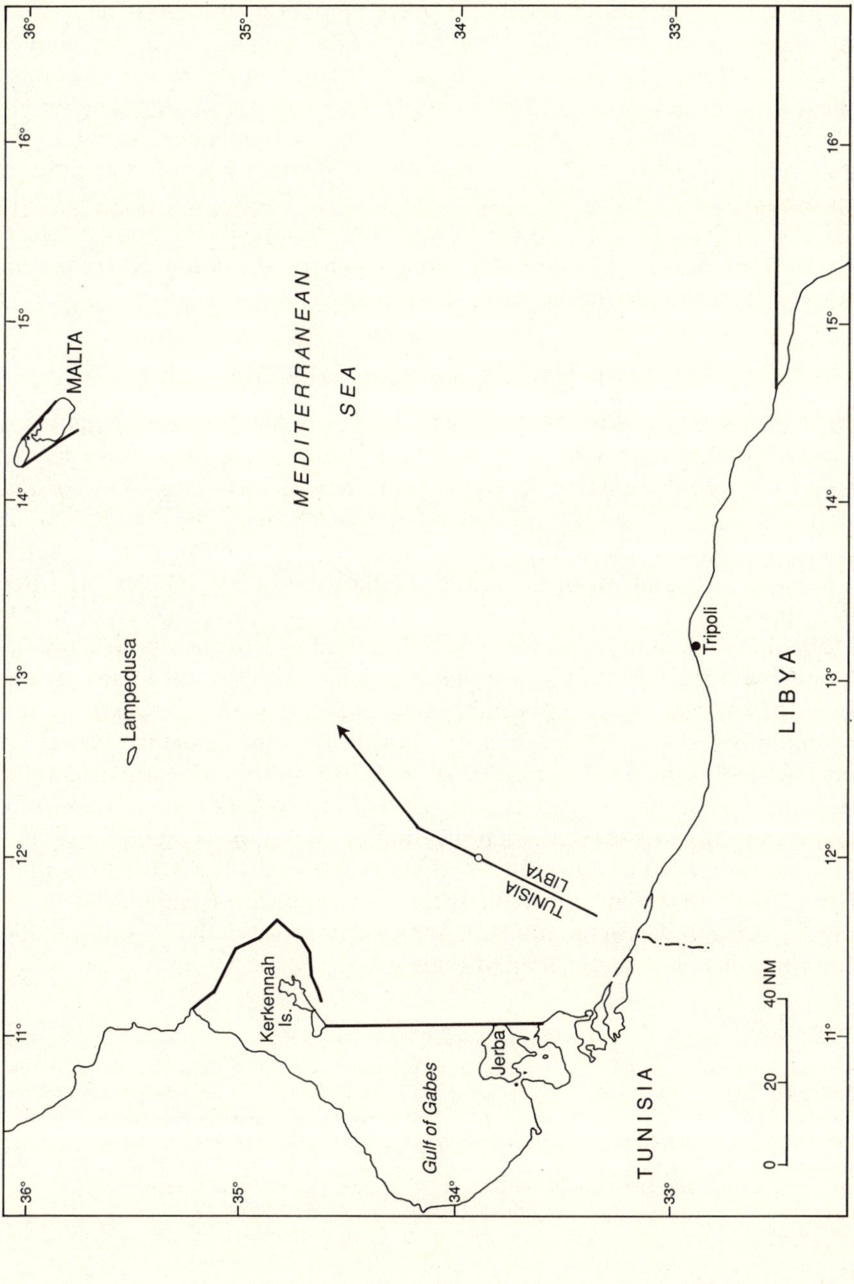

Source: Leanza, U. (ed), *Mediterranean Continental Shelf: Delimitations and Regimes*, Oceana Publications, New York 1988.

MAP 28 Continental shelf boundary: Libya–Tunisia (1982 Judgment)

third States being reserved',[102] the line ended with an arrow pointing in Malta's direction (see Map 28).

This is submitted to be a recommended, practical solution in situations where, due to the geographical limitation of the area involved, the conflict of claims of two parties may overlap with the claims of other States. Thus, the parties agreed in 1988 to implement the Court's decision, and it was provided in the same Agreement that the last segment of the boundary line would extend 'north-eastward until it meets a line of delimitation with third States'.[103]

The practicality of the Court's approach is emphasized even more in the light of the complications resulting from its deviation therefrom in similar circumstances in the *Libya–Malta* case.

(*f*) *The Libya–Malta Judgment*

The Court was asked by the parties to determine the 'principles and rules of international law' and the 'practical application' of these rules to the delimitation of the area of the continental shelf appertaining to Libya and Malta.[104] The proper interpretation of this request, the Court stated, 'necessarily entails the indication by [it] of the method or methods which it considers to result from the proper application of the appropriate rules and principles'.[105]

The fundamental problems which faced the Court in its attempt to comply with the parties' request may be confined to two problems. These were, firstly, the impact of a third-party claim which was exerted by the Italian intervention and the subsequent refusal by the Court to allow it,[106] and, secondly, the extensive reliance by Libya on the principle of natural prolongation which was apparently justified by the Court's assertion (only sixteen months before) that 'identification of natural prolongation may . . . have an important role to play in defining an equitable delimitation'.[107] The Court's pronouncement on these two problems is submitted to be largely associated with the fundamental characteristic of the Mediterranean as a semi-enclosed sea of limited dimension.

[102] Para. 133(*b*)(1).

[103] Agreement between the Libyan Arab Jamahiriya and the Republic of Tunisia to implement the Judgment of the International Court of Justice in the Tunisia–Libya Continental Shelf Case, signed on 8 Aug. 1988 and entered into force on 11 Apr. 1989. Both texts of the agreement and the exhange of notes of ratification (in Arabic) are provided by courtesy of the People's Bureau of Foreign Liaison, Tripoli. The agreement follows the parallels of longitude and latitude and bearings recommended by the Court.

[104] Art. 1 of the *compromis*, text in the *Libya–Malta* case, para. 2.

[105] Para. 19.

[106] See *infra*, nn. 110, 116.

[107] The *Tunisia–Libya* case, para. 44.

Following the promulgation by Italy in the *Intervention* case of shelf claims overlapping with a significant part of the area of dispute between the parties,[108] both Libya and Malta expressed the view that the Court should not feel inhibited from extending its decision to the whole area of dispute.[109] Apparently apprehensive of the Court's intention to limit the scope of delimitation,[110] Libya had suggested, in oral pleading, a 'practical solution'. It had proposed that in areas where there were overlapping claims with Italy the Court could indicate the delimitation line between Libya and Malta by a pecked line (as opposed to a continuous line) so as to indicate clearly that this delimitation was entirely subject to any Italian claims.[111] Furthermore, Libya had pointed out that the caveats and reservations which the Court would include in its Judgment would in any event protect the rights of third States by preventing such a delimitation from being definitive *vis-à-vis* such third States.[112] The parties and the Court were at one that the decision of the Court would, by virtue of Article 59 of the Statute of the Court,[113] have binding effect only between the parties, and not against third States.

Nevertheless, the Court saw fit to limit the jurisdiction granted to it by the parties. It insisted that

> in the sense of permitting the delimitation of the areas of the shelf which 'appertain' to the Parties, as distinct from areas to which one of the Parties has shown a better title than the other, but which might nevertheless prove to 'appertain' to a third State *if the Court had jurisdiction to enquire into the entitlement of that third State, the decision must be limited to a geographical area in which no such claims exist.*[114]

The italicized words highlight the paradox in the Court's reasoning. Although realizing that it had no jurisdiction to entertain the claims of a third State, it nevertheless decided to exclude areas of *actual* dispute

[108] See *infra*, s.2(g)(iii)(1).

[109] Para. 20.

[110] The Court in the *Application by Italy to Intervene* stated, *inter alia*, that 'it will, so far as it may find it necessary to do so, make it clear that it is deciding only between the competing claims of Libya and Malta'. It then went on to observe that 'if, as Italy has suggested, the decision of the Court in the present case, taken without Italy's participation, had for that reason to be more limited in scope between the Parties themselves, and subject to more caveats and reservations in favour of third States, than it might otherwise have been had Italy been present, it is the interests of Libya and Malta which might be said to be affected, not those of Italy. It is material to recall that Libya and Malta, by objecting to the intervention of Italy, have indicated their own preferences.' Para. 43.

[111] Information given by Professor Bowett, Counsel for Libya.

[112] *Application by Italy to Intervene*, para. 20.

[113] Art. 59 provides that 'the decision of the Court has no binding force except between the parties and in respect of that particular case'.

[114] Para. 21 (emphasis added).

between the parties by virtue of that third State's claims.[115] This was exactly the object of Italy's request to intervene which the Court had denied.[116]

Thus, taking into consideration the claims of Italy, the Court decided to confine itself to a short segment of the area disputed by the parties to the case, that is to say, areas where no claims by Italy were advanced, between the meridians 13° 50′ E. and 15° 10′ E. Furthermore, as a consequence of this self-imposed limitation, the Court excluded from its jurisdiction an area of dispute between the parties to which Italy laid no claims; that is the area east of meridian 15° 10′ E. but south of the parallel 34° 30′ N. The Court found that it could not 'pass any judgment on this area so long as national attribution of the continental shelf lying immediately to the north of it (that is, east of the meridian 15° 10′ E. and north of the parallel 34° 30′ N.) has not been settled by agreement between those States concerned [including Italy]'.[117]

It is submitted that this is an unprecedented approach[118] by the Court which exceeded the ordinary, yet sufficient, means of protecting the alleged rights of a third State.[119] In the circumstances of this case in particular one wonders whether the application of the Court's approach would have led it to discard its jurisdiction entirely, had Italy alone or

[115] As Judge Oda pertinently observed, '[b]y doing so, the Court loses sight of the scope of the dispute between the two original Parties, thus falling short of that full exercise of jurisdiction which they are entitled to expect'. His Dissent to the Judgment, 113, para. 11.

[116] The object of Italy's intervention was quoted by the Court as follows: 'Italy is asking the Court . . . to take into consideration the interests of a legal nature which Italy possesses in relation to various areas claimed by the main Parties, on certain parts of those areas, and accordingly *to provide the two Parties with every needful indication to ensure that they do not, when they conclude their delimitation agreement pursuant to the Court's judgment, include any areas which, on account of the existence of rights possessed by Italy, ought to be the subject either of delimitation between Italy and Malta, or of delimitation between Italy and Libya, or of a delimitation agreement as between all three countries*.' *Application by Italy to Intervene*, para. 17 (emphasis added).

[117] Para. 22.

[118] This unprecedented approach by the Court was anticipated by Judge Morozov in *Application by Italy to Intervene*. Citing the Court's remark that 'Libya and Malta, by objecting to the intervention by Italy, have indicated their own preference', he observes that 'In the context of the paragraph, this sentence can only be interpreted to the effect that the Court, before considering the merits of the case between Libya and Malta, appears to have decided that their interests could be affected as a consequence of exercising their right to object to the Italian request.' He further observes that 'This seems like an attempt to predict that Libya and Malta could have made a rod for their own backs. Such an unusual approach is, of course, unprecedented in the history of international jurisdiction.' His Sep. Op. to the Judgment, 30.

[119] e.g. the approach of the Court in the *Tunisia–Libya* case, see *supra*, text at nn. 97–102. In the *Anglo-French* Arbitration the Court felt it sufficient to point out that its 'decision, it scarcely needs to be said, will be binding only as between the Parties to the present arbitration and will neither be binding upon nor create any rights or obligations for any third States, and in particular for the Republic of Ireland, for which the Decision will be *res inter alios acta*'. Para. 28.

together with other neighbouring States, such as Greece and Tunisia, advanced 'ambitious' claims relating to the same area of dispute between the parties. This was a hypothesis raised by the Court in passing but to which it provided no satisfactory answer. The Court simply indicated that neither party in this case suggested that the Italian unilateral claims were particularly unreasonable.[120] Beside the fact that this statement may be taken as implying a premature 'judicial blessing' on a unilateral claim whose full impact on the Judgment neither party in the case was able to predict,[121] the reasonableness of the Italian claims might be questioned in the light of the very criterion put by the Court itself.

As will be seen, the Court put a maximum limit to what might be a reasonable extension of the Libyan shelf northward, described by the Court as the median line which represented 'a delimitation between a portion of the southern littoral and a portion of the northern littoral of the Central Mediterranean'.[122] Applying the same criterion, the reasonableness of the Italian claims could have been questioned since they obviously extended southwards of this same median line hypothesized by the Court.[123] Even if the parties failed to point out this factor, one feels that it was still up to the Court to provide some explanation for this self-evident contradiction in its Judgment.

Furthermore, with regard to the latitudinal length of the Italian claims, between 15° 10′ E. and the Greek–Italian shelf boundary line,[124] the Court should have been able to consider the reasonableness of this length, for it utilized the same Italian coastal front which had already been used in the said shelf boundary with Greece. In other words, the Italian coasts of Calabria and Apulia[125] were used to define Italy's alleged delimitatory rights *vis-à-vis* Greece, Libya, and Malta. The Court was not of course endowed with jurisdiction to judge the merits of the Italian claims. One feels, however, that, in view of the weight given to these claims, the Court would be expected to elaborate on this point which is submitted, moreover, to be unsettled in the law on delimitation.

[120] Para. 23.

[121] The full details of the Italian claims were submitted to the Court after the closing of the oral pleadings in the *Application* case. See Italian Reply to Judge Oda, Reply of the Agent of Italy, Professor Arangio Ruiz, 30 Jan. 1984.

[122] See *infra*, text at n. 153.

[123] This is obvious from the fact that the Court's 'median line' between Sicily and Libya was located at parallel 34° 36′ N., whereas the southward limit of the Italian claims (see *infra*, s.2(*g*)(iii)(1), (2)) was located at parallel 34° 30′ N., that is, 6′ south of the Court's hypothetical line. It is worth mentioning, moreover, that, according to the Italian written reply, this southern extension was not announced as Italy's maximum seabed jurisdiction in the area. Rather, in the area 'south of this line' Italy sought to establish 'the delimitation line' of its continental shelf with Libya.

[124] See *infra*, s.2(*g*)(iii)(2).

[125] See the Italian written reply, *supra*, n. 121; the Judgment, para. 23.

However, having thus delimited its 'area of jurisdiction', the Court proceeded to consider the rules of international law applicable to the dispute. It quoted with approval its statement in *Tunisia–Libya* that it was bound to decide the case on the basis of equitable principles which would guarantee an equitable result.[126] In its effort to identify the applicable equitable criteria or practical methods, the Court considered the delimitation provisions of the 1958 and 1982 Conventions. In respect of Article 6 the Court noted that only Malta was party thereto and thus Article 6 was not as such applicable in the relationship between them.[127] Neither was the Court able to accept the Maltese view that equidistance should be applied as a natural consequence of the principle of distance. It stated that equidistance could not be applied 'even as a preliminary and provisional step towards the drawing of a delimitation line . . . That a coastal State may be entitled to continental shelf rights by reason of distance from the coast, and irrespective of the physical characteristics of the intervening sea-bed and subsoil, does not entail that equidistance is the only appropriate method of delimitation.' The application of equitable principles in the particular relevant circumstances 'may still require the adoption of another method, or combination of methods, of delimitation, even from the outset'.[128]

Turning to the 1982 Convention the Court recalled its statement in *Tunisia–Libya* that the Convention delimitation formula was empty of any specific criterion of delimitation and, instead, emphasis was placed on the equitable solution which had to be achieved.[129] The Court, however, referred to the parties' emphasis on the distinction in law between the rules applicable to the entitlement and definition of the continental shelf and those related to its delimitation.[130] The Court was eager to state that the questions of entitlement and delimitation 'are not only distinct but are also complementary . . . The legal basis of that which is to be delimited, and of entitlement to it, cannot be other than pertinent to that delimitation'.[131]

Within this context the Court observed the development of the law in the 1982 Convention, asserting that the rights enjoyed by a State over its continental shelf would also be possessed over the seabed and subsoil of any exclusive economic zone which it might establish. The Court found that the possibility of extending such rights to the legally permissible extent of the exclusive economic zone should be taken as a relevant circumstance for the delimitation of the continental shelf. The Court emphasized that 'this does not mean that the concept of the continental shelf has been absorbed by that of the exclusive economic zone; it does

[126] Paras. 28, 45–6. [127] Para. 26. [128] Para. 43. [129] Para. 28.
[130] Para. 27. [131] Ibid.

however signify that greater importance must be attributed to elements, such as distance from the coast, which are common to both concepts'.[132] Referring to the practice of States indicating the adoption of the concept of the exclusive zone—with its rule on entitlement by reason of distance—as part of the customary law, the Court reasserted that, for juridical and practical reasons, 'the distance criterion must now apply to the continental shelf as well as to the exclusive economic zone; and this quite apart from the provision as to distance in paragraph 1 of Article 76 [of the 1982 Convention]'.[133]

The Court in this paragraph seemed to attempt to prevent the inevitable conclusion that it extracted delimitation criteria from rules which *define* the continental shelf rather than from rules which are specifically concerned with delimitation. Yet it is obvious that the rule of entitlement by distance embodied in the EEZ regime (to which the Court made an explicit reference) is the same as that embodied in Article 76(1). It is not necessarily incorrect[134] to resort to Article 76 for the purpose of extracting rules on delimitation, especially since, as submitted earlier,[135] the specific rule on delimitation, i.e. Article 83, has no self-sufficient content. The unequivocal reliance by the Court on the 'distance principle' should not, however, be justified by reference to so artificial an argument.

However, still preoccupied with the soundness of its reasoning, the Court was aware that its conclusion would reinforce the position of one party to the dispute, i.e. Malta, against the other. For Malta, the 'distance principle' should govern its delimitation with Libya and the reference in Article 76 of the 1982 Convention represented 'a consecration' of the distance principle.[136] For Libya, 'the delimitation should be accomplished in such a way as to leave as much as possible to each Party all areas of continental shelf that constitute the natural prolongation of its land territory into and under the sea'.[137]

In an attempt, it is submitted, to assert the independence of its approach from that of Malta, the Court tried to compromise two distinct (geographical and geophysical) principles under one, yet unidentified, 'legal' concept. Following up its statement quoted above, the Court continued,

[132] Para. 33.

[133] Para. 34.

[134] Cf. e.g. Professor Bowett, who argues that the Court's emphasis on distance introduces a novelty, for the Court extracts from a criterion established for the definition of the outer limit a principle which is introduced into the delimitation process. See his Comment in U. Leanza (ed.), *The International Legal Regime of the Mediterranean Sea* (Giuffrè Editore, Milan, 1987), 265.

[135] See *supra*, Ch. 5 s.2(*e*).

[136] Para. 34.

[137] Submission No. 3, *Memorial* submitted by the Libyan Arab Jamahiriya, vol. i (1983), 163; Judgment, para. 11.

This is not to suggest that the idea of natural prolongation is now superseded by that of distance. What it does mean is that where the continental margin does not extend as far as 200 miles from the shore, natural prolongation, which in its physical origins has throughout its history become more and more a complex and juridical concept, is in part defined by distance from the shore, irrespective of the physical nature of the intervening sea-bed and subsoil. The concepts of natural prolongation and distance are therefore not opposed but complementary.[138]

How the concept of natural prolongation could be partially defined by distance from the shore, and in what way the two concepts could be considered complementary, was left unanswered.[139] The judicial utilization of 'natural prolongation' does not eliminate its essentially geophysical character any more than it does the essentially geometrical character of the concepts of distance or equidistance. The notion of transforming a geophysical factor into an as yet *undefined* juridical concept is of no value for the clarification of the rules of law.[140] It would have been more acceptable if the Court had simply reiterated the reasonable view that the development of the law since the 1977 *Anglo-French* Award has progressively indicated the relative demise of the concept of natural prolongation, at least within the 400-mile distance.[141] This development, in the Court's view, was related only to 'natural prolongation' *in its geophysical sense*, but not in its alleged, undefined juridical sense:

The Court . . . considers that since the development of the law enables a State to claim that the continental shelf appertaining to it extends up to 200 miles from its coast, whatever the geological characteristics of the corresponding sea-bed and subsoil, there is no reason to ascribe any role to geological or geophysical factors within that distance either in verifying the legal title of the States concerned *or in proceeding to a delimitation as between their claims*.[142]

Primarily on the basis of this reservation the Court discarded the central argument of Libya that there existed 'a fundamental discontinuity'

[138] Para. 34.

[139] Judge Oda states that this statement by the Court 'is surely, at least within the 200-mile context, no more than a method of keeping "natural prolongation" alive by artificial respiration'. His Dissent to the Judgment, 129.

[140] The idea that 'natural prolongation' is a juridical concept was first postulated by the Court of Arbitration in the *Anglo-French* Arbitration. However, what the Court wanted to turn attention to in that case was the appropriate suggestion that the 'physical' factor should be considered within a politico-geographical context, reflecting the correlation between a *State's coast* and the offshore shelf area (including its geographical dimensions), but not the physiography of a whole continent: 'the concept of natural prolongation is a juridical concept which connotes the natural prolongation under the sea not of a continent or geographical land mass but of the land territory of each State. And the very fact that in international law the continental shelf is a juridical concept means that its scope and the conditions for its application are not determined exclusively by the physical facts of geography but also by legal rules. Para. 191.

[141] See *supra*, Ch. 4 s.3(*b*)(ii)(1).

[142] Para. 39 (emphasis added).

between the continental shelf of Libya and Malta identified by the 'Rift Zone' within which the line of delimitation between the two States ought to be effected. The Court responded that 'since the distance between the coasts of the two Parties is less than 400 miles, so that no geophysical feature can lie more than 200 miles from each coast, the feature referred to as the "rift zone" cannot constitute a fundamental discontinuity terminating the southward extension of the Maltese shelf and the northward extension of the Libyan shelf as if it were some natural boundary'.[143]

In the light of the above enumeration of the 'legal principles' governing the case, the Court proceeded to what it called the 'first stage' in the process of applying an appropriate criterion and method. The ultimate equity of the resulting delimitation would be tested against the other relevant circumstances of the case. The kind of criteria and methods that ought to reflect the legal title to claims less than 200 miles apart was not difficult to predict. In the Court's view, within this distance State entitlement was based 'on the principle of adjacency from the coast, or, to use the traditional term, on the principle of adjacency as measured by distance'.[144] Consequently, the Court considered that 'the tracing of a median line between [the opposite coasts of the parties], by way of a provisional step in a process to be continued by other operations, is the achievement of an equitable result'.[145]

The Court then went on to check the equity of this provisional line in the light of the relevant circumstances.[146] With regard to the basepoints from which the equidistance line would be considered, the Court found it appropriate to disregard the straight baseline system of Malta, which connected, *inter alia*, the island of Malta to the uninhabited Rock of Filfla. The decision to ignore Filfla, the Court explained, was made within the context 'of eliminating the disproportionate effect of certain islets, rocks and minor coastal projections'.[147]

The Court also found that there was a considerable disparity in the lengths of the relevant coasts of the parties, to a degree constituting 'a relevant circumstance which should be reflected in the drawing of the delimitation line'.[148] This disparity between their lengths, the Court considered, was a pertinent circumstance which should not be confused with the use of the same factor in assessing ratios of proportionality. This is a significant development from the case-law, which hitherto examined the

[143] Ibid.
[144] Para. 61.
[145] Para. 62.
[146] The Court's pronouncements on certain considerations viewed by the parties as relevant circumstances, namely the relevant economic position of the States, the size of their land mass, and the issue of security, were considered, *supra*, Ch. 4 s.3.
[147] Para. 64.
[148] Para. 68.

role of this factor within the context of proportionality.[149] The Court pointed out that

It is . . . one thing to employ proportionality calculations to check a result; it is another thing to take note, in the course of the delimitation process, of the existence of a very marked difference in coastal lengths, and to attribute the appropriate significance to that coastal relationship, without seeking to define it in quantitative terms which are only suited to the *ex post* assessment of relationships of coast to area. The two operations are neither mutually exclusive, nor so closely identified with each other that the one would necessarily render the other supererogatory.[150]

Within the context of the 'restricted' area of dispute, the Court took the figures of 24 miles' length and 192 miles' length, respectively representing the relevant coasts of Malta and Libya, as a basis 'to justify the adjustment of the median line so as to attribute a larger shelf area to Libya'.[151] As to how far northward the median line should be 'transposed', the Court explained that this should be decided within the wider geographical context in which the delimitation would have to be effected. In this respect the Court regarded the geographical location of the Maltese islands, appearing as a minor feature of the northern seaboard of the central Mediterranean, as a relevant circumstance, especially pertinent to the issue of adjustment.[152] The Court obviously had in mind the relatively limited dimension of the Maltese islands and their proximity to the much larger Italian island of Sicily. The Court saw fit to establish a second notional median line representing 'a delimitation between a portion of the southern littoral and a portion of the northern littoral of the Central Mediterranean'.[153] Thus, the Judgment considered 'a notional median line between Libya and Sicily' to represent what might be the extreme limit of such an adjustment.[154] In this respect the Court could indeed be criticized for self-contradiction; for the same factor which justified in its view the adjustment of the line northward, that is, the disparity between the parties' coastlines, was also manifested between the short coast of Sicily if measured against that of Libya.

However, as mentioned earlier, the Court took the 15° 10′ E. meridian, corresponding as it did to the Italian claim, as the eastern limit of the relevant area. The median line between Malta and Libya intersected the 15° 10′ E. meridian at about 34° 12′ N., and the notional median line between Libya and Sicily intersected it at about 34° 36′ N. The Court suggested that a transposition 'northwards through 24′ of latitude of the

[149] In *Tunisia–Libya*, para. 131, and in *Gulf of Maine*, ICJ Reports (1984), paras. 218 and 221–2, the comparable lengths of the coasts of the parties were used for applying the factor of proportionality. See *supra*, Ch. 4 s.3(*b*)(iii)(6).

[150] Para. 66. [151] Para. 68. [152] Ibid. [153] Para. 69. [154] Para. 72.

Libya–Malta median line would therefore be the extreme limit of such northward adjustment'.[155]

Weighing up all the relevant circumstances of the case the Court concluded that a shift northward 'of around three-quarters of the distance . . . achieves an equitable result in all the circumstances'.[156] Thus a northward shift of 18′ E. meridian as a reference line was suggested, resulting in a delimitation intersecting the reference line at 34° 30′ N. 'By a mere coincidence', Judge Oda observes, '—a fact which the Judgment does not mention this—this point is the same as the southwestern corner of the Italian claim in the Ionian Sea.'[157]

With regard to proportionality, in the form explained earlier by the Court, it was inevitable that the Court found 'such a test [of proportionality] would be meaningless in the absence of a precise definition of the "relevant coasts" and the "relevant area"'.[158] The Court stated two 'practical difficulties' rendering the application of proportionality inappropriate. First, 'the geographical context is such that the identification of the relevant coasts and the relevant areas is so much at large that virtually any variant could be chosen, leading to widely different results' and, second, 'the area to which the Judgment will in fact apply is limited by reasons of the existence of claims of third States'. Any application to the proportionality test within this limited area would be unrealistic and would produce ratios that could perhaps be 'overthrown' by possible future delimitations with third States.[159]

Yet, apparently becoming aware of the difficulty inherent in its deliberately restricted jurisdiction, the Court sought a 'cosmetic' application of the test of proportionality. Turning to the areas of shelf lying on each side of the line, the Court said, it was possible 'to make a broad assessment of the result' and, without giving the result of this assessment even in broad terms, it was satisfied that there was 'certainly no evident disproportion in the area of shelf attributed to each of the Parties respectively such that it could be said that the requirements of the test of proportionality as an aspect of the equity were not satisfied'.[160]

(g) *The Judgment's impact on delimitation problems in the area*

Many of the majority judgment's pronouncements on controversial issues, such as the elevation of the distance principle as a basis of State entitlement into customary law, as well as the hypothesis of a delimitation between Libya and Sicily, have been analysed and criticized in the separate

[155] Ibid. [156] Para. 73. [157] His Dissent to the Judgment, 130.
[158] Para. 67. [159] Para. 74. [160] Para. 75.

and dissenting opinions attached to the case.[161] What is pertinent to this section is the discussion of the complexities ensuing from the Court's self-imposed restriction of the area of dispute between the parties. Complexities have already been reported between the parties to the case as well as other neighbouring States. Before embarking on these specific issues, it might be appropriate to speculate upon the Judgment's implications on potential bilateral delimitations in the region in general.

(i) The general picture of 'third States'' interests in the Mediterranean

The approach of the Court will certainly bring confusion to the prospect of bilateral delimitations which none of the States of the region, especially in the light of the Court's approach in *Tunisia–Libya*,[162] could have expected. One needs only, by way of example, to ask how Turkey and Syria can press ahead with their bilateral delimitation if Cyprus promulgates extensive claims to the north and east.[163] How can Libya and Greece define their 'relevant area' without the prior agreement of Egypt, given that the Greek island of Crete lies opposite the Libyan–Egyptian frontier?[164] In the western Mediterranean, the complexities associated with Corsica, Sardinia, and the Balearic Islands, which were discussed earlier, will be intensified further if Algeria denies full effect to the Balearic Islands, a possibility that can be inferred from Algeria's position at UNCLOS III.[165] In the Alboran basin, how might a delimitation between Spain, Morocco, and Gibraltar (UK) proceed on a bilateral basis given the location of the Spanish islands of Alboran (situated almost half-way between the coasts of Spain and Morocco), Chafarinas (lying off the coast of Morocco),[166] and the intrusion of the British enclave into the Strait of Gibraltar?

It might be argued that such situations are best resolved on the basis of a tripartite or even quadripartite settlement. It takes no stretch of imagi-

[161] See e.g. Sep. Op. of Judge Sette-Camara, 65–6, 68–9; Joint Sep. Op., 82; the Diss. Opinion of Judge Schwebel, 181–2.

[162] See *supra*, text at nn. 97–102.

[163] Art. 2 of Law No. 8 of 15 Apr. 1974 states that the term continental shelf includes the 'sea-bed and subsoil of the Republic [of Cyprus], but beyond the territorial waters, where the depth of the superjacent waters admits of the exploitation of the natural resources of the said areas; provided that, in relation to any States whose coasts lie opposite those of the Republic, the outer boundary of the continental shelf shall, unless otherwise agreed between the Republic and that State, in no case extend beyond the median line'. Cyprus is about 36 and 54 n.m. from the nearest points on the coasts of Turkey and Syria, respectively.

[164] Crete lies about 184 n.m. from the Libyan–Egyptian coast. For the legal position of Greece on shelf delimitation, see *infra*, Ch. 10 s.3(*b*).

[165] See *supra*, n. 24.

[166] Alboran Island is 1 sq.km. in area with a total coastline of under 1 km. For the Spanish–Moroccan dispute over, *inter alia*, the Chafarinas, see *supra*, Ch. 8 s.4(*b*)(iii).

nation, however, to contemplate the difficulties inherent in such a solution. Leaving aside the possibilities of a direct, negotiated settlement, the prospect of which, as State practice seems to indicate, is still limited and rudimentary,[167] the submission of such disputes to a third-party settlement incurs several problems. It requires a special agreement binding on all the States concerned and defining the 'area of dispute' between the parties. It would be extremely difficult for the parties to reach a definition that does not prejudice the position of either one or the other. Even if such a problem is overcome, there still remain the difficulties not only of handling three or four sets of pleadings rather than two, but also of devising appropriate criteria and methods, as well as examining a wider range of 'relevant circumstances'. For instance, the decision of the Court on the 'equity' of a certain criterion, method, and 'relevant circumstances' between two parties might be affected by adverse considerations relating to the interest of the remaining parties. In the *North Sea* cases the Court was fortunate to find that the interests of The Netherlands and Denmark *vis-à-vis* West Germany were so similar that the cases were treated in effect as a two-party dispute.[168]

The heavy burden on the Court as well as on the parties themselves is not only related to these *de facto* problems. It is indeed a problem of finding applicable principles of law and devising relevant techniques in a system which has so far developed within the context of two adversaries, not three or four. A tripartite or quadripartite boundary settlement by adjudication is still unprecedented. The language as well as the *travaux préparatoires* of the 1958 and 1982 Conventions' delimitation provisions generally presuppose bilateral disputes.[169] State practice is still, as just observed, rudimentary and only partially relevant.[170]

Having, therefore, to await future developments, one would wish to express a concern arising by analogy with decisions in the 'single boundary' cases. The ICJ or an arbitration tribunal may be tempted to apply, *mutatis mutandis*, the 'neutral criterion' approach, whereby all criteria

[167] State practice on multilateral boundaries appears limited in two respects. First, so far, there seems to be no one example of a direct negotiated agreement providing for the delimitation of the *entire* boundary line between more than two States. The practice shows that States prefer to negotiate their boundaries on a bilateral basis and then collectively agree on establishing the *point of convergence* of their boundary lines. Such limited purpose agreements adumbrate very few, if any, rules or techniques that might be relevant to the problem under consideration. Second, the number of these latter types of agreements is still limited. Out of perhaps over 100 potential trijunction or quadrijunction settlements only 3 are reported to have been agreed upon, namely Indonesia–Malaysia–Thailand, 1971; India–Malaysia–Sri Lanka, 1976; India–Indonesia–Thailand, 1978. See *LIS*, 'Maritime Boundaries of the World' (20 July 1988), table 3, 12 ff.

[168] See *supra*, Ch. 3 s.3(*a*).

[169] See *supra*, Ch. 2 s.3; Ch 3 s.1; Ch. 4 s.1.

[170] See *supra*, n. 167.

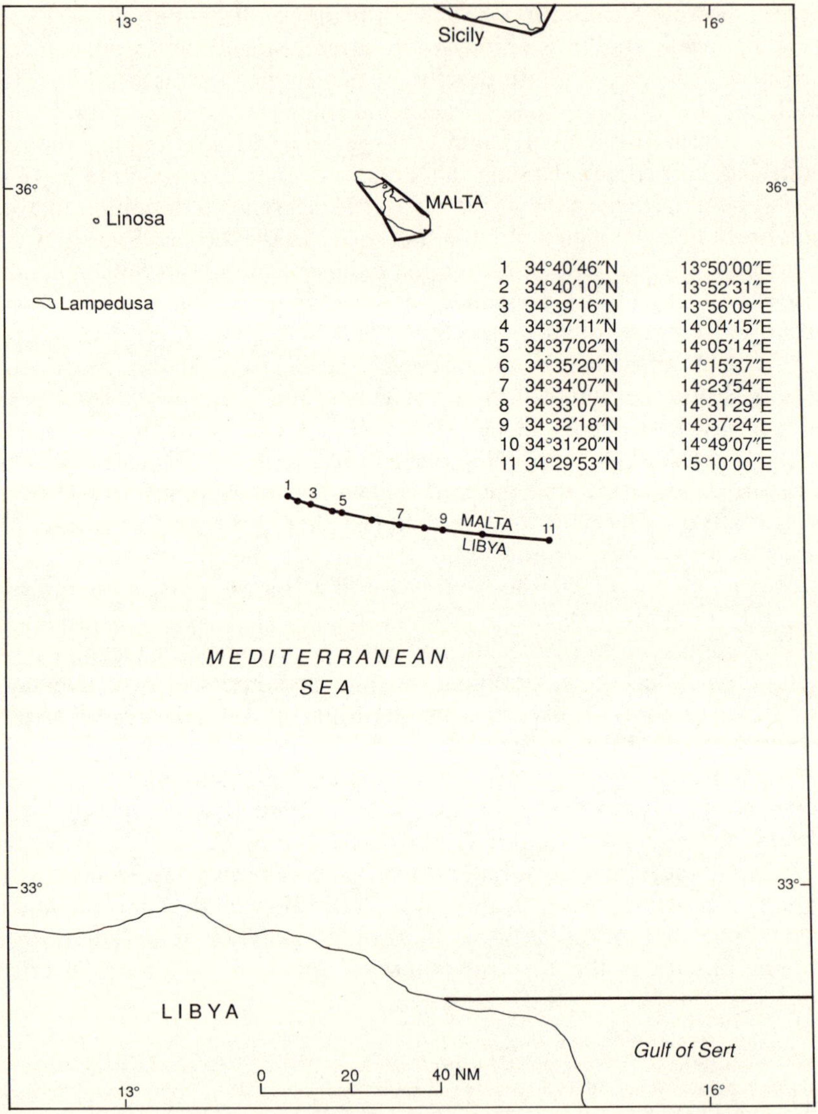

MAP 29 Geographical co-ordinates of the Libya–Malta continental shelf boundary

and relevant circumstances that are only related to some parties to the disputes but not to the other(s) would be excluded. This is of course an easy way of 'escaping' the difficulties inherent in these kinds of disputes. The adverse consequences, however, on the development of the law on delimitation will be similar, it is submitted, to those already noted with regard to the problem of a single boundary delimitation.[171]

(ii) Complexities between the parties to the case: the continuity of the dispute

The sole aim of submitting the dispute to the Court was to enable the parties to reach an agreement which would solve their boundary dispute once and for all. As a consequence of the Court's refusal event to point out the principles of law applicable between the parties in the *whole* area of dispute, the parties found themselves substantially in the same position that they were in before the submission of the dispute to the Court. More precisely, Italy's claims in the east, beyond the 15° 10′ line of longitude, included large parts of the Medina Bank, the area principally in dispute between Malta and Libya. The Court did not seem to be disturbed by this awkward result. It merely repeated its remark in *Application by Italy to Intervene* that the parties should bear the consequences of their negative view on the Italian application to intervene.[172]

On the one hand, neither Malta nor Libya seems to have been deterred by the probability of the Court's Judgment being restricted in scope as a consequence of the Italian claims. The prospect of such a restriction did not persuade these countries to abandon their opposition to Italy's application to intervene.[173]

The discussion of the *Intervention* case is outside the scope of this study, but it is pertinent to recall that it was the Court, not the parties, that refused to grant Italy permission to intervene.[174]

However, as a result of this situation the parties were unable in their negotiations following the Judgment to reach agreement as to which States were entitled to delimit the shelf boundary east and west of meridians 15° 10′ and 13° 50′, respectively. It was reported that Malta invited

[171] See *supra*, Ch. 5 s.3(*d*).

[172] See *supra*, n. 110.

[173] Para. 23.

[174] Contrary to the position of Malta in its *Application to Intervene* in *Tunisia–Libya*, in *Application by Italy to Intervene*, the Court viewed the Italian application as practically requesting it to recognize Italy's alleged rights, and hence, for the purpose of being able to do so, to make a finding, at least in part, on disputes between Italy and one or both of the parties (see paras. 29–34). Reviewing this objective against Art. 62 of the Statute of the Court (establishing the basis of jurisdiction by way of intervention), the Court concluded that the scope of that Article would not permit the acceptance of such intervention, 'since the only cases of intervention afforded by that Article would be those in which the intervener was only seeking the preservation of its rights, without attempting to have them recognized, the latter objective appertaining rather to a direct action' (para. 37). Having reached this conclusion, the Court affirmed that it was for 'the Court to decide upon a request for permission to intervene, and the opposition of the Parties to a case is, though very important, no more than one element to be taken into account by the Court' (para. 46).

Libya to extend the boundary line both eastward and westward of the limits placed by the Court. The position of Libya was that the parties should confine themselves to the Judgment. It was also reported that Libya reiterated its position that in the area east of longitude 15° 10′ the dispute remaining was exclusively between Libya and Italy. Malta for its part contended that the area relevant to delimitation with Libya should be extended as far east as a point corresponding with Benghazi, that is, approximately longitude 20° E.,[175] a proposition that some dissenting opinions appear to have supported.[176] It was therefore natural that the parties reserved their differences and were only able to concur in their Agreement of 1986 to draw a boundary line which followed, in effect,[177] the line suggested in the Court's Judgment.

(iii) Complexities involving a third State

The bilateral dispute between Libya and Malta has now been further complicated by a third 'dispute' concerning Italy *vis-à-vis* Malta and Libya. It is exactly the situation that Libya sought to avoid by introducing the 'pecked-line' solution.[178] The two disputes are considered below.

1. *The Italy–Malta dispute*

As a party to the 1958 Shelf Convention Italy has asserted in its continental shelf law of 1967 that, in the absence of agreement regarding the boundary with opposite States, concessions extending beyond the median line may not be granted.[179] However, Italy's unilateral claims in the central Mediterranean were never made known before its intervention in the *Libya–*

[175] A written interview with Prof. El-Murtadi Suleiman, Agent of Libya in the *Libya–Malta* case and delegate in the Libya–Malta Boundary Agreement of 1986 (see *infra*, n. 177).

[176] See Judges Oda's and Schwebel's dissenting opinions, respectively, 171, 178.

[177] Agreement between the Republic of Malta and the Socialist People's Libyan Arab Jamahiriya Implementing Article III of the Special Agreement and the Judgment of the International Court of Justice, signed on 10 Nov. 1986. (Text of the Agreement is provided by courtesy of the People's Bureau of Foreign Liaison, Tripoli.) It should be noted that, whilst the negotiated boundary between the parties is effected within the Court's proposed meridians of 13° 50′ E. and 15° 10′ E., the northern extension of the boundary exceeds in some points not only the Court's 'equitable' shifting of latitude 34° 30′ N. but also the 'maximum' northward shift prescribed by the Court at latitude 34° 36′ N. The difference was due to the two parties' proper application of the median line (adjusted 18′ northward), every point of which is equidistant from two points on each coast. The Court's initial median line was apparently drawn only from two points on the Maltese and Libyan coasts, namely Delimara and the Cape of Homs. For the geographical co-ordinates of the negotiated line see Map 29.

[178] See *supra*, text at n. 111.

[179] Art. 1 of Law No. 613, 21 July 1967 on Research and Development of Hydrocarbons in the Territorial Sea and the Continental Shelf. Text in *WESDLS* (1983) (Binder II) (Italy), 5.

Malta case. In that case Italy specified, by geographical co-ordinates, its claims in areas mainly east of Malta but also, to a smaller extent, west of the island of Malta, which extended to a region lying between opposite coasts of Libya and Malta. With regard to its boundary with Malta east and south of the Maltese islands, Italy denies any Maltese extension of continental shelf east of the meridian 15° 10′ E. and south of the parallel 34° 30′ N.

According to the unilateral claim of Malta of 1981,[180] the boundary line with Italy in the north and in the north-east is a median line. The initial segment of this line (exactly between the south-western coast of Sicily and Malta) corresponds with the provisional line negotiated between the two States in 1965/70.[181] Eastwards, the line extends in a south-easterly direction (equidistant from the south-eastern coast of Sicily and the southern coast of the Italian peninsula and the Maltese islands) and reaches its south-eastern extremity at a point situated about 52 n.m. south-south-west of the Italian–Greek boundary terminal point of 1977. It is this eastward extension of the Maltese claim which overlaps with the Italian unilateral claim described earlier (see Map 24).

Malta's rationale for the extension of this line is obvious. The terminal point of this line is equidistant from relevant points situated on the coasts of Malta, Italy, and Libya, respectively. Italy argues, on the other hand, that in this area Malta is not entitled to a full extension of the median line so far eastwards as to approach the provisional terminal point of the Italian–Greek boundary line. Italy takes the view that the prolongation of the Maltese median line eastwards should end at a point equidistant from the south-eastern extremity of the coast of Sicily and Malta and then 'turn clockwise [in a southerly direction] leaving Maltese areas to its west and Italian areas to the east, for both geographical and [morphological] reasons'.[182]

[180] See *supra*, n. 14.

[181] See *supra*, n. 17.

[182] *Application by Italy to Intervene*, argument of Counsel for Italy, *Oral Pleading*, CR 84/1, p. 37. This view of Italy was challenged by Counsel for Malta in that case as being contrary to Italy's prior adherence to the median line solution with Malta. In June 1975 the two sides held their only meeting that dealt with the problem of delimitation. According to the Maltese notes of that meeting, 'the Italian representative stressed that [Italy was] willing only to settle the boundaries of the continental shelf [with] Malta comprehensively and on the basis of equidistances'. The Maltese note also recorded that Italy expressed the view that it 'had no objection to signing a protocol based on a median line demarcation'. The note then recorded an extended discussion and sharp disagreement about the effect to be given to the Pelagian Islands (the Notes of the meeting, recorded by the Maltese Government, are reproduced in annexe 14 to Malta's Observations on the Italian Application for Permission to Intervene and quoted by Counsel for Malta, Mr Lauterpacht, ibid. *Oral Pleadings*, CR 84/6. pp. 23–4). In the same oral argument, however, Counsel for Italy disputed these submissions and stated that, on the contrary, 'very serious difficulties arose on that occasion with regard to the delimitation of the continental shelf between the

Geographically, this deflection of the equidistant line southwards is justified, according to Italy, by the disparity in the lengths of the relevant coasts of the two States as well as by the disparity in the size of their land masses. Geomorphologically, Italy seems to take the view that the Maltese claim should not stretch so far eastwards as to exceed the region's natural prolongation which belongs to the 'Ragusa–Malta' Plateau, 'which constitutes a natural prolongation, in a south-easterly direction, of the Italian landmass'.[183]

In the light of the 'general rules' decided by adjudication, only the first factor adverted to by Italy seems to be legally relevant to the dispute. In the case most relevant to the area, *Libya–Malta*, the Court stated in the *dispositif* that 'the disparity in the lengths of the relevant coasts of the Parties' was a factor relevant for achieving an equitable delimitation.[184] In respect of the second 'factor', Italy's position runs counter to the firm assertion of the cases that the 'size of the landmass' is irrelevant for boundary delimitation purposes.[185]

With regard to the third 'factor' of geomorphology, it is well known that the Ragusa–Malta Plateau constitutes the geomorphological link between the 'Ragusa shelf on which Malta is situated and the Ragusa area of Sicily'.[186] It seems, therefore, unwarranted for Italy to argue for a correction of the median line on the basis of submarine morphology which constitutes as much the natural prolongation of the Maltese islands north and eastwards as that of Sicily southwards.

A more reasonable interpretation of Italy's argument on geomorphology, however, may suggest that Malta's claim eastwards should not exceed the eastern extremity of the Ragusa–Malta Plateau, which is separated from the Ionian Sea and Ionian Abyssal Plain by a remarkably steep, north–south escarpment known as the Sicily–Malta Escarpment.[187] However,

[Pelagian Islands] on one side, and Malta on the other. This was an obstacle not merely to agreement on that delimitation but also to the commencement of examination of any other problem of delimitation between the two countries' (Counsel for Italy, Mr Sperduti, CR 84/2 (trans.), 24). Italy further argued that, according to its own *aide-mémoire* of that meeting, Italy proposed the median line only in the area between Malta and the Pelagian Islands, but not in the area north and east of Malta (CR 84/7 (trans.), 14).

[183] CR 84/1, p. 37.

[184] *Libya–Malta* case, para. 79.

[185] See *supra*, Ch. 4 s.3(*b*)(ii)(2).

[186] See e.g. I. Finetti, 'Structure, Stratigraphy and Evolution of Central Mediterranean', 34 *Bolletino di geofisica teorica ed applicata* (1982), 247; 'The Geology and Geophysics of the African Central Continental Margin South of Sicily', a summary paper in *Memorial* submitted by the Libyan Arab Jamahiriya (1983), vol. i, *Technical Annex*, pt. ii.

[187] In the *Libya–Malta* case, Libya dealt with the geomorphological extension of this feature. The Escarpment was submitted to form the north-eastern boundary of the 'Rift Zone' which separated in the physical sense the natural prolongation of the Libyan land mass northwards from the natural prolongation of Malta's land mass southwards. The Escarpment was described 'as running from approximately the 37° 30′ N. parallel just off

this view (also advanced by Libya in *Libya–Malta*) seems to be equally faced with the difficulties mentioned above in relation to Italy's argument over the entitlement of the Pelagian Islands.[188] More specifically, although the Court in *Libya–Malta* declined to express any view as to the legal importance of the 'Escarpment-Fault Zone'[189] which was described as bounded in the north by the Sicily–Malta Escarpment,[190] the reasons it stated for rejecting Libya's argument with regard to the 'Rift Zone'[191] seem to cast an *a fortiori* doubt as to the validity in law of the Sicily–Malta Escarpment as a fundamental discontinuity terminating the eastward extension of the Maltese shelf.

However, although it is now almost certain that, as a result of the Libya–Malta delimitation Agreement, Malta will no longer have valid claims *vis-à-vis* Italy and Libya as far southwards as parallel 34° 29′ 53″,[192] it is also obvious that any final demarcation of the maritime boundary between Italy and Libya is contingent upon the settlement of this pending Italo-Maltese dispute. With this factor in mind, other possible 'relevant factors' to a prospective Libyan–Italian maritime boundary are considered below.

2. *The prospect of an Italian–Libyan shelf boundary*

In its unilateral claims made before the Court in 1984,[193] Italy effectively postulated a boundary line, *inter alia*, with Libya. It considered itself to have rights over an area delimited on the west by the meridian 15° 10′ E., to the south by the parallel 34° 30′ N., to the east by the delimitation line agreed between Italy and Greece and its projection southward, and to the north by the Italian coasts of Calabria and Apulia. The latitudinal length of this portion of the Italian coast (roughly between 15° 10′ E. and 18° 55′ E.) is about 191 n.m. If the 1977 Italy–Greece delimitation is projected south it meets the Libyan coast between approximately the 18° 45′ E. and 19° E. meridians. The relevant Libyan coastline (measured following its general direction) is approximately 165 n.m. The Italians' reference to

the east-facing coast of Sicily south-eastward to the Heron Valley'. In its southern extremity it is linked up with a second escarpment—the Medina Escarpment—which extends south-west along the eastern edge of the Medina Bank down to the Melita Valley, approximately the 33° 30′ N. parallel. The Sicily–Malta Escarpment is said to comprise one of the steepest, most pronounced features in the entire Mediterranean—statistically, its dimensions are thought to be as follows: maximum vertical drop 3,200 metres; length approximately 120–130 n.m.; inclination of slope 1:4–1:10. To the east of this Escarpment lies the Ionian Sea with its Abyssal Plain. See Libyan *Memorial* (1983), i. 33–4.

188 See *supra*, s.2(*c*)(i).

189 Para. 34.

190 See Libyan *Memorial* (1983), 33.

191 *Supra*, text at nn. 20–1.

192 This corresponds with the most southerly point (Point 11) on the line agreed upon by Libya and Malta. See *supra*, n. 177, and Map 29.

193 *Supra*, n. 121.

latitude 34° 30′ N. clearly indicates the utilization of the equidistance line. But, in view of the general way in which this was determined, the line does not seem to reflect precisely any relevant points on the coastlines.

The Libyan view on this unilateral delimitation by Italy is not clear. Whilst the general determination by Italy of 'the relevant area' seems to accord with the view of Libya (at least in so far as denying Malta any right *vis-à-vis* both Libya and Italy east of longitude 15° 10′),[194] the precise determination of the delimitation line and the relevant coasts might lead to disagreement between the two States. In the oral pleading in *Application by Italy to Intervene*, Counsel for Libya questioned the soundness of the Italian claims, 'since, so far as Libya is concerned, they have never before been formulated, and indeed have not been clearly formulated even up to [the time of the proceedings]'.[195]

However, within this context it is possible to define certain factors which may constitute points of disagreement between the two States. The first factor will perhaps be related to the basepoints from which the dividing line will have to be effected. The extension southward of the Italian–Greek boundary line (which is considered by Italy to constitute the eastern limit of its delimitation area with Libya) intersects the Libyan coast at approximately longitude 18° 55′, that is, almost at the innermost part of the coast of the Gulf of Sert. In other words, a considerable part of the Gulf area will be relevant to the boundary with Italy. Identifying Italy's 'legal interest' in *Application by Italy to Intervene*, Counsel for Italy submitted that a point of

> difference [with Libya] derives from the position taken by Libya with regard to the starting line of any northern delimitation of its submarine areas. I refer to the well-known Libyan claim concerning the status of a considerable part of the Gulf of Sert. Unable as it is to accept that claim, Italy has a precise legal interest in the sense that a decision might be taken in favour of the Libyan contention before Italy had a chance to defend its position.[196]

However, in the light of the approach of the Decisions regarding States' claimed baselines,[197] Libya might be advised to argue its case on the basis of the principle of non-encroachment, which involves, as stated earlier, the factor of security,[198] rather than on its 'historic rights' *per se*. The 'cutting-off' effect of the Gulf of Sert area is substantial. If the Gulf's closing line is entirely ignored for delimitation purposes, an equidistant line with Italy will be shifted as far southward as 103 n.m. However, the

[194] See *Counter-Memorial* submitted by Libyan Arab Jamahiriya (1983), vol. i, paras. 2.26–2.27. See also the *Libya–Malta* case, para. 22.

[195] Counsel for Libya, Sir Francis Vallat, CR 84/4, p. 62.

[196] *Oral Pleadings*, CR 84/1, pp. 42–3.

[197] See *supra*, Ch. 1 s.3(*b*)(ii).

[198] See *supra*, Ch. 4 s.3(*b*)(iii)(3); Ch. 7 s.4(*c*)(iv).

success of the Libyan argument regarding this point seems to depend largely on the 'method' of settling the boundary. While the Decisions are unanimous in principle on the relevance of the principle of non-encroachment and the related factor of security,[199] the limited scope they give to these elements[200] makes it very doubtful that a court or arbitration tribunal would accept full exclusion of the Gulf of Sert region from the area of delimitation. Libya, however, might be in a better position in a direct, negotiated agreement. A 'trade-off' can be reached in the light of the enclosure by Italy of the Gulf of Taranto. The Gulf, though of much smaller size than that of Sert, is located opposite the Gulf of Sert[201] and the parties could, therefore, reach some exchangeable concessions.

Another factor of possible contention is the determination of the 'relevant coast' of Italy. The final segment of the agreed shelf boundary between Italy and Greece is effectively a line equidistant between the western coast of Greece and its islands on the one hand, and the south-western coast of the Italian peninsula and the eastern coast of Sicily, on the other. As seen above, this same section of the Italian peninsula is also claimed by Italy for a second shelf zone, on the basis of comparative lengths, as against Libya. Although the exact extension latitudinally of the prospective boundary between Italy and Libya depends largely on the agreement between the two States and Malta on a tripoint, Libya might challenge this 'dual' use by Italy of its coastline. The ultimate objective of Libya in this regard would be to limit the relevant opposite coast of Italy to the 'sole' of the Italian 'boot', that is, between Pt. della Allice and Capo Spartivento. This would result in a considerable disparity between the lengths of the two coasts in favour of Libya.

With regard to the geomorphological characteristics of the area, there seems to exist (in the light of the scientific evidence available and the Court's decision in *Libya–Malta*[202]) no physical feature which might be said to constitute an east–west 'natural boundary' separating the shelves of the two States. The characteristics of the 'relevant area' seem, therefore, to warrant the application of a median line, modified in order to reflect some or all of the following 'relevant' circumstances:

1. the position of the trijunction point of the boundary lines of Italy with both Libya and Malta;
2. the relationship between the lengths of the relevant coasts of the two States; and
3. the configuration of the two States' coastlines and the encroachment

[199] See *supra*, Ch. 4, s.3(*b*)(iii)(3).
[200] See ibid.
[201] See *supra*, Ch. 7 n. 236.
[202] See *supra*, nn. 186–91 and accompanying text.

and security effects relating thereto, as clearly displayed by the two States' claims to 'historic rights' in the area.

One fact which is bound to determine the magnitude of the parties' arguments in respect of these circumstances is the importance of the area for resources. So far, there is no evidence suggesting that the area could be of great potential for hydrocarbon or fishing resources. If future development proves otherwise, the legal arguments of the parties will perhaps witness an attempt by either side to utilize to the utmost extent the circumstances suggested above.

3. FUTURE DEVELOPMENT: THE IMPACT OF PROSPECTIVE 200-MILE CLAIMS

(*a*) *A provisional abstention on 200-mile claims*

Claims to 200-mile exclusive economic or fishery zones are almost non-existent in the Mediterranean. To date, only four States—France, Spain, Morocco, and Egypt, which also front on other, non-Mediterranean coasts—have proclaimed a 200-mile EEZ; in the case of France and Spain, their EEZ only pertains to the Atlantic Ocean, although they did not exclude the possibility of instituting such zones off their Mediterranean coasts.[203] As for Morocco, although the *Dahir* of 1980 did not distinguish between the Atlantic and Mediterranean coasts, it does not appear to have implemented the 200-mile zone off Morocco's Mediterranean coast. A map published in July 1981 by the Department of Cartography did not indicate the existence of a Moroccan EEZ on the Mediterranean side of the Strait of Gibralter.[204]

To date, Egypt is the only Mediterranean State which exercises EEZ jurisdiction in the Mediterranean. When depositing its instrument of

[203] The French Law of 16 July 1976 created an 'economic zone' extending to 188 n.m. beyond the outer limit of the territorial sea. According to Art. 5 of that Law, the exact promulgation of the economic zone is left to 'decrees of Council of State'. So far, several decrees have implemented the said law over various continental and overseas coasts, including the Atlantic coast (the Law and executive Decrees are reprinted in *WEDLS* (Binder I) (France), 161–9). The policy of France in this regard seems to be to adopt a 'reciprocity' principle, which responds to the attitude of the neighbouring States in respect of unilateral claims (see F. Lacharriere, 'La Zone économique française de 200 miles', 22 *AFDI* (1976), 641, at 649–50). Art. 1 of the Spanish Law of 20 Feb. 1978 establishes an EEZ of 200 miles 'from the base lines used to measure the breadth of the territorial sea'. The same Law, however, limits the 200-mile EEZ to the Atlantic coast of Spain, but declares the Government's right to extend the EEZ to 'other coasts of Spain' (repr. in *WEDLS* (Binder IV) (Spain), 47).

[204] See argument of Mr Quéneudec, Counsel for Libya in *Libya–Malta* case, CR 84/32 (trans.), 8.

ratification of the 1982 Convention on 20 August 1983, Egypt declared that it

will exercise as from this day the rights attributed to it by the provisions of parts V and VI of [the 1982 Convention] in the exclusive economic zone situated beyond and adjacent to its territorial sea in the Mediterranean Sea and in the Red Sea . . . the outer limit of its exclusive economic zone [is established] in accordance with the rules, criteria and modalities laid down in the Convention.[205]

The absence of EEZ claims by other States bordering the Mediterranean is not indicative, however, of their opposition in principle to the EEZ concept.[206] Rather, it seems to represent a recognition that the Mediterranean does not at present lend itself to the kind of far-reaching claims to ocean space associated with the EEZ concept. Such a recognition is obviously related to a twofold, economico-geographic reason. This was voiced by several Mediterranean States at UNCLOS III. For instance, while expressing his State's support in principle of the EEZ concept, the delegate of Turkey added

Although Turkey [is] . . . surrounded on three sides by seas, it [stands] to receive no direct benefit from the establishment of economic zones since the narrow and semi-enclosed nature of those seas would prevent it from extending its jurisdiction to more than a fraction of the intended maximum breadth of the economic zone. Furthermore, as the Fishery Country Profiles prepared by the [FAO] demonstrated, those seas [are] not well endowed in living resources.[207]

Similar views were expressed by the delegates of Tunisia, Egypt, Spain, Lebanon, and Algeria.[208]

However, whilst the geographical reason, that is, the Mediterranean's limited space, is a permanent one, the economic reason might not have the same perpetuity. Changes in the economic-legal order associated with, and advancement of technology available to, the Mediterranean littorals may provide the incentive necessary to assert their 200-mile jurisdictions there.

In respect of living resources (the essential reason for the 200-mile claims), it is generally assumed that the Mediterranean is a fairly unproductive sea.[209] This general assumption, however, does not minimize the

[205] *LSB* (Mar. 1987), 4.

[206] Most of the Mediterranean States endorsed the EEZ concept at UNCLOS III. See e.g. statements of the delegates of Yugoslavia, Cyprus, Greece, Italy, Tunisia, Egypt, Albania, Turkey, Libya, Spain, Lebanon, and Algeria, in UNCLOS III, *Official Records*, Plenary Meetings (24th mtg.), 173–222, respectively.

[207] Statement of Mr Akyamac (Turkey), ibid. 213.

[208] See respectively ibid. (27th–28th mtgs.), 203, 204, 216, 221, 222.

[209] See A. Zei, 'Prospectives for Mediterranean Fisheries and Aquaculture', 3 *Ocean Management* (1976), 219, at 220; G. Blake, 'Mediterranean Non-energy Resources: Scope for Cooperation and Dangers of Conflict', in G. Luciani (ed.), *The Mediterranean Region* (Croom Helm, London, 1984), 41, at 47.

high value attributed to the Mediterranean fisheries, nor does it rule out the possibility of increasing fish catches.

The Mediterranean has been estimated to be able to support a fishery which exceeds in value that of much more productive areas such as the north-east Atlantic and south-east Pacific.[210] This remarkably high value is related to the fact that the States of the region are high consumers of fish, with a large proportion of the fish catch being utilized for direct human consumption. Mediterranean fisheries are also remarkable in the sense that an abnormally high proportion of the species is economically important; about 120 out of 500 species are fished commercially.[211]

In terms of fish stocks, while some demersal fish stocks along the northern coast are overfished,[212] there are some areas in the eastern and southern coasts of the Mediterranean which are under-exploited.[213] Furthermore, the comparision of quantities of pelagic species captured in the past suggests that catches in the eastern Mediterranean, extending from Greece to the Libya–Tunisia border, might be increased by two or three times, if the economic obstacles to their marketing could be overcome.[214]

One cannot, therefore, consider the Mediterranean as one entity as regards the productivity of its living resources. In fact, the recent statistical study of the GFCM splits the Mediterranean into several regions and subregions, each reflecting different species and yearly catches.[215] Whilst the western region reflects a year-to-year fluctuation in demersal fish catch, the eastern and southern regions have a yearly increase in pelagic catch. All in all, it has been forecast that the Mediterranean catches could be doubled in volume,[216] and this prospect is sufficiently attractive to make investment in the fishing industry worthwhile for several Mediterranean States. It is already reported that some Mediterranean States, particularly along the southern and eastern coasts, have ambitious plans for the development of their fishing industry,[217] which will add up to a considerable increase in fishing productivity.[218]

[210] UN, FAO, GFCM, *Report of the Twelfth Session: Prospectives for Fisheries, Development to 1985* (Rome, 1974), 38.

[211] See M. Beazley, *Atlas of the Oceans: Mediterranean: Living and Mineral Resources* (M. Beazley Publisher, London, 1977), 140.

[212] UN, FAO, GFCM, *Report of the Fifteenth Session* (Rome, 1980), 76.

[213] UN, FAO, GFCM, *Report of the Twelfth Session: Prospectives for Fisheries*, 7; UN, FAO, *Atlas of Living Resources of the Seas* (Rome, 1981), sheet G.1.

[214] UN, FAO, GFCM, *Report of the Twelfth Session: Prospectives for Fisheries*, 7; UN, FAO, *Atlas*, FAO, sheet G.2.

[215] UN, FAO, GFCM, *Regional Bulletin of Statistics* (Rome, 1976), 49–53, esp. annexe 1.

[216] UN, FAO, *Fish Resources in the Mediterranean and Black Seas: Consultation on the Protection of Living Resources and Fisheries from Pollution in the Mediterranean* (Rome, 1974), 1.

[217] UN, FAO, GFCM, *Report of the Twelfth Session: Plans and Politics for Fisheries Development* (Rome, 1974), 39.

[218] It was estimated that by 1985 the Mediterranean littorals' demand for fish could reach

This relatively promising picture of the Mediterranean fishery might prove more attractive in the light of the increasing adoption of 200-mile exclusive zones all over the world. Mediterranean States, especially major maritime States, are expected to increase their catch from the Mediterranean, as they are progressively evicted from these zones in important areas such as the North Sea and the coast of West Africa. For instance, although Italy is a member of the EEC, thus enjoying the non-discriminatory fishery policy of that community,[219] it still takes about 87 per cent of its catch from the Mediterranean.[220] Spain is also expected to increase its fishing activity in the Mediterranean in order to make up for its exclusion from important fishing grounds in the North Sea and the Atlantic Ocean.

The fact that Spain became a member of the EEC did not seem to reduce the importance of its catch from the Mediterranean. On the contrary, the EEC itself has substantially reduced Spanish activity in the existing EEC common EFZ in the North Sea and the Atlantic Ocean. By 1982, the Community was granting Spain a mere 106 licences to operate in waters where Spain claimed to have had some 600 boats fishing in the early 1970s.[221] Soon after the accession of Spain to the EEC in 1986, further measures to restrict Spanish fishermen were taken. For instance, the number of Spanish vessels able to operate in EEC waters has been limited to 300, only 150 of which are permitted to fish at any one time.

anything from 4,350,000 to 6,100,000 tons. A great proportion of this demand would have to come from the Mediterranean Sea itself. See UN, FAO, GFCM, *Report of the Twelfth Session: Prospectives for Fisheries*, 74.

[219] The general principle governing the EEC common fishery policy is the obligation of each member State to guarantee freedom of access to fishing in waters placed under its own jurisdiction. An important exception to this principle is to allow the coastal State to restrict fishing within areas up to 12 miles to vessels which traditionally fish in those waters and operate from ports in that geographical coastal area. For instance, according to the Act of Accession of Greece to the EEC, Italy is entitled to adopt measures restricting the freedom of others to fish in waters within the 6-mile limit, with an extension *vis-à-vis* Greece of the limit to 12 miles in specific areas mentioned in Art. 111 of the same Act. See Act Concerning the Conditions of Accession of the Hellenic Republic and the Adjustments to the Treaties, 10 *ECCP* (1979) (Cmnd. 7650), 46.

[220] Italy's interest in the Mediterranean fishery is well documented. Ever since the 1960s, Italy has been engaging in bilateral arrangements, especially with neighbouring Yugoslavia and Tunisia, aimed at facilitating access of its fishermen to waters adjacent to these States. The last arrangement was the creation with Yugoslavia of a 'fishery zone' in the Gulf of Trieste. The zone includes waters on both sides of the boundary line established in the Osimo Treaty of 1975. The Agreement was signed in Feb. 1983 and entered into force in June 1987. See *Gazzetta ufficiale della Repubblica Italiana*, 188 (13 Aug. 1987). Talks with Tunisia to agree on a similar arrangement have not yet proved successful. See *Quaderni del Centro di Ricerche Giuridiche per la Pesca e la Navigazione da Diporto* (Le Società Internazionali Misti di Pesca, 1987).

[221] See M. Wise, 'European, National and Regional Concepts of Fishing Limits in the European Community', in G. Blake (ed.), *Maritime Boundaries and Ocean Resources* (Croom Helm, London, 1987), 117, at 128.

These vessels can only operate in specific regions of the EEC's EFZ. Access to the large 'Irish Box' has been denied to Spanish fishermen until 1995.[222]

The likely outcome of this global/regional development is that many Mediterranean States will turn to the Mediterranean fishery in order to meet their increasing demand for fish, taking utmost advantage of the present abstention from exercising EEZ/EFZ rights. This would in turn compel many of those States to change this 'policy'. In fact (in addition to the Egyptian claim to a full 200-mile EEZ), rather conservative steps are beginning to be taken to assert fishery claims beyond 12 miles; Malta established a 24-mile fishery zone in 1978[223] and Syria followed by establishing a 35-mile territorial sea in 1981.[224] However, in order properly to evaluate the potential of the Mediterranean's EEZ economy, non-living resources should also be included.

The potential of the Mediterranean for natural resources varies from already developed or highly promising hydrocarbon areas to areas of unknown potential in very deep waters and to the much less promising basins. Already developed or promising areas are mainly restricted to relatively shallow waters in the Gulf of Valencia, in the Adriatic, the Tunisian Plateau, the Sicily–Malta Plateau, and in the plateau on which Corsica and Sardinia lie.[225] Current offshore production is restricted to oil and gas fields in these areas.[226] The impact of these resources on maritime boundaries, if any, is already achieved, since it is in these areas that present shelf boundaries, as seen earlier, have been resolved. It is still, however, relevant to see the potential of the relatively inaccessible deep waters of the Mediterranean for hydrocarbon and other mineral resources.

Many factors are responsible for the transformation of organic materials deposited millions of years ago into oil through long, slow, and complex changes.[227]

Two natural phenomena are regarded, however, as indicative of hydrocarbon potential. First, their occurrence in economic quantity depends on

[222] Act Concerning the Conditions of Accession of the Kingdom of Spain and the Portuguese Republic and the Adjustments to the Treaty, 15 *ECCP* (1985) (Cmnd. 9634), esp. Arts. 158–9, pp. 88–9.

[223] Territorial Waters and Contiguous Zone (Amendment) Act, 1978, *UNLS* (ST./LEG/SER.B/19) (1980), 63.

[224] Decree of 1981, *LSB* (1983), 61.

[225] See G. Luciani, 'The Mediterranean and the Energy Picture', in *The Mediterranean Region*, 1; F. Bastianelli, 'Boundary Delimitation in the Mediterranean Sea', 5 *MPR* (1983), 1.

[226] See Luciani, 'The Mediterranean and the Energy Picture', 6–7, for location and annual production of these fields.

[227] For a brief discussion, see R. Kent, 'The Sea's Non-living Resources, Present and Future,' in Greenich Forum IX (ed.), *Britain and the Sea* (Scottish Academic Press, Edinburgh, 1984), 100, at 109–11; Ying-jeou Ma, *Legal Problems of Seabed Boundary Delimitation in the East China Sea* (University of Maryland, School of Law, Baltimore, 1984), 14.

the availability of a rich source rock (usually a shale deposited under anoxic conditions) and the accumulation of many natural 'traps' which are sealed by impervious rocks (such as igneous rocks) as a result of crustal movement. The thicker the sedimentary rock, the greater the potential yield of oil. The most favourable habitat of hydrocarbons offshore is sedimentary basins where layers of sediment acquire a thickness of over 1 kilometre. Secondly, the passage of history adequate to have released mobile oil or gas from the original solid source is required. Past drilling data indicate that rocks of certain geological ages are of greater potential than those of other ages. More than half of the oil in the world, in fact, is found in rocks of the Tertiary period (65 million to 2 million years ago). In this period, rocks of the Neogene Epoch (25 million to 2 million years ago) account for a large proportion of world hydrocarbon production.

Considering these two phenomena alone, some deep basins of the Mediterranean seem to be very promising. In deep areas situated between the Balearic Islands, the French coast, and Sardinia, a huge area has been found of potential oil interest that may be valuable when exploration and production technology allow exploitation out to the 2,500–3,000-metre isobath.[228] In the Tertiary sedimentary sequences of the Adriatic and Ionian Seas mostly natural gas fields have been found, while in the Mesozoic calcareous reservoirs of the central Adriatic and the area surrounding Sicily, mostly oilfields have been discovered.[229] Exploration in the Italian seabed in waters deeper than 200 metres will become more and more important, especially in the lower Adriatic and the Ionian Sea where positive results have already been obtained.[230]

The continental shelf of Yugoslavia and Albania seems to have an interesting oil potential, both in the Tertiary clastic sequences and in the Mesozoic carbonates of the Dalmatian series. The continental shelf of Albania is a Mio-Pliocene (25 million–13 million year ago) sedimentary extension of the Albanian land mass. About two million tons of oil a year are produced on the Albanian land mass from similar terrain.[231]

[228] F. Bastianelli, 'Border Delimitation in the Mediterranean Sea', 17 *SI* (1982), 319, at 332.

[229] Italy has developed offshore operations technology more than any other State in the Mediterranean. Italian offshore technology has already been successful in manipulating difficult weather and depth conditions for installing the first deep-water platform in the Mediterranean Sea for the Libyan field of Bouri. The technology involves an offshore permanent loading system, comprising a biarticulated column which is anchored to the seabed by means of a piled foundation. At the top of the column is a triaxial joint which supports one end of the yoke, the other end of which is attached to a permanently moored storage tanker. The Bouri Field has a water depth of 169 metres. It is targeted to produce 150,000 barrels per day. The first stage of installation began in 1986. Information given by Agip NAME, Tripoli, Dec. 1986.

[230] Bastianelli, 'Border Delimitation', 323.

[231] Bastianelli, 'Boundary Delimitation', 2.

In terms of accessibility to these deep basins, the development of offshore drilling technology suggests that the obstacle of depth, the critical factor in the Mediterranean, will soon be removed. In the field of exploration, recent data show that the French have already developed a drillship which is capable of drilling in waters of more than 900 metres in depth.[232] The Italian firms Agip and Eniepsa are reported to employ drillships capable of drilling at similar great depths, although currently being used in shallower waters.[233] World-wide, more sophisticated semi-submersible drilling vessel technique is developed as operations move into deeper and deeper waters. Exploration wells can now be drilled in 6,000 feet of water, but the deepest-water fields in production are still in the range of 600–900 feet.[234] On the other hand, the technological limit for exploitation is more stringent than for exploration. It faces the two-sided problem of laying pipes and of setting up a central production unit. However, with regard to pipelines, technology employed in the Mediterranean has achieved a breakthrough with the construction of the Transmediterranean gas pipeline which has been laid down at depths of up to 600 metres and traverses a fairly rugged seabed area, between the Tunisian coast and Sicily, over a distance of about 155 km.[235] The same technology could be used for somewhat greater depths, up to 900 metres.

As for production units, present offshore technology has not only improved in fixing production units based on platforms above the water,[236] but also presented an alternative technique in the form of underwater production systems. Such a technique is already operating in the North Sea, but in waters of only 147 metres in depth. It is reported, however, that 'similar units can be used in waters several thousand feet deep'.[237]

Deep drilling studies in 1970 revealed extensive evaporite formations almost everywhere in the deep Mediterranean basins, which may reach up to 3,000 feet thick in places and consist mainly of rock salt, sulphur, and potash salts—basic raw materials of the chemical industry.[238] These deposits are currently only mined on Sicily and other Mediterranean islands where they reach the surface. Exploitation of such deposits is still beyond current techniques since they are confined to deep basins. Evaporite deposits are indeed considered a huge Mediterranean reserve, estimated to be at least 240,000 cubic miles.[239] Future development of

[232] *Total Information*, 78: 33.
[233] *Offshore* (5 June 1981).
[234] Kent, 'The Sea's Non-living Resources', 110.
[235] Luciani, 'The Mediterranean and the Energy Picture', 27–8.
[236] The tallest existing production platform was installed by Shell over the Cognac Field in Louisiana. It operates in more than 1,000 feet of water. Ibid. 3.
[237] *Petroleum Intelligence Weekly* (9 Apr. 1981), 12; *Business Week* (15 Mar. 1982), 36.
[238] Beazley, *Atlas of the Oceans*, 140.
[239] Ibid.

suitable techniques as well as possible diminishing of land-based supplies may render such deposits of commercial value and hence of political importance.

The Mediterranean basins do not seem to offer great potential for deep seabed mining; rates of sedimentation in this semi-enclosed sea are generally too high to allow the formation of manganese nodule fields.[240] However, hydrothermal solutions emanating from the current-swept crests of recently active submarine volcanoes have resulted in the formation of small manganese and iron-rich deposits that are both near to land and in relatively shallow and accessible areas. Moreover, preliminary data suggest that, due to river and other land-based discharges, coastal areas in the Mediterranean appear to have promising locations for placer mineral deposits (such as chromite, rutile, zircon, and monanzite) which probably continue to the adjacent continental shelf. The data also indicate that, given the geological and oceanographic conditions occurring in the Mediterranean, there is 'no doubt that more and probably larger placer deposits should exist in this region'.[241]

Whilst more detailed studies are of course needed for pinpointing the Mediterranean's natural resources potential, technological advancements will make exploitation of such resources feasible. The proximity of the Mediterranean basin to present and future markets for mineral resources in Europe, Asia, and Africa will provide the necessary economic incentive. This will certainly change the present picture of a few inter-State boundary agreements in the Mediterranean. The incentive provided by the Mediterranean's future natural resources, combined, more importantly, with its fishery resources will no doubt force the Mediterranean States to incorporate the regime of the 200-mile EEZ.

(*b*) *Prospective impact on the type of boundaries established*

Having concluded that the 200-mile jurisdiction in the Mediterranean will soon be claimed by the coastal States, it is important to examine the likely impact of this imminent development on the law applicable to maritime boundaries. There is little doubt that it is the multitude of the 'relevant factors' that such claims may produce which will almost certainly determine the 'politics' of future boundary-making in the Mediterranean.

[240] On the process of formation of deep-sea manganese nodules see D. Cronan, *Underwater Minerals* (Academic Press, London, 1980), esp. 62 ff.

[241] S. P. Varnavas, 'An Fe-Ti-Cr Placer Deposit in a Cyprus Beach Associated with the Troodos Ophiolite Complex: Implications for Offshore Mineral Exploitation', 5 *Marine Mining* (1986), 405, at 406.

(i) Single boundaries as 'equitable solutions'

Looking at the Mediterranean from a resource-management point of view, the Mediterranean States would of course be interested in delimiting their seabed as well as water-column jurisdictions by one single boundary. Most certainly this would be most favourable for direct, negotiated agreements.

As mentioned earlier, it is not difficult to envisage the practical advantages of a single maritime boundary, especially with regard to States' regulatory powers over resource-exploitation in a semi-enclosed sea like the Mediterranean.[242] Such considerations seem to be *a fortiori* applicable to States which already have shelf boundaries. Such boundaries would at least be looked upon as 'a special circumstance' in the negotiation of a new boundary. More important is the fact that none of the six negotiated agreements (including that between Libya and Malta putting into effect the Court's Judgment) is based on factors exclusively related to the seabed; and they can, therefore, be related to the water column as well. As seen earlier, geomorphological features, such as the Ionian Abyssal Plain and the abyssal depths between Sardinia and the Balearic island of Minorca, were totally ignored. The 'boundary lines' produced by the Court between Tunisia and Libya, and Libya and Malta, also followed suit. In effect, all the boundaries employed equidistant or geometrical lines, adjusted to reflect practical convenience, displacement of islands, and disparity between coastal lengths. The effect of these factors cannot be said to be different in new boundaries involving the water column. In fact, as between Libya and Malta, it was reported that the two States had already considered the application of the 1986 shelf boundary to their future, overlapping EEZs.[243]

Having reached this general conclusion, one should, however, specify certain areas in which the prospect of new single boundaries may not be automatically resolved by the existing shelf, or simply by a strict, equidistant/median line. The dispute between Spain and France over the delimitation of their EEZs in the Bay of Biscay is a good example. Although both States appear to be in favour of a single boundary line, their positions as to the principles applicable are diametrically opposed. France maintains that this can be done by extending the existing shelf boundary to a distance of 200 miles from points on the two States' respective coastlines. Spain, reiterating its initial disagreement in 1974 on a boundary line other than equidistance,[244] argues that the line of the

[242] *Supra*, Ch. 5 s.2(*g*).

[243] Information given by Mr El-Murtadi Suleiman and Mr Mizzi, respectively the Agents of Libya and Malta in the *Libya–Malta* case, 6 May (written interview) and 8 June (Valletta) 1988.

[244] *Supra*, Ch. 4 s.4(*b*)(i).

1974 Shelf Boundary Agreement should be revised. This diversity of opinion caused the breakdown of negotiations in February 1979. There is little doubt that Spain's interest in the fishing of the area lies behind its argument for negotiating a new boundary line.[245]

Inside the Mediterranean, the existing controversy between France and neighbouring Spain and Italy on the principles of law applicable to their respective shelf boundaries has already been discussed. The dispute might be aggravated further if the water column were added.

Another possible caveat may emerge in the form of a 'stress zone' in the Alboran and Balearic Seas along the median line between Europe and Africa. The possible application of the EEZ regime by Morocco, Algeria, and Tunisia will deprive, *inter alia*, Spain and Italy of the right to fish in waters near the coasts of these States. There is little doubt that this will raise the difficult question of giving effect to islands lying 'half-way' or 'on the wrong side', including Sardinia, Alboran, the Chafarinas, and the Balearic Islands. The prospect of delimitation in this area is already affected by the two sensitive political issues of sovereignty over the 'Spanish *presidios*' and Gibraltar.[246]

It is also possible that Tunisia will argue for a different water-column boundary with Libya. The recent shelf boundary suggested by the Court did not reflect the position of Tunisia, certainly that which related to its alleged long-standing fishery rights as far east as the ZV 45° line.[247] Although the two States were recently reported to have taken steps towards the implementation of the Court's Judgment,[248] Tunisia's argument for negotiating a new single boundary remains a possibility. Such an argument, however, would be difficult to sustain, given the Court's finding in *Tunisia–Libya* that the line of 26° represented the only *de facto* fishery line between the two States.[249]

Finally, a preference for a separate EEZ/EFZ boundary may also be expressed by States of established fishing activity such as Italy and Greece. If a case were taken to adjudication such States would be hesitant to submit a single boundary settlement which might fail to take account of established fishery interests. This is a strong possibility in the light of States' awareness of the 'neutral criterion' adopted by adjudication.[250]

[245] E. Orihuela, 'La delimitación de espacios maritimos', ii, doctoral thesis (University of Murcia, 1986), 453–554.

[246] *Supra*, Ch. 8 s.4(*b*)(iii); s.5(*c*)(ii).

[247] See *supra*, Ch. 8 s.3(*b*)(ii)(1).

[248] On 5 Apr. 1988 the two Governments agreed to press ahead with the implementation of the 1982 *Tunisia–Libya* Judgment and to establish a joint venture for the exploitation of the resources of their continental shelf. 'Joint Communiqué' issued (in Arabic) in Tripoli on 5 Apr. 1988.

[249] See *supra*, Ch. 8 s.4(*b*)(ii)(2) and (3).

[250] See *supra*, Ch. 5 s.3(*d*).

(ii) The relevance of future Mediterranean resources

The picture of the Mediterranean deep-sea hydrocarbons and other minerals reveals that their effect as relevant factors lies largely in their very long-term potential. As maritime boundaries tend, however, to be permanent, it is possible that negotiators of inter-State boundaries would bear in mind the fast development in technology and future depletion of land-based resources which could turn an offshore reserve into an actual source. Thus the potential effect of these future sources on boundary-making should not be excluded.

The discovery of promising hydrocarbon resources in the Tunisia–Malta Plateau, the Medina Bank, and the shallow continental shelf of the Adriatic may indicate the extension of deposits in these areas to much deeper zones in the east (the Ionian Abyssal Plain and adjacent area) and in the west (the Balearic Abyssal Plain). In such areas central zones of seabed would be involved and their hydrocarbon potential may thus become relevant to the negotiation of future boundaries.

Evaporite deposits appear to be located across central areas of seabed[251] where maritime boundaries will run, and thus could provide one reason among many why coastal States would seek to negotiate maximum allocations of seabed.

The economic potential of placer deposits is not yet clear. At a future date some well-known potential mining sites, such as the Nile Delta and the Gulf of Corinth, may prove commercially attractive.[252] In any event, their relevance to boundary delimitation appears to be minimal. As they tend to be close to the coast, on the continental shelf, the only possible dispute might be on the boundary between adjacent States in geographically limited areas such as the Aegean Sea.

The main characteristics of the Mediterranean fishery are a wide variety of species (though they are not particularly abundant), a large number of coastal States, and fisheries which are very largely coastal in nature.[253] Rich fishery grounds across central areas similar, for instance, to Georges Bank in the Gulf of Maine are practically non-existent.[254] Apart from the

[251] See illustrative map in *Beazley*, *Atlas of the Oceans*, facing p. 140.

[252] Blake, 'Mediterranean Non-energy Resources', 57–8; S. Varnavas and P. Papatheodorou, 'Marine Mineral Resources in the Eastern Mediterranean I', 6 *Marine Mining* (1987), 37.

[253] UN, FAO, GFCM, *Report of the Thirteenth Session: The Organization of Fisheries Research* (Rome, 1976), 64.

[254] Only two small fishing grounds are reported by the GFCM, one showing an interesting concentration of sardines off the Nile Delta which appear to be limited to coastal areas and another in the western Mediterranean, showing a rich area for shrimps located around the Balearic Islands, though nowhere near a notional equidistance boundary in the area. See UN, FAO, GFCM, *Report of the Twelfth Session: Prospectives for Fisheries*, 7–8; UN, FAO, GCFM, *Rapport de la consultation technique pour l'évaluation des stocks dans les divisions statistiques Baléares et Golfe du Lion* (Rome, 1980).

two instances mentioned above, the impact of the Mediterranean's future fishery will, therefore, be demonstrated in the establishment by the coastal States of exclusive economic or fishery zones (and thus generating more boundary settlements) rather than in considering it as a direct, 'relevant factor' for delimitation purposes.

(iii) Conclusion

The economic potential of the Mediterranean's living and non-living resources seems to indicate that overlapping zones of 200-mile claims will soon be a reality. This development will almost certainly have its impact on the Mediterranean littorals' policy regarding suitable approaches for future maritime boundary delimitation.

In most cases, there seem to exist no reasons for deciding against the adoption of a single boundary, most certainly when EEZs—rather than EFZs—are involved. To the establishment of these single boundaries two groups of facts seem to be particularly relevant. First, there appears to be little evidence in most cases to suggest the 'inequity' of these existing boundaries for determining the boundaries of the water column as well. Whilst a Tunisian argument *vis-à-vis* Libya for renegotiating the existing shelf boundary to reflect the new element of the water column is not excluded, this argument would run against the Court's findings in *Tunisia–Libya* regarding the 26° line.[255] An alternative solution might be to keep the same boundary and agree on some sort of co-operative fishery arrangement.

There is, finally, the possible impact of the yet unknown mineral reserves of the Mediterranean deep basins. The rather rudimentary studies conducted so far indicate that these could be turned into important deposits. As they are likely to straddle central areas of overlapping claims their potential impact on future maritime boundary delimitation will be contingent to a large extent upon their economic importance to the States involved.

[255] *Supra*, text at n. 249. One has to recall, however, the comment of the Court in *Tunisia–Libya* regarding the argument of Tunisia that its established 'fishing rights' in the area of delimitation should be regarded. The Court stated that this claim was 'more nearly related to the concept of the exclusive economic zone' (para. 100).

10

A Case-Study: The Aegean Sea Continental Shelf Dispute

1. INTRODUCTION

The Aegean is a semi-enclosed sea of the Mediterranean basin. It is bounded in the south by the chain of the Cretan islands, in the east by the coast of Turkey, and in the west and north by mainland Greece. It comprises an area of 70,000 sq.m. and is 297 to 323 n.m. in length (north–south) and 160–303 n.m. in width (east–west). Within this sea basin there are over 2,000 islands, islets, and rocks, mainly arranged in a series of chains. With the exception of Imyros, Tenedos, and the Rabbit Islands which lie off the entrance to the Dardanelles, the Aegean islands, with an area of approximately 17,500 sq. km., belong to Greece.

It is this geo-political factor that underlies the contending claims of Greece and Turkey regarding the principle of delimitation applicable to the Aegean Sea. This geo-political factor also renders difficult the possibility of introducing relevant solutions in the light of the 'applicable law'. In fact, it has been remarked that, because of the geo-political peculiarity of the Aegean, analogy to State and judicial practice is of little, if any, value.[1] This chapter aims, after reviewing the contending positions of the parties in the two following sections, to seek a geographical approach whereby the complexity of Aegean geography could be resolved, thus allowing the question of its delimitation to be analysed in the light of 'comparable' State and judicial practice. The exclusion of geology or geomorphology is dictated by at least two factors. First, the geophysical characteristics of the Aegean basin seem to reveal no fundamental discontinuity which may allow the consideration of a 'natural prolongation' criterion for the purpose of delimitation.[2] Secondly, but most importantly,

[1] See e.g. Professor Bowett, *The Legal Regime of Islands in International Law* (Oceana Publications, New York, 1979); B. Kwiatkowska, 'Maritime Boundary Delimitation between Opposite and Adjacent States in the New Law of the Sea: Some Implications for the Aegean', paper given at the International Symposium on Aegean Issues (Cesme, İzmir, 15–17 Oct. 1987).

[2] It should be recalled, however, that in the discussions that took place between the two States in the 1970s Turkey argued that, geomorphologically, the continental shelf upon which the eastern Aegean islands are situated constitutes 'the natural prolongation of the Anatolian peninsula, whereas the Greek islands . . . do not possess a shelf of their own'.

is the fact that the dispute between the two States ought to be governed by customary law.[3] This law seems now to indicate, as acknowledged in the Decisions,[4] that in the case of a semi-enclosed sea like the Aegean, where the distance between the coasts of States is less than 400 miles, no criterion for delimitation of the shelf areas can be derived from the notion of natural prolongation. Nor is there any 'reason why the factor which has no part to play in the establishment of title should be taken into account as a relevant circumstance for the purposes of delimitation'.[5] This also means that, in principle, the frame of analogy suggested below, although primarily concerned with the existing shelf dispute between the two parties, can hardly be immaterial for an EEZ or comprehensive, multi-purpose boundary.

The purpose of this approach is not to suggest certain specific methods of delimitation. Rather, the aim is to examine, in the light of 'comparable' State and judicial practice, some factors that might be deemed relevant for devising suitable criteria and methods for delimiting the maritime space of the Aegean. This attempt is specifically made in realization of the two States' agreement in the 1976 Berne Protocol to, *inter alia*, establish a standing committee 'to study State practice and international

(See Turkey's *Note verbale*, 27 Feb. 1974, quoted in *Aegean Sea Continental Shelf*, *Pleadings*, annexe II (1980), 23). Geologically, Turkey argued that the Anatolian islands, including Rhodes and the rest of the Dodecanese islands, constitute a 'geological continuation of the Turkish mainland'. An island such as Rhodes was argued to be 'part of the Anatolian mainland until upper Pleistocene'. (Minutes of the Meetings of the Greek and Turkish Continental Shelf Experts, Section of Geological Data presented by Turkish delegation, 2 Feb. 1976, ibid. 169). Scientific evidence, however, suggests that neither the geomorphology nor the geology of the Aegean could provide a proper criterion for delimitation. The major geomorphological characteristics of the Aegean are two troughs, respectively called the Anatolian Trough (extending in the north in a south-west–north-east direction from the Gulf of Saros, north-east of Imyros, to the Greek Megnesia peninsula, with a maximum depth of 1,468 metres) and the Cretan Trough (extending in the south from the Gulf of Argolis (Greece) eastwards and then to the north and ending at the western slope of Rhodes, with a maximum depth of 2,509 metres). Both Troughs are considered an incidental break in the otherwise continuous island-shelf slope around them (see J. C. Whitney, 'The Aegean and Related Seas (Physiography)', in D. Stanley (ed.), *The Mediterranean Sea: A Natural Laboratory* (Hutchinson & Ross, Stroudsburg, 1971), 16–19). Geological evidence suggests that the Aegean land mass of Anatolia and Greece as well as the Aegean seabed belong to the same Alpine geosyncline, with similar geotectonical zones stretching in great arcs through Greece, the Aegean basin (in a south-easterly direction), and into the Anatolian peninsula. See Whitney, 'The Aegean and Related Seas'; P. L. Ryan, 'The Tectonics and Geology of the Mediterranean Sea', in P. Maxwell (ed.), *The Sea*, iv (M. N. Hill, New York and London, 1970), 387–484; Bowett, *The Legal Regime of Islands*, 262–5.

[3] See *infra*, s.3.

[4] See *supra*, Ch. 4 s.3(*b*)(ii)(1); Ch. 9 s.2(*f*).

[5] *Libya–Malta* case, ICJ Reports (1985), para. 40. It has to be noted, however, that in the *Aegean Sea Continental Shelf* case, the Court put emphasis on 'natural prolongation' in order to relate the shelf dispute between Greece and Turkey to 'the Territorial status' of the coastal States and was, therefore, able to conclude that the dispute was excluded from its jurisdiction according to the Greek instrument of accession to the General Act. See *Aegean Sea Continental Shelf* case ICJ Reports (1978), paras. 86–90.

law rules to identify certain principles and practical criteria' which might be applicable to the Aegean.[6] More recently, Greece, an adamant supporter of strict equidistance, signed the 1982 Convention which provides, *inter alia*, that the delimitation of the continental shelf/EEZ 'shall be effected by agreement . . . in order to achieve an equitable solution', without expressing any interpretative observation on this proviso.[7]

2. CLAIMS AND CONCESSION AREAS

In a strictly legal sense, only Greece has made a positive claim to continental shelf jurisdiction.[8] However, both States have announced certain activities for surveys, exploration, and (for Greece) exploitation of hydrocarbon deposits in the Aegean. Presented below is a summary of these activities.

(*a*) *Turkey*

In November 1973 and June and July 1974 Turkey issued exploration concessions to the Turkish Petroleum Company (TRAO) in respect of sections of seabed in the eastern half of the Aegean. According to maps published in the Turkish *Official Gazette* the western limits of the concession areas largely centre around a notional median line between the Turkish straight baseline of the Anatolian coast and the Greek mainland.[9] The maps also show the Greek islands in the eastern Aegean to be credited with 'enclave' zones corresponding presumably to the Greek-claimed territorial sea limit of 6 miles.

In May 1974 and February 1976 Turkey asserted its claim by sending survey ships to conduct magnetometric and seismic tests along the western limits of the concession areas. These activities were declared by Turkey to have been held in the 'Turkish continental Shelf'.[10] In March 1987 Turkey

[6] Para. 8 of the Berne Protocol, 11 Nov. 1976, in 16 *ILM* (1977), 13.

[7] Greece signed the 1982 Convention on 10 Dec. 1982 with the following declaration: 'In areas where there are numerous spread out islands that form a great number of alternative straits which serve in fact one and the same route of international navigation, it is the understanding of Greece, that the coastal State concerned has the responsibility to designate the route or routes, in the said alternative Straits, through which ships and aircrafts of third countries could pass under transit passage regime, in such a way as on the one hand the requirements of international navigation and overflight are satisfied, and on the other hand the minimum security requirements of both the ships and aircrafts in transit as well as those of the coastal State are fulfilled.' *LSB* 5, (July 1985), 13.

[8] Decree-Law No. 142/1969 Concerning Exploration for and Exploitation of the Mineral Resources in the Seabed and the Beds of Lakes, *UNLS* (ST/LEG/SER.B/16) (1974), 87. Art. 1 adopts the depth/exploitability criterion of the 1958 Shelf Convention.

[9] *Resmi Gazete* (the *Official Gazette* of Turkey) (1 Nov. 1973; 6 June, 18 July 1974).

[10] Statement by Turkish Minister for Power and National Resources. For the development

renewed its intention to conduct seismic surveys west of the island of Samothrace, apparently to counter possible Greek moves to search for oil in the disputed continental shelf. The Turkish plan was, however, called off following diplomatic efforts to ease the tension aroused between the two States.[11] So far there are no published records of the result of Turkish prospecting for oil in the Aegean.

(*b*) *Greece*

The practice of Greece with regard to granting oil concessions in the Aegean started in the early 1960s. This practice, however, does not reflect Greece's view on the shelf boundary with Turkey, for, unlike the latter, Greece has not so far published concession areas coinciding with its claims in the Aegean.[12]

However, in terms of oil production Greece is the only offshore producer in the Aegean. In 1974 commercially significant deposits were discovered south-west of Thasos. In 1975 the production capacity from this area (now called the Prinos Field) was estimated at a rate of 20,000 barrels a day. Following the sinking of two exploration wells, it was first estimated that output from the Prinos Field would rise to a maximum of 50,000 barrels per day. But the sinking of further exploration wells revealed a fault in the field and, consequently, estimates were put down to 25,000 barrels per day. In 1987 it was reported that the Prinos production capacity is 28,000 barrels per day.[13] In addition to the Prinos Field, drilling has established the existence of a small natural gas field in the Thasos area, which has still to be delineated and which could be commercially exploitable.

However, these incidents of faulting are of interest in terms of the hydrocarbon potential in the Aegean. They might be related to a large-scale earth movement which may explain the poor rate of sedimentation in the eastern Levantine, including the Aegean. The suggestion is that the evidence of small scarps in some of the Aegean basins and of tilted and relic plain surfaces indicates that reorganization may still be in progress and that the normal sedimentary levelling processes have been unable to

of tension between the two States following the Turkish claim, see A. Wilson, *The Aegean Sea Dispute*, Adelphi Papers No. 155 (Institute for Strategic Studies, London, 1979–80).

[11] See *The Times* (27, 28 Mar. 1987).

[12] Greece appears to have granted only five limited concessions in the Aegean. See *Basic Oil Laws and Concession Contracts: Europe*, supp. 15 A-O (Contract with Texaco 1968); Contract with Ada Oil Exploration Corporation 1969, supp. 27 A-O; Contract with An-Car, 1970, supp. 30; Contract with Oceanic (the Prinos Concession), 1974, supp. 26.

[13] *The Times* (28 Mar. 1987).

keep pace with the tectonic deformation.[14] More accurate evidence is of course needed to substantiate this suggestion, which, if proved true, will certainly cast doubt on the potentiality of the Aegean for hydrocarbon exploitation. Seabed reorganizations may cause significant deformations to the natural gas and oil 'traps' and consequently the faulting of potential gas or oilfields.

3. PRINCIPLES OF DELIMITATION CLAIMED

(*a*) *Turkey*

Turkey is neither a party to the 1958 Shelf Convention, nor a signatory to the 1982 Convention, though it was an adamant supporter in UNCLOS III of the 'equitable' principles. In application of these principles, Turkey argues, the particular situation of the islands

> is a second major factor in the problem . . . the rules established by international practice, as shown by several agreements which have already been concluded, deny the granting of an equal value to all the islands without taking into account their characteristics and their particular situation when it is a matter of continental shelf delimitations. Both the [eastern Aegean islands] and the whole of the Aegean Sea . . . constitute a typical example of 'special circumstances' and for that reason should be appropriately treated with a view to the application of the rules of international law.[15]

At UNCLOS III Turkey submitted draft articles which clearly reflected its position regarding the Aegean Sea dispute:

> 1. Where the coasts of two or more adjacent States are adjacent and/or opposite, the continental shelf areas appertaining to each State shall be determined by agreement among them, in accordance with equitable principles.
> 2. In the course of negotiations, the States shall take into account all the relevant factors, including, *inter alia*, the geomorphological and geological features of the shelf . . . and special circumstances such as the general configuration of the respective coasts, the existence of islands, islets or rocks of one State on the continental shelf of the other.[16]

[14] Ryan ('The Tectonics and Geology of the Mediterranean Sea', 415) wrote in 1970 that the 14 largest shocks recorded in the 20th century in the Aegean region occurred around the perimeter of the 'aseismic zone' of the central Aegean.

[15] Turkey's *Note verbale* (1974), 24.

[16] UNCLOS III, *Official Records*, iii. 201. See also Turkish draft article on the regime of islands, which proposed that 'an island situated in the economic zone or on the continental shelf of other States shall have no economic zone or continental shelf of its own if it does not contain at least one tenth of the land area and population of the State to which it belongs'. Ibid. iii. 230.

In his comment on the Article the Turkish delegate added that islands, islets, and rocks within the area to be delimited deserved 'special attention' since their presence could have such a distorting effect 'as to deprive a State of any meaningful economic zone' and that a State should 'not be made to suffer from that situation'.[17]

(*b*) *Greece*

As a party to the 1958 Shelf Convention, Greece[18] has declared the principle of international law on which it based its seabed claims in the Aegean. Thus, following the announcement of the Turkish unilateral boundary-concessionary line, Greece in its 1974 *Note verbale* stressed two points: (1) the territorial sovereignty over the Greek islands 'extends according to international law to their sea-bed and subsoil as well as to their continental shelf'; and (2) the delimitation of the continental shelf is based both in the theory and practice of international law on the principle of equidistance as provided in Article 6(1) of the Geneva Convention.[19]

In its unilateral application instituting proceedings against Turkey before the ICJ,[20] Greece stated that the delimitation of the continental shelf between the Greek islands and the coast of Turkey 'is to be effected by reference to the points at which respectively natural prolongations of the Greek and Turkish territories, whether insular or continental, meet and overlap; and that that point can only be the median line'.[21] The legal position of Greece is therefore designed to deny Turkey any seabed area beyond the coasts of the Aegean islands situated off the coast of Anatolia. Thus at UNCLOS III Greece strongly advocated an equal treatment of islands and islets with continents regarding their entitlement to seabed rights. In its draft articles submitted to the Conference Greece suggested the deletion of the phrase 'special circumstances' and declared that

> no State is entitled to extend its sovereignty beyond the median line every point of which is equidistant from the nearest points of the baselines, *continental or insular*, from which the continental shelf of each of the two States is measured.[22]

These are the contrasting claims of the two States regarding the principles of law applicable to the delimitation in the Aegean. These claims, however, are obviously influenced by factors of a rather political charac-

[17] UNCLOS III, *Official Records*, ii. 213, 214.
[18] Greece ratified the 1958 Shelf Convention on 6 Nov. 1972. For the accompanying reservation see *supra*, Ch. 6, text at n. 16.
[19] Greek *Note verbale* (1974), 21.
[20] *Aegean Sea Continental Shelf*, Judgment, 3.
[21] See *Aegean Sea Continental Shelf*, *Pleadings*, 10.
[22] UNCLOS III, *Official Records*, ii. 202 (emphasis added).

ter.[23] In making such claims, Turkey is anxious to break through the 'barrier' of the Greek islands, from Samothrace to Rhodes in the eastern Aegean, not only to participate in the potential oil resources, but also to secure Turkish free access in the eastern Aegean to and from the Dardanelles and major ports of Anatolia, such as İzmir. Air traffic is also involved, especially when Greece banned flights across the Aegean in 1974 when Turkey challenged the 1951 rules of the International Civil Aviation Organization which had given Greece responsibility for all air traffic control over the sea.[24]

Turkey feels that existing legal regimes such as innocent passage, freedom of navigation, and overflight in the high seas might be in jeopardy at times of crisis with Greece. Such fears were prominent when Greece in the mid-1970s remilitarized Limnos and the adjacent islands in the north-east and the Dodecanese group in the south-east.[25]

For Greece, Turkey's equidistant line in the middle of the Aegean is seen as part of a Turkish move to restructure the political frontier decided in the 1923 and 1947 Agreements.[26] These agreements are viewed by Athens as ensuring that the political interface—though not a *de jure* maritime boundary—lies along a line equidistant from the Anatolian coast and the Greek-owned islands closest to that coast so as to preserve the islands' political continuum with the Greek mainland.[27] Any attempt to cut off this continuity between the islands and mainland Greece is regarded as threatening the national integrity of the islands. In the mid-1970s, for instance, Greece blamed some of the population decline in the eastern islands on apprehension at the possibility that the Greece–Turkey

[23] See N. Deloukas, 'The Controversy between Greece and Turkey in the Aegean Sea', *Issues and Studies* (Jan. 1980), 70, at 76.

[24] See C. Sazanidis, 'The Greco-Turkish Dispute over the Aegean Airspace', unpublished paper translated from Greek based on part IV of his book, *The Greco-Turkish Relations 1973–1978* (in Greek) (Thessaloniki, 1979), 144–58.

[25] Limnos and the adjacent islands were demilitarized in the 1923 Lausanne Peace Treaty and the Lausanne Convention on the straits of the same year. Also according to Art. 12 of the 1947 Peace Treaty with Italy the Dodecanese island group was transferred to Greece with a proviso stipulating for the demilitarization of the group and the duty to keep them demilitarized. After Turkey's invasion of Cyprus in 1974, the group was the object of massive militarization by Greece. For Greek argument regarding the 'tacit' abrogation of the demilitarization clauses, see N. Ronzitti, 'Demilitarization and Neutralization in the Mediterranean', 6 *IYBIL* (1985), 35, at 39–44. See also counter-claims by the two parties in letters to the UN Secretary-General, UN Docs. S/11668, 9 Apr. 1975; S/11627, 14 Apr. 1975; S/12176, 13 Aug. 1976.

[26] See *supra*, Ch. 8 s.4(*b*)(i)(1). Turkey has expressed, though indirectly, its dissatisfaction with the consequences of these political arrangements which turned the Aegean into a 'Greek preserve'. See N. Deloukas, 'The Controversy between Greece and Turkey in the Aegean Sea', *Issues and Studies* (Jan. 1980), 70–80.

[27] Therefore, before the Court, Greece expressly rested its claims to continental shelf in respect of the eastern islands upon, *inter alia*, the territorial and political unity of Greece. *Aegean Sea Continental Shelf*, Judgment, para. 87.

frontier might move westwards across part of the Aegean, thereby 'isolating' some Greek islands.[28]

The issue of political security has been, therefore, a central factor in the formulation of each party's legal position. The two States' dispute over the air traffic in the Aegean as well as their conflict in Cyprus[29] has further transformed the issue of delimitation into a political one. An attempt to suggest a 'legal solution', such as the one suggested below, should not, therefore, lose sight of this factor.

4. GEOGRAPHICAL FRAMEWORK OF ANALOGY

(*a*) *Introduction*

Neither of the rigidly opposed positions of Greece and Turkey would serve as a basis for legal consideration of the boundary delimitation in the Aegean. Giving effect to all the Aegean islands, as Greece argues, entails that Turkey is virtually deprived of any Aegean shelf or EEZ of its own. The inequity in this claim appears from two factual statements: (1) in the Aegean basin Turkey has a coastal length of about 427 n.m. (that is over half of its Mediterranean façade) along which about 10 million (about 22 per cent of the total population of Turkey) are settled. This is compared with a Greek population of about 145,071 living on the Aegean islands. The Aegean mainland coast of Greece is about 384 n.m. long. (2) The other factor is Turkey's geographically disadvantaged situation on the rest of its Mediterranean coast. A hypothetical median/equidistant line between the Turkish coast on the Levantine and the rest of the neighbouring coasts, including (once again) the Greek islands of Rhodes, Karpathos, Kasos, and Crete, allots to Turkey a very limited continental shelf, not comparable with a coastal length ranking as the sixth longest coast in the Mediterranean (about 546 n.m.).[30]

Considering these two factors alone, the potential unfairness apparent in denying the Aegean-facing coast of Turkey a reasonable 'proportion' of the Aegean shelf or EEZ is reinforced by the fact that this would result in cutting off Turkey from maritime areas adjacent to its coast (between the Greek islands or west thereof) and would eventually result in a maritime boundary 'so near' to its coast that the other party might exercise rights opposite that coast in a manner that 'could prevent the exercise of its own right to development or compromise its security'.[31]

[28] *The Times* (18 Nov. 1976).

[29] See *supra*, Ch. 8, n. 3.

[30] See Table 1.

[31] See *Libya–Malta* Decision, para. 73, and *Guinea–Guinea Bissau* Decision, 25 *ILM* (1986), para. 124.

The adjustment of the above situation in order to reach an equitable shelf or EEZ delimitation in the Aegean cannot be seen as an attempt to redress 'natural inequity' or, as the ICJ said, 'refashion geography'. The situation in the *North Sea* cases was one in which there were

> three States whose North Sea coastlines are in fact comparable in length and which therefore, have been given broadly equal treatment by nature except that the configuration of one of the coastlines would, if the equidistance method is used, deny to one of these States treatment equal or comparable to that given the other two.

In such a situation, the Court concluded, there was 'a case where, in a theoretical situation of equality within the same order, an inequality is created'.[32]

The presence of numerous islands in the northern Aegean (where, as will be seen shortly, a shelf or EEZ boundary may be envisaged) does not totally prevent the existence of 'comparable coastal equality' between the respective coastlines of Greece and Turkey.[33] What withholds this 'comparable equality' is a factor of *islands location*, and to consider this factor as a relevant circumstance, creative of inequity, in the light of the equality between the respective coastlines, is not a question of totally refashioning geography. If the Court's emphasis on equitable principles is viewed as a refusal to give full weight to the *mainland because of the overall coastal configuration*, then it would seem to follow that *islands location*, which has the potential to produce equally inequitable results, should be treated in a similar fashion and such islands should be given some reduced weight in shelf or EEZ delimitation.

On the other hand, to ignore all the Aegean islands for the purposes of boundary delimitation, as Turkey seems to argue, is an argument which flies in the face of the established rule affirming that, in principle, islands have their own maritime jurisdiction.[34] Indeed, it is the recognition of this rule that often complicates the process of inter-State delimitation.

In practice the actual delimitation is usually affected by a combination of factors such as the location of the islands, their size, distance from the mainland, and their distorting effect on the delimitation. Measuring the relative weight of each factor is a function of the particular context and it is difficult, if not impossible, to make generalizations. However, it can be said that the two main factors are location and size.

In terms of location, that is, in the order of proximity to the mainland coast of the sovereign State: (1) islands located within the territorial sea

[32] *North Sea* cases, ICJ Reports (1969), para. 91 (emphasis added).

[33] See *infra*, s.4(*b*), (*d*).

[34] See Art. 1(b) of the 1958 Shelf Convention. See also *supra*, Ch. 2 s.3(*d*).

of the mainland of the sovereign State are usually granted full effect in generating their own shelves;[35] (2) islands located close to but outside the territorial sea of the sovereign State's mainland coast are accorded partial effect;[36] (3) islands located on or near the equidistance line (constructed in disregard of their presence) between the sovereign State and the opposite/adjacent State are given full effect,[37] partial effect,[38] or no effect,[39] depending on other considerations; (4) islands located close to the mainland coast of the opposite State are usually accorded limited effect or no effect.[40] In terms of size, small islets or rocks are usually ignored or given limited effect in shelf delimitation, since they are too small to support a permanent population, and they are often uninhabited and have little or no economic value. Table 5 and Map 18 illustrate only the most important islands of the Aegean in terms of size, population, and location.

An equally relevant factor is the whole geographical context of the area subject to delimitation. The *Libya–Malta* case stands, as shown earlier,[41] for the proposition that the extent of an island's maritime jurisdiction in a semi-enclosed sea is also determined in the entire geographical context. This, as will be seen from the analysis provided below, is also reflected in judicial and State practice, where in several cases the narrowness of the area subject to delimitation is taken as another important 'restrictive' factor in the decision to deal with islands.

It is submitted that the above generalizations might be addressed to the Aegean only if the entirety of the Aegean, that is, a sea dominated by islands throughout its length, is divided into two sectors (northern and southern), each being carefully examined so as to reflect the principal relationship between the Aegean mainland coasts and the islands on the one hand, and the existence of corridors of high seas (within which the shelf/EEZ boundary between Greece and Turkey is envisaged) on the other. Factors influencing islands' effect on shelf/EEZ boundaries, such as size or economic importance, are considered together with location as the connecting factor.

35 See e.g. agreements mentioned *infra*, n. 57, but also see agreements analysed in text accompanying nn. 64–6, *infra*.

36 See e.g. Saudi Arabia–Iran, *LIS* 24, where the Iranian island of Kharg, 17 miles offshore, was given half-effect in delimiting the median line; Italy–Yugoslavia, *supra*, Ch. 9 s.2(*a*), where the Yugoslav islands of Jabuka and Andrija were granted partial effects in influencing the median line.

37 See *infra*, s.4(*b*)(i)(1).

38 See *infra*, s.4(*b*)(i)(2).

39 See *infra*, s.4(*b*)(i)(3).

40 See e.g. Australia–Papua New Guinea, *infra*, text at n. 83.

41 See *supra*, Ch. 9 s.2(*f*).

Table 5. *Islands in the Aegean Sea*

Island group	Area in sq. metres	Population (1985)
Sporadhes		
Skiros	209,499	2,757
Skopelos	95,479	4,451
Alonnisos	64,468	1,528
Skiathous	47,941	4,127
Pelagos	24,766	7
Peristera	14,155	15
Kasos	65,975	1,184
Tilos	62,825	301
Simi	58,100	2,273
Cyclades		
Andros	379,668	9,020
Naxos	428,125	14,037
Tinos	194,206	7,730
Amorcos	120,665	1,718
Mikonos	85,481	5,503
Psara	39,770	460
Crete	8,261,183	502,082
Lesvos		
Lesvos	1,630,378	88,601
Limnos	475,606	15,721
Thasos	378,845	13,111
Samothrace	177,977	2,871
Ag. Evstratios	43,230	296
Dodecanese		
Rhodes	1,398,075	87,831
Karpathos	301,175	4,645
Kos	290,275	20,350
Kalimnos	110,875	14,295
Astipalaia	96,850	1,030
Yioura	10,945	2
Skantzoura	6,000	—
Piperi	4,250	—
Leros	52,950	8,127
Nisiros	41,400	916
Patmos	34,050	2,534
Khalki	28,125	334
Kastellorizo	9,000	222
Chios		
Chios	341,583	48,700
Antipsara	4,443	—
Samos		
Samos	476,198	31,629
Ikaria	255,257	7,559

Source: *Statistical Yearbook of Greece (1986)* (National Statistical Service of Greece, Athens, 1986), 20–1.

(*b*) *The northern sector*

This refers to the part of the Aegean bounded in the south by the almost overlapping belts of the 6-mile territorial seas allotted to Euboea, Andros, Tinos, Mikonos, Ikaria, Samos, and from there up to the Turkish mainland coast (see Map 18). From these belts northwards lies over 69 per cent of the total area of the Aegean high seas (measured from the outer limits of 6-mile territorial sea belts of the insular and mainland coasts).[42] This is the most controversial area in the Aegean. Offshore hydrocarbon deposits have so far been discovered only in this sector. It is in this sector only that the problem of delimitation can be seriously considered, since, as will be seen, nowhere else in the Aegean could Turkey establish a reasonable shelf or EEZ jurisdiction. In this northern sector the following island–mainland situations may be considered.

(i) Midway islands

These are the islands of Samothrace, Limnos, and Ag. Evstratios. They are all inhabited and lie from north to south in a south-western arrangement which straddles a notional equidistant/median line starting at the Greek–Turkish land boundary terminus and extending southwards between the coasts of Turkey (measured from the straight baselines system of 1964)[43] and Greece (measured from a hypothetical straight line encompassing Thasos and the 'headlands' of the Gulfs of Singitikos, Toronaios, and Thermaikos). Islands in the following shelf boundary agreements have at least one common factor. They lie on or near median lines drawn from the respective mainland coasts.

1. *Full effect to midway islands*

Two cases are of some interest: the Shetlands in the North Sea between the United Kingdom and Norway and Tsushima Island in the Korea Strait between Japan and South Korea. In the former case, the 1965 Agreement between Britain and Norway gave full effect to the Shetland Islands in drawing the median line boundary.[44] In the latter case, selected basepoints on the coast of Tsushima Island were used in drawing an equidistant shelf boundary between Japan and South Korea in the Korean Strait.[45]

A close look at the factors of the two cases reveals significant differences from the Aegean example. First, in both cases the islands involved are

[42] See *supra*, Ch. 8 s.4(*b*)(i)(1).
[43] See *supra*, Ch. 6 s.3(*g*).
[44] Agreement of 1965, *LIS* 10: 2.
[45] Agreement of 2 Sept. 1977, *LIS* 75.

fairly large (Shetlands: 552 sq.m.; Tsushima: 271 sq.m.) and populous (Shetlands: 17,298; Tsushima: 58,672). Islands of such magnitude are far more important than the small and unpopulous islands of Samothrace, Limnos, and Ag. Evstratios (see Table 5). Secondly, although utilizing the Shetlands and Tsushima Islands as basepoints certainly displaced the median line to varying degrees, thus creating distorting effects, the geography of the areas involved still allowed enough room for not only reducing or eliminating these distorting effects, but also making reciprocal exchanges. The Shetlands lie outside the British straight baseline system and are 96 and 173 miles from the nearest British and Norwegian mainland coasts, respectively. The distance between the two coasts in that part of the North Sea is about 270 miles.

Significantly, the Norwegian Trough lying only a few miles off Norway could have caused Britain to deny the continuity of the Norwegian shelf west of this trough. The median line was drawn in disregard of geomorphology as a potential limiting factor. Whether this concession of the United Kingdom was linked to granting full effect to the Shetlands is speculative. But such a possibility obviously exists.

Tsushima Island is situated 29 and 45 miles off the coasts of Korea and Japan, respectively. The distance between the two coasts facing the island is 98 miles. In the mouth of the Korea Strait lies the Korean island of Cheju Do, 45 miles offshore. To make up for the full effect given to Tsushima, Cheju Do was used as a basepoint in constructing the southern portion of the median line shelf boundary between Japan and South Korea. In the northern sector in the Aegean, the distance between the two States' respective baselines (as determined above) varies dramatically between 117 and 76 miles. Reciprocal concessions are non-existent, since there are no geomorphological features or Turkish outlying islands which could be used by Turkey as a 'trade-off' with Greece. The Shetlands and Tsushima are therefore not comparable examples for the Aegean midway islands.

2. *Partial effect of islands*

Two cases are particularly interesting—the Yugoslav islets of Pelagruz and Kajula and the small Italian islands of Pantelleria, Linosa, Lampione, and Lampedusa. According to the Italy–Yugoslavia Shelf Agreement,[46] Pelagruz and Kajula, midway islets in the Adriatic Sea, were accorded a 12-mile zone, which created a bulge in the otherwise smooth median line between the Italian and Yugoslav mainland coasts. As seen earlier, this solution is understood in the light of factors not only related to the islands' location. The Yugoslav concession of the area outside the 12-

[46] See *supra*, Ch. 9 s.2(*a*).

mile limit was partially in response to concessions made by Italy when she accepted similar treatment for the island of Pianosa, as well as allowing Yugoslavia to utilize as basepoints numerous offshore islands.[47]

The Italy–Tunisia Shelf Boundary Agreement is of particular interest, because, just as in the Aegean islands discussed here, each of the four islands—straddling or near a mainland-to-mainland median line—belongs to Italy. In order to offset the potential disadvantage to Tunisia, the parties agreed on a median line which disregarded these four islands. In the area of the islands the boundary deviated, so as to grant these islands semicircles of 12–13 miles' radius.[48] The major consideration behind this solution was the fact that a strict median line (utilizing these islands as basepoints) would bring the line to a distance up to 20 miles off the Tunisian mainland coast, thus causing a considerable inequity to that State. Similar solutions to midway islands in relatively narrow sea areas are found in the Iran–Saudi Arabia[49] and Abu Dhabi–Qatar[50] Agreements.

3. *No effect (islands ignored)*

Three delimitation agreements are of particular interest here. In delimiting their modified median line in the Gulf, Bahrain and Saudi Arabia[51] chose to ignore the islands of Greater Lubainah and Smaller Lubainah. The boundary passes through turning-points, two of which are located on the easternmost tip of the former, leaving it to Saudi Arabia, and the westernmost tip of the latter, leaving it to Bahrain. The width of the water in the area of the islands is less than 12 miles.

The second case is Kachchativu Island, lying 1.2 miles from the median line drawn between the respective coasts of India and Sri Lanka in the 45-mile wide Palk Strait.[52] The 1974 Agreement between the two States allocated the island to Sri Lanka, but the line was drawn without real regard to this island. The island was denied even a 12-mile territorial sea which Sri Lanka claims.

The final example is the 1978 Torres Strait Agreement between

[47] See ibid.

[48] See *supra*, Ch. 9 s.2(*c*).

[49] Continental Shelf Agreement (1968), *LIS* 24, which determines a median line between the two States' mainland coasts, ignoring the Iranian island Farsi and the Saudi island Al-Arabiya, both lying within 12 miles of the median line. The islands are both given a belt of 12 miles of territorial sea drawn between them according to equidistance.

[50] Continental Shelf Agreement (1969), *LIS* 18. The boundary, which is not strictly equidistant, deviates as it approaches the small island Dayinah which belongs to Abu Dhabi. The deviation produces a semicircle around the island marking the 3-mile territorial sea of the island.

[51] Agreement of 22 Feb. 1958, *LIS* 12.

[52] Agreement of 26–8 June 1974, *LIS* 66.

Australia and Papua New Guinea,[53] whereby several Australian insular formations lying within 12 miles of the agreed seabed boundary line[54] were granted enclaves of territorial waters of 3 miles. It was further provided that these territorial seas should not be enlarged or reduced. The width of the Torres Strait in the area of the islands is about 90 miles.

A primary factor in all these three cases is that the question of sovereignty over these islands was undetermined. This is normally suggested as a relevant factor in ignoring the island, where a delimitation settlement also resolves the question of its status. What is certainly more relevant is that these islands are small and situated in narrow areas. In such areas, had the islands been granted full effect, enormous distortions would have resulted; and, because the areas are limited, solutions to adjust such distortions would have been impractical.[55] In the Aegean example, if the islands of Samothrace, Limnos, and Ag. Evstratios were to be used as basepoints a median line would run no further than 12.8–24 n.m. off the coast of Turkey.

(ii) Offshore islands (Skiros and the Sporadhes group)

The Sporadhes group comprises eight principal islands (see Table 5) and other uninhabited tiny islets and rocks. The eight islands are about 93.85 sq.m. in area and have a population of 10,130 inhabitants. The furthest islands in the group, that is, Pelagos, Yioura and Piperi, lie about 48 miles offshore and are almost uninhabited (only nine people live on Yioura and Pelagos and none on Piperi). Skiros is more important in terms of size (80.7 sq.m.) and population (2,757), but lies approximately 18 miles offshore; thus allowing, unlike the series of the Sporadhes Islands,[56] a corridor of high seas between its 6-mile belt of territorial sea and that of the mainland. If the easternmost islands of the Sporadhes and Skiros were used as basepoints in delimiting a land-to-land median line, they would deflect the line about 37 miles towards Turkey, thus resulting in Greece receiving a major proportion of the high-seas area available east of these islands.

[53] Agreement of 18 Dec. 1978, 18 *ILM* (1979), 291.

[54] e.g. Turnagain Island, Pearce Cay, Black Rocks, Anchor Cay, and Ecost Cay.

[55] This also appears to be the major reason behind ignoring the midway island of Hans in drawing the 1973 Canada–Denmark (Greenland) shelf boundary in the Nares Strait. The parties have apparently avoided the resolution of the issue of the island's ownership. But the narrowness of the water body would apparently preclude the attribution of a zone of water, should the issue of ownership be resolved. See Boundary Agreement of 17 Dec. 1973 and analysis of *LIS* 72.

[56] The islands extend from the mainland in an east-north-easterly direction, one after another, and, together with their 6-mile belts, produce a continuous overlap for about 48 miles eastwards.

There is little doubt that the general practice relating to offshore islands is their inclusion as basepoints for the purposes of boundary delimitation.[57] This general practice, however, is not absolute. State and judicial practice reveals that in narrow and enclosed or semi-enclosed sea areas the application of this general rule may lead to inequity. The treatment of offshore islands in the following instances is comparable to the situation under consideration.

1. *Adjudication*

The Court of Arbitration's treatment of the Scilly Isles in the 1977 Arbitration and the ICJ's treatment of the Kerkennah Islands in the 1982 Judgment are *a fortiori* comparable to the Aegean offshore islands.

1. The Scillies, which lie at their nearest point some 21 miles from the British Land's End, consist of a group of some forty-eight islands of which six are inhabited. They have a sizeable population and for centuries have been considered an integral part of the United Kingdom. The islands were regarded by the Court as 'unquestionably islands offshore of the United Kingdom'. The Court also noted that the existing 12-mile fishing zones of the mainland and of the Scillies merge into one and, if the United Kingdom were to establish a 12-mile territorial sea, the same would be the case with their territorial waters.[58]

The main issue with regard to the Scillies was whether they should be treated normally, so as to have a full effect on the equidistance boundary as against the corresponding French basepoint on Ushant. Having considered the location, size, and importance of the Scillies, the Court pointed out that the additional projection of the islands into the Atlantic, when superadded to the westward projection of the Cornish peninsula, constituted an element of distortion which justified the delimitation of a boundary other than by the strict median line (utilizing the islands). Instead, the Court gave the Scillies only half-effect and considered this 'to be an appropriate and practical method of abating the disproportion and inequity which otherwise results from giving full effect to the Scilly Islands as a basepoint for delimiting the course of the boundary'.[59]

2. The Kerkennah Islands are situated 11–14 miles off the mainland of Tunisia. They are comparatively large (65 sq.m.) and populous (more than 15,000). The economic importance of the shallow waters around

[57] See, for instance, the following agreements: Indonesia–Malaysia, *LIS* 50; Finland–Soviet Union, *LIS* 56; India–Indonesia, *LIS* 62; Australia (Papua New Guinea)–Indonesia, *LIS* 87.

[58] *Anglo-French* Arbitration, 18 RIAA, para. 248. The UK, of which the Scillies are a part, extended its territorial sea to 12 n.m. in 1987. See Territorial Sea Act 1987, *The Law Reports Status* (c.49) (1987).

[59] *Anglo-French* Arbitration, para. 251.

them for mobile and sedentary fishing has been recognized for centuries.[60] Nevertheless, alluding to State practice,[61] the Court resolved that granting the Kerkennahs full effect would 'amount to giving them excessive weight'[62] and, instead, 'the Court considers that *it should go so far* as to attribute to the islands "a half-effect"'.[63]

While the geographical setting of the area within which the Scillies and Kerkennahs are situated is different from that of the Aegean, one cannot help but conclude that if islands of the Scillies' and Kerkennahs' magnitude may be given only half-effect in the name of equity, it is all the more logical to grant the tiny islands of Skiros and Sporadhes lesser weight or no weight at all, given, furthermore, the limited geography of the sea area around them.

2. *State practice*

In the Iran–Saudi Arabia Shelf Agreement,[64] the Saudi offshore islets Janah, al-Jurayd, al-Qurrayyin, and Qiran, all lying within 12 miles of the Saudi coast and each other, were ignored when constructing the shelf boundary, which was essentially an equidistant line.[65] If used as base-points, the islets would have deflected the line several miles towards Iran. The segment of the boundary within the sphere of influence of the islands has been determined from the respective mainland coasts, including the Iranian offshore islet Nakhilu. The reasons for including this islet and excluding the Saudi islets were obviously related to differences in size and location. The Saudi islets are smaller than Nakhilu and, more significantly, Nakhilu is closer to the coast of Iran in comparison to the Saudi islets, which are at some distance from the mainland coast. The a priori analogy of these Saudi islets to the Greek island Skiros, which is 18 miles offshore, is self-evident.

The second example is the Iran–Qatar Boundary Agreement establishing a median line, ignoring, *inter alia*, the offshore Iranian islands Sheikh Shoeyb, Hendorabi, and Qeys. All these islands lie within 12 miles of the Iranian coast and of each other. One major factor behind ignoring these islands was the narrowness of the seabed area subject to delimitation between the two States.[66] This also seems to be the reason

[60] See *supra*, Ch. 6 s.3(*j*).

[61] *Tunisia–Libya* case, ICJ Reports (1982), para. 128.

[62] Ibid.

[63] Para. 129 (emphasis added).

[64] Continental Shelf Agreement of 24 Oct. 1968, *LIS* 24.

[65] See A. El-Hakim, *The Middle Eastern States and the Law of the Sea* (Manchester University Press, Manchester, 1979), 95.

[66] Continental Shelf Agreement of 20 Sept. 1969, *LIS* 25. A second factor was the presence of small islands off the coast of Qatar, the sovereignty of which had been in dispute. As these could not be included for delimitation purposes, the two parties seemed to

for giving the three Iranian islands Ye Four, Bani Four, and Sirri no effect in the seabed boundary between Iran and the United Arab Emirates.[67] The islands lie within 12 miles of the Iranian coast and of each other, stretching out towards the median line. Opposite Sirri the line curves southwards so as to coincide with Sirri's 12-mile territorial sea. This third example is comparable to the arrangement of the Sporadhes group.

The final example of comparable value is the Shelf Agreement between Canada and Greenland (Denmark). In the central sector of the boundary passing through the Nares Strait there lie several offshore islands, the biggest of which are the Danish Crozier and Franklin Islands. Apparently because of the narrowness of the water body, the islands were not even accorded belts of territorial sea.[68]

(iii) Islands clustering against the Turkish coast

The extension of the above-described median line south of Ag. Evstratios, thus allocating to Turkey a meaningful shelf/EEZ zone, would be possible only if the outlying islands of Lesvos, Psara, Chios, and Samos were 'enclaved' within their belts of territorial waters. Such a solution in the Aegean raises the question of its equitableness in the light of State and judicial practice.

1. *The relevance of the* Anglo-French *Award*

In this case the Court, it will be recalled, reviewed the hypothetical view that if the Channel Islands (clustering against the coastline of France to a marked degree) were 'a projection' of the British mainland, and so constituted the 'opposite coast' *vis-à-vis* France in the area, the mid-channel median line would automatically deviate southwards in a long loop around the Channel Islands.[69] This, in the Court's opinion, would have been 'as extravagant legally as it manifestly is geographically'.[70] Applying the principle that 'the delimitation must be in accordance with equitable principles',[71] the Court awarded these islands only a 12-mile enclave in the French continental shelf.

Certain elements of the situation in the Channel Islands offer an analogy to the case under consideration. With regard to the applicable law, the Court of Arbitration found that, in the circumstances of the case, the

have reached a trade-off agreement, each ignoring the other's offshore islands. See El-Hakim, *The Middle Eastern States*, 100.

[67] Continental Shelf Agreement of 31 Aug. 1974, *LIS* 63.

[68] The Boundary Agreement of 17 Dec. 1973. See analysis in *LIS* 72.

[69] *Anglo-French* Arbitration, para. 189.

[70] Para. 194.

[71] Para. 97.

rules of customary law were applicable.[72] This means that the Court's interpretation of these rules is relevant to the Aegean situation where the rules of customary international law (only Greece is party to the 1958 Shelf Convention) apply. Geophysically, like the continental shelf between Britain and France,[73] the Aegean seabed is, as noted earlier, characterized by its fundamental geological and geomorphological continuity. Therefore, the 'natural prolongation' argument raised by Turkey is just as irrelevant in this case as the Court decided it was when it rejected a similar argument by France.[74] Geographically, the parties for the greater part of the Aegean are opposite States, as are the United Kingdom and France.

As for the two parties' coastal relations in the area where the above-mentioned islands are expected to have influence, one finds more related elements for the purpose of analogy. East of these islands runs the mainland coast of Turkey between Bab Burmu and Kusadasi, with two major curvatures, east of Lesvos and Samos. Opposite to this coast, in approximate equality (Skiros and Ag. Evstratios being ignored), runs the coast of Greece (represented by the mainland coast of Volos, Euboea, and the eastern coasts of the almost continuous land bridge of the islands of Andros, Tinos, and Mikonos (see Map 18)). The Greek islands under consideration are not only situated on 'the wrong side' of a median line drawn between these two opposite coasts, but literally cluster against the Turkish coast. In fact Lesvos and Samos lie within arms of Turkish coastal curvatures.

In the *Anglo-French* Arbitration, it was this equality between the opposite coasts of France and the United Kingdom on the one hand and the presence of the Channel Islands upsetting this equality on the other that impelled the Court to consider the enclave solution.[75] Finally, as was the case in the Channel Islands region,[76] the extent of the continental shelf in the northern Aegean is remarkably modest and the scope for adjusting the equities is correspondingly small.

One difficulty in accepting the analogy of the Channel Islands is said to be the Aegean islands' fairly uniform density from the Greek to the Turkish mainland.[77] The Channel Islands' position, 'close to the French coast' and 'wholly detached' from the United Kingdom, '*was quite dif-*

[72] Paras. 61, 65.
[73] Para. 108.
[74] See *supra*, Ch. 3 s.3(*b*).
[75] Ibid. para. 183.
[76] Ibid. para. 201.
[77] See e.g. Bowett, *The Legal Regime of Islands*, 267–71, 277–81; Kwiatkowska, 'Maritime Boundary Delimitation between Opposite and Adjacent States in the New Law of the Sea', 22–3; J. Georgacopoulos, 'The Aegean Sea Continental Shelf Problem: Presentation of the Greek Case', 6 *IBL* (1978), 479, at 493.

ferent from . . . the situation where numerous islands stretch out one after another long distances from the mainland'.[78] It will be recalled, however, that this italicized description by the Court appears prominently in the Aegean *only* when it is seen as one, geographical, i.e. undivided unit. For delimitation purposes, the equity in studying the geography of the Aegean in sectors would seem, as attempted here, to yield 'comparable pictures' that could certainly permit reasonable analogy with situations such as the Channel Islands. On the other hand, although the Aegean example is unique in terms of its unbalanced geo-political distribution, it must not be overlooked that the legal framework within which the Court decided the course of the boundary in the Channel Islands region was not confined to the factor of the islands' 'geographic isolation' only, but

> [t]o state this conclusion is not, however, to deny all relevance to the size and importance of the Channel Islands which, on the contrary may properly be taken into account in balancing the equities in the region. Put also of this framework are the limits of the territorial seas and coastal fisheries of the United Kingdom and France . . . Other elements of the framework are the various equitable considerations invoked by the parties regarding their respective navigational defence and security interests in the region.[79]

An interesting factor of analogy to the Aegean (where Turkey is the only coastal State along the whole of the Aegean's east coast) is the Court's conclusion that

> it is sufficient to say that, in its view, they [the above-mentioned factors] tend to evidence the predominant interest of the French Republic in the southern areas of the English Channel, a predominance which is also strongly indicated by its position as a riparian State along the whole of the Channel's south coast.[80]

2. *State practice*

Instances of inter-State agreements which have dealt with situations approximate to the Greek 'detached' island are not common. However, the 1978 Agreement between Australia and Papua New Guinea[81] delimiting their respective fishing rights, seabed rights, and territorial sea jurisdiction in, *inter alia*, the Torres Strait is a good case for analogy to the Aegean. The principal boundary line is the shelf line which runs through the approximate 'middle' of the Torres Strait, separating the two States, thus ignoring the numerous Australian islands, islets, and reefs which dot the area.

The treatment of the two Australian island groups (Abusi and Saibi)

[78] *Anglo-French* Arbitration, para. 199 (emphasis added).
[79] Paras. 187, 188.
[80] Ibid.
[81] See *supra*, n. 53.

situated off the coast of Papua New Guinea is relevant to the Aegean. The islands, being ignored for seabed boundary purposes, were granted only territorial sea enclaves of 3 miles. A median line between these islands and the coast of Papua New Guinea was also delimited. Like the Aegean 'outlying' islands, these two groups lie in the narrow Strait of Torres, approximately 90 miles wide. They are all inhabited and of considerable size; lie in the vicinity of a foreign coast which is opposite their owning State's mainland; and south of them are situated numerous other insular formations belonging to Australia. The fact that the question of sovereignty over these islands was resolved in the same agreement did not seem to have a decisive impact on the delimitation arrangement.[82]

One further striking and original feature of the Torres Settlement is the fishery jurisdiction arrangement. The two States have agreed on delimiting the line between their respective fishery jurisdictions, which for a great part coincides with the seabed line. But at the point of its intersection with meridians corresponding to the western and eastern outer limits of the Abusis and Saibis respectively, the line extends northwards so as to grant these two groups (and the Australian islands to the south) full effect for fishing jurisdiction purposes. This fishing jurisdiction area forms, however, part of a greater 'Protected Zone' designated primarily to protect the traditional way of life and livelihood of the indigenous inhabitants, including their traditional fishing and free movement.[83]

Unfortunately, it has not been possible to establish how essential mobile and sedentary fishing activities in the Aegean are. It may well be the case that such activities have been maintained as traditional activities of the inhabitants of the Anatolian coast and/or the inhabitants of the islands. If this is so, solutions similar to the Torres arrangement may well serve as a good precedent to obviate difficulties in reaching a comprehensive boundary settlement in the Aegean.

(*c*) *The southern sector*

In the southern sector, that is, the area bounded in the north by the almost overlapping territorial sea belts of Samos, Ikaria, Mikonos, Tinos, and Andros (see Map 18), the Aegean turns into a 'Greek lake'. The Turkish coast is literally 'cut off' from the Aegean basin by the barrier of Greek islands: from the Dodecanese in the east, the Cretan islands in the south, the Cyclades in the centre, and the Peloponnese peninsula in the

[82] See K. Ryan and M. White, 'The Torres Strait Treaty', 7 *AYBIL* (1976–7), 87, at 92–8. Art. 2 of the Agreement provides, *inter alia*, for the recognition by Papua New Guinea of the Australian sovereignty over these islands.

[83] Arts. 4–11 of the Agreement. See also Ryan and White, 'The Torres Strait Treaty', 103 ff.

west. These Greek territories, together with their surrounding belts of territorial sea (even the 6-mile limit), effectively screen the Turkish coast to exclude extension of Turkish jurisdiction throughout the southern Aegean. In contrast to the northern sector, even an enclave solution to the Dodecanese would only allow negligible 'fingers'[84] of Turkish seabed jurisdiction west of these islands, for these 'fingers' would have to terminate a short distance east of the continuous territorial sea belts of the Cyclades Islands.

One possible accommodation, however, may be envisaged in the area (south of the Aegean Sea proper) around the most easterly Greek island of Kastellorizo (also known as Megisti). The island is 3.5 sq.m. in area, with only 222 inhabitants. It is situated some 60 miles east of Rhodes (the nearest major Greek island) but only 1 mile off the coast of Turkey. The island was Turkish until occupied by Italy in 1912 and handed over to Greece in 1947. If allowed to generate a full maritime jurisdiction, Kastellorizo would effectively cut off Turkey's access to the area of high seas available south-east and north-east of the Cretan islands and Cyprus, respectively.

(*d*) *Resultant guidelines*

(i) Introduction

The approach to analysing the Aegean case in the context of geographical sectors—each reflecting the mainlands–islands relationship—is submitted to be the first step that must be taken for the construction of a conceptual framework for the delimitation of the Aegean basin. This approach has mitigated the Aegean's geographical complexity and allowed constructive analogy to State and judicial practice.

A review of this 'comparable' practice certainly reflects the view that, whereas islands in large sea situations are usually granted full or considerable weight, subject, prima facie, to their size and location, in narrow and enclosed sea areas, where chances of redressing inequity are correspondingly small, the effect of islands in extending the coastlines of the riparian States is seen as an important element in the decision to assign to them reduced value, or no value at all. This is a factor of great relevance to the Aegean Sea, where the area available for delimitation is very limited.

A decision by the parties on the procedure by which the boundary problem is to be settled may significantly influence the nature of the

[84] For the 'fingers Approach' suggested for the delimitation in the Aegean, see Wilson, *The Aegean Sea Dispute*, 38; D. Karl, 'Islands and the Delimitation of the Continental Shelf: A Framework for Analysis', 71 *AJIL* (1977), 642, at 671–2.

'factors' which may be deemed relevant.[85] This, however, does not prevent one from observing some factors which are thought relevant for the purpose of providing general guidelines for the resolution of the boundary dispute in the Aegean.

(ii) The continuity of the Aegean shelf

The situation in the northern Aegean is one of a common shelf with an overlap of the two States' coastal projection. This is related to the fact that the whole Aegean constitutes one geophysical continuation, as well as to the fact that the delimitation would be effected between two States which are much less than 400 miles apart. The presence of numerous islands might be said to have prevented the existence of one clear overlap. This, however, causes only superficial difficulty in the way of determining equitable delimitation in the Aegean. A delimitation-orientated analysis of the geography of the Aegean may overcome this problem.

(iii) Redressing the inequitable effect of the Aegean islands

A comparison of hypothetical boundaries in the northern Aegean drawn with and without reference to the islands demonstrates that the use of all the islands as basepoints would certainly have a 'disproportionately distorting effect'[86] on the seabed delimitation. Analysed against 'comparable' State and judicial practice, the situation seems quite amenable to treatment either under the judicial principle of equity or by application of the more common rule of 'equidistance–special circumstances'.

(iv) Inclusion of the territorial sea for evaluation of proportionality

The argument for including the territorial belts of 6 miles around the islands[87] for calculation of proportionality seems to be reasonable in the light of the fact that, as seen earlier,[88] approximately 45 per cent of the Aegean basin is territorial sea, with 35 per cent belonging to Greece and 8.8 per cent belonging to Turkey. Considerations relating to the conducting of an equitable test of proportionality may override in this instance the theoretical, albeit legal, criterion of the definition of the continental

[85] For instance, if the parties decided to resolve the boundary dispute by direct agreement, issues such as their dispute over the Aegean air traffic, the militarization of the eastern Aegean islands by Greece, and the Cyprus crisis would probably be relevant or 'trade-off' factors in the agreed seabed boundaries. Such 'factors' have no direct legal impact before a judicial body such as the ICJ.

[86] *North Sea* cases, para. 36.

[87] *Supra*, Ch. 8 s.4(*b*)(i)(1).

[88] Ibid.

shelf doctrine. The decision of the Court in *Tunisia–Libya* to include the territorial sea of the parties in the calculation of proportionality provides a priori analogy to the situation in the Aegean. The Court took this decision because, as pointed out by Judge Jiménez de Aréchaga,

> it would seem that in a case such as the present, which is different . . . from the *North Sea Continental Shelf* cases in that there is an enormous difference between the waters claimed as internal and territorial by each Party, it would be inequitable not to take into account, for the overall evaluation of fairness and proportionality, the whole expanse of water, on the sole ground that legally the continental shelf begins at the outer limit of the territorial sea. To do so would be . . . to allow . . . an exaggerated adherence to the strict letter of the law when equity demands a broader approach for the purposes of compensation.[89]

Such a definition of the continental shelf seems to have lost, furthermore, much of its legal significance in the light of the development of the law, reflected in the 1982 Convention, which defines both the shelf and the EEZ in terms of distance 'from the baseline of the territorial sea' and not from the outer limit.

(v) Identification of proportionality in the Aegean

The factor of proportionality is best effected by comparing the ratio of the lengths of the two States' respective coastlines to the ratio of the seabed/water-column areas that a boundary line in the Aegean might assign to each State. It should be recalled, however, that a 'nice, arithmetical' calculation of proportionality cannot afford a general principle, providing an independent source of rights to maritime jurisdiction.[90] The calculations detailed below are suggested only to ascertain the applicability of 'proportionality' to the Aegean.

Admittedly, the ratio factor is obviously difficult to apply in a situation like the Aegean, where many scattered Greek islands have to be included in the delineation of the 'coastal front' of Greece. However, if one concentrates on the northern sector, where a meaningful shelf or EEZ boundary might be envisaged, an appropriate device for calculating the islands may mitigate the problem. In this northern sector the calculation of the east, north, and west coastlines produces no problem.

Measured 'in the general direction of the coastline',[91] the relevant coast

[89] His Sep. Op. in *Tunisia–Libya* case. 139.

[90] See *supra*, Ch. 4 s.3(*b*)(iii)(6).

[91] See *supra*, Ch. 4 s.3(*b*)(iii)(1). It must be recognized that there is a certain amount of subjectivity in a measurement 'in the general direction of the coastline' which depends upon the degree to which one ignores coastal irregularities. This will not make a marked difference in this case, because, more noticeably in the northern sector, Greece and Turkey have coastlines with similar degrees of irregularity and thus a more exacting measurement would not destroy the ratio measured below.

of Turkey, between Kusadasi and the Greek–Turkish land boundary terminus, is about 243 miles. The Greek coastal front from there up to the southern tip of Euboea (Thasos and the Sporadhes group—except Piperi, Yioura, and Pelagos—are included in a hypothetical straight line)[92] is about 248 miles. The southern 'coastal front' of this northern sector comprises the relevant coasts of Andros, Tinos, Mikonos, Ikaria, and Samos. The allocation to these islands of 6-mile territorial sea belts produces an almost continuous overlap. The length of these island coasts (including their overlapping territorial sea belts) is about 135 n.m.

The problem, however, still arises with regard to the rest of the Greek islands[93] situated seaward of the above identified coastal front. These islands constitute a substantial portion of Greece's territory and should be given full effect in the calculation of the *proportionality test*. This, it must be emphasized, is not the same as using them as basepoints for depicting a boundary line since such treatment would produce, as demonstrated earlier, a distorted and disproportionate delimitation.

The criterion for measuring the 'size' of these islands for proportionality test purposes is not the total area but rather the length of the coastline. Since the coastline for this purpose should be measured in the general direction of the coast, the maximum length of an island, that is, the distance between the two points on an island's coastline which are furthest from each other,[94] would be an appropriate measure of the length of the islands' coastline. By doing this the location of an island, i.e. offshore, midway, or off-lying, would be irrelevant for *proportionality purposes*, because the 'furthest points distance' would be at any rate the *maximum* measure of the amount of coastline which could be counted had all the islands been situated 'offshore', thus forming part of the general direction of the mainland coast of Greece. Applying this to the above listed Greek islands would add to the coastal front of Greece some 178 miles, representing the total maximum lengths of these islands.

The application of the criterion of proportionality dictates that the seabed area of the northern sector should be divided between Greece and Turkey as follows: the operative coastline of Greece is 248 + 135 + 179 = 562 n.m; the operative coastline of Turkey is 243 n.m. The total length of the coastlines is 562 + 243 = 805 miles. The ratio of each

[92] See Map 18.

[93] These are Chios, Lesvos, Samothrace, Limnos, Ag. Evstratios, Skiros, Psara, and Antipsara.

[94] Karl ('Islands and the Delimitation of the Continental Shelf') uses this measure also to identify islands of 'substantial' portions of a State's territory which should be accorded some effect in the delimitation of maritime boundaries. He used a subjective ratio between the island's 'maximum length' and the total length of the State's coast so as to allow a 'cut-off' percentage that may be ignored for delimitation purposes.

State's coastline to the total length of the coastlines will determine that State's proportional share of the area to be delimited. Thus the shares are as follows: Greece 562/805 = approximately 70 per cent, Turkey 243/805 = approximately 30 per cent.

The application of this proportionality criterion to the southern sector does not produce a similarly accurate result. But even in this sector the criterion can serve as one of the 'balancing factors' which help establish the overall evaluation of an equitable seabed or EEZ boundary in the Aegean.[95]

(vi) Resource interests

Because the function of extended maritime jurisdictions is directly related to economic interests, the location of mineral resources in the Aegean might be the central factor underlying the two States' legal argument.[96] This is certainly the case if oil and gas fields are discovered in areas outside the existing 6-mile territorial sea belts in the Aegean. Although it was recently reported that the international consortium exploiting the Prinos Field announced plans to drill a promising new well east of Thasos outside its 6-mile territorial sea,[97] no official data exist to confirm the exact hydrocarbon potential of the Aegean seabed. The presence of more accurate data will, however, be read in two ways: as supporting each side's claim to an equitable share of the resources of the region, or as an indicator of each side's ability to exert pressure for a favourable settlement.

One possible measure that might encourage the parties to resolve the boundary issue is for them to agree on some arrangement in the event of any common deposit of hydrocarbon resources extending across a prospective boundary line. State practice reveals that in such cases States commit themselves to consulting with each other with a view to reaching agreement on the manner in which the accumulation or deposit may be most effectively exploited and on the equitable sharing of the benefits from such exploitation.[98] Perhaps an alternative solution is to postpone the ultimate decision on delimitation in certain areas and

[95] Utilizing the Aegean Islands specified by Greece in its Application before the Court (*Aegean Sea Continental Shelf* case, ICJ Reports (1976), 15) and the other (largest) islands and applying the above method of calculation, the ratio model for the southern sector should produce the following: the Greek coast (northern sector 562 + southern sector 482) = 1,044; the Turkish coast (northern sector 243 + southern sector 184) = 427. Thus the ratio in the whole Aegean should be as follows: Greece 1,044 ÷ 1,471 = approx. 70%; Turkey 427 ÷ 1,471 = approx. 30%.

[96] See *supra*, Ch. 4 s.3(*b*)(iii)(2)(*a*).

[97] *The Times* (28 Mar. 1987).

[98] See e.g. the continental shelf agreements examined in *supra*, Ch. 4 s.4(*b*).

to endeavour instead to exploit the resources by a 'joint development zone'.[99] Joint zones have been created in several bilateral agreements, such as those between Saudi Arabia and the Sudan, Japan and Korea, Malaysia and Thailand, Norway and Iceland, and, most recently, Australia and Indonesia.[100]

(vii) Preservation of traditional activities

The importance of this factor is related to the narrowness of the Aegean basin and the role of this basin as a source of livelihood to the inhabitants of the islands and the coastal areas alike. It might be the case that a certain community maintains long-standing activities associated with the resources of the sea, such as fishing or offshore tunnelling for natural resources. Some of these activities are of course only related to the discussion of the delimitation of the 'water column'. As mentioned earlier, however, it will not normally be feasible to contemplate two different seabed/water-column boundaries in such a limited area as the Aegean. Therefore, it may be possible for Greece to argue, for instance, that the enclave solution to certain islands in the eastern Aegean would harm the economic activities of the inhabitants of the islands. Turkey might, on the other hand, establish that a full weight solution would affect established fishery or mineral activities associated with the inhabitants of the coastline. In such a situation, the rights of the inhabitants can be preserved under a proper scheme of delimitation. The solution adopted in the Torres Strait Agreement to preserve the rights of the islanders is a striking precedent that might be considered for the Aegean.[101]

(viii) The factor of security

The factor of security, though not critical to any of the Decisions,[102] seems to be a more important issue in the Aegean, given the course of the relations between the two States, the various conventions and treaties relating to the land frontier, sovereignty over the eastern islands, their demilitarization, passage through the straits, and the geographical configuration of the area.[103]

[99] Turkey proposed the 'joint venture' solution in the negotiations between the two States which took place in the mid-1970s. Greece did not reject the proposal in principle. See e.g. Minutes of the Meetings of the Greek and Turkish Experts, Berne, 2 Feb. 1976, *Aegean Sea Contienental Shelf*, *Pleadings* (1986), 156–7.

[100] See e.g. *supra*, Ch. 5, text at n. 65.

[101] See *supra*, text at nn. 78–80.

[102] See *supra*, Ch. 4 s.3(*b*)(iii)(3).

[103] See *supra*, s.1 and text at nn. 25–8. For a brief account of the historical development of the conflict between the two countries, see B. Beeley, 'The Greek–Turkish Boundary: Conflicts at the Interface', 3 *IBGT* (1978), 351 ff.

It is relevant, for instance, to consider the implication of a boundary line extending from the land frontier at the mouth of the Martisa River. The direction of the land frontier at this point is perpendicular to the general directon of the coastline. Solutions such as drawing a boundary line between the adjacent coasts perpendicular to their general direction, or a lateral line, or the prolongation of the direction of the land frontier, are considered by State practice and the Decisions to be relevant circumstances.[104]

The choice of one of these lines might be influenced by security considerations. A strict equidistant line, for instance, would tend toward the Dardanelles, thus raising a problem of security for the Turkish straits, recognized by the restoration to Turkey in the 1923 Treaty of Lausanne of Tenedos, Imyros, and the Rabbit Islands which lie off the entrance to the Dardanelles.

More important are the overall implications of the boundary line in the rest of the Aegean. Neither Greece nor Turkey wants its seabed/EEZ areas to be disjoined or separated, thus putting vital interests, such as political unity, and unimpeded air and sea access to and from the Aegean, at stake. A proper solution might be that both States agree on a joint jurisdictional rather than delimitatory arrangement in certain areas sensitive to their security.

It is possible, for instance, that both parties might agree to the mutual exercise of certain shelf/EEZ rights which do not relate directly to security, such as marine scientific research and the protection and preservation of the marine environment. Conversely, neither party would exercise any of the rights which involves significant economic or security implications, such as exploration and exploitation of natural resources, establishment of artificial installations, or military exercises, without the express consent of the other.[105]

Such arrangements assume of course a political willingness on the part of the two States to tackle the politico-economic and technical issues involved. But having overcome such difficulties, these solutions would facilitate the willingness of the two sides to reach a delimitatory arrangement in other parts of the Aegean basin as may be determined in the future by direct negotiations or through adjudication, and, if the mutual zones were retained in such a delimitation, would also reduce the tension which at present underlies the two States' conflicting claims over this limited geographical basin.

[104] See *supra*, Ch. 2 s.2 and Ch. 4 s.3(*b*)(iii)(1).

[105] For the same view see D. Karl, 'The Delimitaton of the Aegean Continental Shelf: Equitable Principles and the Problem of Islands', paper given at the International Symposium on Aegean Issuer (Cesme, İzmir, 15–17 Oct. 1987), 27.

Conclusions

The post-Geneva Conventions law on maritime boundaries has witnessed a remarkable increase of materials, including the prolonged discussions at UNCLOS III, the conclusion of scores of delimitation agreements, and, most importantly, the adjudication of six boundary cases within a five-year span.

Part I of this study attempted to consider whether this body of materials denotes a continuation or otherwise of the rules embodied in the 1958 Geneva Conventions. The review of these materials does not always provide a clear answer. Whilst it established with some certainty *the very general approach* to solving a maritime boundary,[1] it left much less certain the identification of rules and principles that put such an approach into effect. This uncertainty in the law can be demonstrated by the following statements by the International Court. In the 1982 *Tunisia–Libya* decision the Court was confident in stating that 'It is . . . the result which is predominant; the principles are subordinate to the goal. The equitableness of a principle must be assessed in the light of its usefulness for the purpose of achieving an equitable result.'[2] Yet only three years later the same Court was seen to be retreating from such a strongly result-orientated approach:

> Thus the justice of which equity is an emanation, is not abstract justice but justice according to the rule of law; which is to say that its application should display consistency and a degree of predictability; even though it looks with particularity to the peculiar circumstances of an instant case, it also looks beyond it to principles of more general application.[3]

The conclusion arrived at in this research—that the 'equidistance–special circumstances' rule of the 1958 Conventions embodies a formula which has crystallized into a rule of law—emerges from the analysis of the various sources on the contemporary law of maritime boundaries.

The fact that the rule was adopted at UNCLOS I (by sixty-three votes to none, with only two abstentions) after careful preparation and discussion by the ILC since 1951,[4] and that the rule has continued as a conventional rule since 1964 (when the 1958 Shelf Convention entered into force; it is now binding for fifty-four States), cannot be ignored in the consideration of the existing law on delimitation.

[1] *Supra*, Ch. 4 s.3(*a*).
[2] ICJ Reports (1982), para. 70.
[3] The *Libya–Malta* case, ICJ Reports (1985), para. 45.
[4] *Supra*, Ch. 2 s.3(*a*); Ch. 3 s.2.

The decision of the ICJ in the 1969 Judgment to make only a casual connection between Article 6 of the 1958 Shelf Convention and general international law was dictated, as seen earlier, by considerations peculiar to both the *geography* (the parties' coastal configuration)[5] and *time* (when evidence on positive law on maritime boundaries was rudimentary)[6] of the *North Sea* cases. The attack by that Judgment on the argument that the rule in Article 6 embodied a substantive standard of equity should, therefore, be evaluated within this context.

This conclusion affects, *a fortiori*, two other matters associated with the general application of the equidistance–special circumstances rule. First, there is considerable evidence that the origin and early application of 'equidistance' was never envisaged within the context of a purely geometrical method of one restricted form of application.[7] Secondly, the more recent practice of States, relating, significantly, to shelf, EEZ, and single boundaries, revealed an outstanding flexibility in the application of that rule in order to adapt the method to any 'relevant' or 'special' circumstance.[8] It was this flexible quality which made reasonable the argument—convincingly advanced by the Court of Arbitration in the *Anglo-French* Arbitration—that 'equidistance–special circumstances' was a substantive rule of equity. The significance of the reasoning given by the Court stems from the fact that this is the only case considered juridically which has been conducted, at least in part, on the basis of Article 6 of the Shelf Convention. The Court's demonstration of its meaning, application, and relationship with general international law cannot be ignored.[9]

The rule remained a basic formula in the negotiations throughout UNCLOS III. Its omission from the final draft was dictated by psycho-political considerations,[10] which eventually led to the hasty introduction of the present formula of Articles 74 and 83 of the 1982 Convention.[11]

The recognition by the International Court of the significance of the 1982 Convention's detachment from the Geneva Conventions[12] might be put forward as a strong argument against the validity of 'equidistance–special circumstances' embodying a rule of law. But the proponents of

[5] This was more compelling in the light of the fact that both Denmark and The Netherlands argued for the application of unqualified equidistance as a necessary expression of the doctrine of the exclusive appurtenance of the continental shelf to the nearby coastal States. See *supra*, Ch. 3 s.3(*a*).

[6] See *supra*, Ch. 4 s.3(*c*).

[7] Supra, n. 4.

[8] *Supra*, Ch. 4 s.4; Ch. 5 s.4.

[9] Supra, Ch. 3 s.3(*b*).

[10] *Supra*, Ch. 4 s.1.

[11] *Supra*, Ch. 4 s.2(*a*).

[12] *Supra*, Ch. 4 s.2(*d*).

such an argument soon face a dilemma in that they are left with a lacuna, the consequences of which are not difficult to identify. As one recent commentator has put it, in discussing the state of the law on boundaries and its real or alleged evolution between 1958 and the present time

> it should be recalled that not a few of those who deal with [boundary] delimitation wonder whether the rule so modified in its formulation, notably by the substitution of a general reference to international law for the rule of 'equidistance–special circumstances' and the mere indication of the 'equitable results' as the aim to be pursued—does really constitute a substantive rule on delimitation . . . or is simply an instrumental or 'blanket' rule into which any content can be poured not only by . . . States' direct agreement but also by judicial or arbitral pronouncement.[13]

The legal anguish which the 1982 Convention formula engenders is well illustrated by the judicial failure, since the 1969 *North Sea* decisions, to attribute to 'equity' an acceptable, substantive content.[14] The mere assurance that equity operates strictly within the framework of law, as the Decisions state,[15] does not explain the nature of the relationship between equity and the method, or choice of methods, resulting from the determination of that 'principle' applicable in a concrete case. While raising the question of this relationship,[16] the Decisions failed in many respects to reflect any logical or consistent trend,[17] simply because reliance on equity (as the only 'primordial' rule of law) meant reliance on an elusive factor not capable of prediction.

Only in the most recent case has the Court attempted to approach this issue with some clarity when it stated:

> That the question of entitlement and of definition of continental shelf on the one hand, and of delimitation of continental shelf on the other, are not only distinct but are also complementary is self-evident. The legal basis of that which is to be delimited, and of entitlement to it, cannot be other than pertinent to that delimitation.[18]

Moreover, the Court proceeded to explain how this correlation between title and delimitation would influence its decision-making process.[19] In particular the Court indicated that it would initially employ the law

[13] G. Arangio Ruiz, 'The Italian Shelf Delimitation Agreements and the General Law on Shelf Delimitation', in U. Leanza (ed.), *The International Legal Regime of the Mediterranean Sea* (Giuffrè Editore, Milan, 1987), 33, at 39.

[14] *Supra*, Ch. 4 s.2(*d*).

[15] Ibid.

[16] See e.g. the *North Sea* cases, ICJ Reports (1969), para. 46; *Tunisia–Libya*, para. 44; *Libya–Malta*, para. 27.

[17] See e.g. some outcomes criticized *supra*, Ch. 4 s.5.

[18] *Libya–Malta* case, para. 27.

[19] *Supra*, Ch. 9 s.2(*f*).

relating to a State's legal title to determine the basic criterion and method for achieving a provisional solution to the dispute:

> the choice of criterion and the method which it is to employ in the first place to arrive at a provisional result should be made in a manner consistent with the concepts underlying the attribution of legal title.[20]

Such a correlation could produce the necessary basis not only for deriving the relevant criteria and methods in a given dispute, but also for identifying principles of more general application.

Accordingly, it became possible to establish that the principle of natural prolongation—which implied the relevance of the geological and geomorphological characteristics of the seabed and subsoil—is irrelevant to most delimitation cases world-wide, where the geophysical continuity of the area of delimitation is usually the characteristic feature, and geographical adjacency measured in terms of distance is the sole basis of title within the 200-mile limit.[21] Alternatively, the correlative principle of non-encroachment by one party on the 'natural prolongation' of the other plays a more dominant role.[22]

The identification of the 'distance principle' as the rule of law governing the entitlement to EEZ/continental shelf within 200 miles[23] certainly enhances the role of 'equidistance–special circumstances' as the rule most relevant to the application of the distance principle.[24] Such a conclusion is supported not only by its overwhelming application in delimitation agreements, but also by numerous references to it in the recent EEZ legislation, which often formulates the delimitation rules jointly for 200-mile zones and the continental shelf.[25]

More indicative in this respect is the actual process of delimitation taken by the Decisions in application of the 'fundamental norm of equitable principles'. Since the 1982 *Tunisia–Libya* case, the Decisions have tended towards narrowing it down to the identification of the relevant factors, with the aim of not only *ameliorating* the strict application of a chosen method but rather *indicating* what that method is to be.[26] In this process the factor of geography has been used almost exclusively in effecting this indicatory function.[27] Other factors are given a secondary role of modifying the criteria and methods reflecting coastal geography in order to reach an 'equitable solution'.[28] This approach has not only produced in reality boundary lines reflecting at least a broad similarity to an 'equidistance–special circumstances' line,[29] but also confirmed its validity reflecting the judicial doctrine of equity.

[20] Para. 61. [21] *Supra*, Ch. 4 s.3(*b*)(ii)(1). [22] *Supra*, Ch. 4 s.3(*b*)(iii)(3).
[23] *Supra*, Ch. 5 s.2(*e*). [24] *Supra*, Ch. 5 s.3(*a*). [25] *Supra*, Ch. 4 s.4(*b*)(ii).
[26] *Supra*, Ch. 4 s.3(*a*). [27] *Supra*, Ch. 4 s.3(*b*)(i), (iii)(1).
[28] Ibid. [29] *Supra*, Ch. 4 s.3(*c*).

The concept of a single boundary delimitation is a logical consequence of introducing 'distance' as the sole basis of seabed and water-column entitlement within 200 miles of the coast. An analysis of the relevant provisions of the 1982 Convention suggests that the relationship between the EEZ and shelf regimes within 200 miles is such that, while the 200-mile zone reflects integrated functions, each regime retains some measure of legal autonomy and distinctiveness within the general framework of the Convention.[30] Such a delicate relationship seems to enhance an initial legal presumption in favour of a single, multi-purpose maritime boundary to divide all jurisdictions within 200 miles.[31] This is further supported by the desire to avoid 'the disadvantages inherent in a plurality of separate delimitations'.[32] These indicia, however, can be disregarded if States have special reasons for having divergent boundaries for seabed and superjacent waters.[33]

The above developments further emphasize the contemporary validity of 'equidistance–special circumstances' as a rule of law. The element of anachronism in the jurisprudence of the Court relates to this point. Notwithstanding its reliance on geography as the *indicative* factor, its recognition of the 'distance principle' as the basis of entitlement, and of the pertinence to delimitation of the close relationship between the regimes of the EEZ and continental shelf, the Court has remained faithful to its view (expressed in the 1969 Judgment and strongly objected to by the dissenting judges) that 'equidistance' is no more than one method having no special status, and not having the character of a 'rule'.[34]

So much for Part I of this study. The study of the Mediterranean maritime boundaries in Part II demonstrated a wide range of relevant issues, owing to the politico-geographical character of a confined basin with many nations sharing its shores. The aim was to point out these issues and to analyse their profound implications for potential maritime boundary delimitation between the Mediterranean States in respect of territorial sea, continental shelf, EEZ, or single boundaries.

Among the issues which figured prominently were the treatment of islands and the interests of third States. The former was discussed in

[30] *Supra*, Ch. 5 s.2(*h*).

[31] Ibid.

[32] *Supra*, Ch. 5 s.2(*g*).

[33] e.g. the present French–Spanish controversy in the Bay of Biscay (*supra*, Ch. 9, text at nn. 245–6) or the Libya–Malta negotiations that preceded the submission of their dispute to the Court (*supra*, Ch. 5, text at n. 175) demonstrated that States which had yet to conclude boundaries for either the shelf or (when EEZs are claimed) the EEZ would have to weigh carefully whether a single boundary best served their interests or whether they would be prepared to contemplate separate boundaries, and whether the uncertainties of litigation constituted a better risk than the compromise necessary for a negotiated solution of either a single boundary or separate boundaries.

[34] See *supra*, text accompanying nn. 5–6.

relation to the current negotiations pending between France, Italy, and Spain,[35] and—in more detail—with respect to the Aegean Sea shelf dispute.[36] The analysis of this dispute in the light of 'comparable' State and judicial practice reveals that innovative solutions could be found, provided that the political will exists on the side of the States concerned.[37]

The problem of third-State interests was discussed in the light of the Court's decision in *Libya–Malta* that it would not deal with a dispute between two parties in an area in which a third, non-party, State had claims.[38] The discussion criticized the effect of the decision in causing the dispute between the parties to continue and complicating the prospect of boundary settlements in the region as a whole. The discussion also highlighted the rudimentary nature of the law on tripartite and quadripartite settlements.[39]

The Mediterranean boundaries also raise other related issues such as unilateral claims to baselines, pending disputes over international boundary termini, and—more significantly—disputes concerning sovereignty over territories.

The relevance of baselines to boundary delimitation is perhaps more manifest where an area enclosed by one State as internal waters is part of a narrow overlapping jurisdiction with another State. The controversy between France and Italy over, *inter alia*, the area east of Corsica is a relevant example.[40] However, the possible impact of the unilateral claims on baselines in general was discussed in Chapter 6. A separate chapter was dedicated to Libya's claim over the Gulf of Sert as a historic bay in the belief that it has profound implications not simply for the problems of inter-State boundaries but also for the distinct, albeit related, question of the ability of the existing rules of law to provide fair solutions. The discussion of the Libyan claim suggests that a legal approach to the assessment of the concept of 'historic bays' in the light of community policy interests and of contemporary State practice and jurisprudence is not difficult to obtain.[41]

With regard to disputed territories, the 'Spanish enclaves' and Gibraltar examples revealed that, because of the limited geography of the Mediterranean, the settlement of their political status was a prerequisite not only for the settlement of the territorial sea delimitation but also for the delimitation of the more extended jurisdictions of continental shelf and (future) EEZ between the States concerned as well as their immediate neighbours. However, the historical, political, geographical, and socio-economic characteristics of these territories were outlined in

[35] *Supra*, Ch. 9 s.2(*d*)(i) and (ii).
[36] *Supra*, Ch. 10.
[37] *Supra*, Ch. 10 s.4.
[38] *Supra*, Ch. 9 s.2(*f*).
[39] *Supra*, Ch. 9 s.2(*g*).
[40] *Supra*, Ch. 9 s.2(*d*)(ii).
[41] *Supra*, Ch. 7 s.5.

order to assess the question of their entitlement to, and impact on, the delimitation of the territorial sea.[42]

Part II also dealt with the important issue of the pattern of Mediterranean maritime boundary delimitation in the future. Given the Mediterranean's limited area and the present economic activity related to its mineral resources, the problem of shelf delimitation was found to be more important today than that of delimitation of EEZ/EFZ jurisdictions. While this explained the present abstention by States from claiming 200-mile zones, a number of factors were thought to be likely to increase the economic importance of the Mediterranean fisheries, which would induce States to exercise their 200-mile jurisdictions, and consequently bring the prospect of EEZ/single boundaries to the fore.[43]

The primary factors that could influence the final solution of the Mediterranean's pending boundaries can be reduced to two groups.

First, there was the impact that the post-Geneva Conventions law (which introduced general reference to 'international law' and 'equitable principles') might have on future maritime settlements in the Mediterranean. It was established that, in so far as third-party settlement was concerned, although both the *Tunisia–Libya* and *Libya–Malta* decisions applied that law, in the result their lines differed very little from what would have resulted from application of the 1958 Shelf Convention.[44] In fact, in *Libya–Malta* the Court moved very close to an equidistance–special circumstances approach when it affirmed that within 200 miles of the coast (or 400 miles between opposite States) the criterion of distance predominated over natural prolongation in the sense of the physical nature of the seabed and subsoil.[45] This obviously implied a judicial preference for 'equidistance', at least as a basis for the method of delimitation, and a future third-party settlement in the Mediterranean would probably find little reason for departure from this approach.

Second, with regard to future, directly negotiated boundaries, the possibility of determining a boundary *ex aequo et bono* would make it possible to depart from the 'distance principle' and 'equidistance' as its basic method of approach. Yet there is little evidence which suggests a wide application of this possibility in the Mediterranean.

In respect of the existing shelf boundaries (including the latest between Libya and Malta),[46] given that there exist no important fishery grounds across these existing lines, there would be little point in arguing for a modification of such arrangements in a future EEZ or single boundary delimitation. Such a possibility, however, might not be excluded with regard to Italy, which might argue, especially *vis-à-vis* Greece and Tunisia, for a separate water-column boundary in areas where it could prove

[42] *Supra*, Ch. 8 s.4(*b*)(iii); s.5(*c*)(ii). [43] *Supra*, Ch. 9 s.3(*a*). [44] *Supra*, Ch. 4 s.3(*c*).
[45] *Supra*, Ch. 5 s.3(*a*); Ch. 9 s.2(*f*). [46] *Supra*, Ch. 9 s.2(*a*)–(*f*).

established fishing rights.[47] In respect of future boundaries, it was established that the Mediterranean's economic/geographical characteristics revealed no special reason for rejecting the single boundary solution, most certainly where EEZs—rather than EFZs—are involved.[48] However, the impact of evolving technology for exploiting Mediterranean resources on the solutions adopted for such boundaries was pointed out. The mineral resources of the Mediterranean's deep basins would straddle central areas of overlapping claims. Their impact on future boundaries would be contingent to a large extent on their economic potential for the States involved. But the possibility could not be excluded that a State might attempt to argue its case on the basis of marine geomorphology or geology in order to extend its seabed 'portion' to areas beyond the median/equidistant line. Libya's attempt to do so with regard to the Medina Bank[49] would reaffirm such a possibility.

In respect of fishery resources, an argument for a water-column boundary reflecting established fishing activity was mentioned above. A similar argument might be advanced by Tunisia *vis-à-vis* Libya in the light of the argument by the former for its traditional fishing banks across the shelf boundary with Libya.[50] Finally, it was shown that the introduction of EEZ claims in the Mediterranean could create 'stress zones' along central areas between North Africa and Europe; and this might give rise to the need for co-operative—rather than delimitative—arrangements.[51]

The Mediterranean Sea is perhaps the only essential factor uniting the States abutting its shores, which are otherwise characterized by diversity, especially in ethnic, religious, cultural, and political perceptions. This contributes, among other factors, to creating situations of potential instability, latent crises, and endemic conflicts.[52] The importance of the peaceful completion of maritime boundary delimitation for reducing tension and thus enhancing the process of peace and security in the Mediterranean is beyond doubt. It is a prerequisite to joint action aimed at the protection, conservation, and rational exploitation of the sea's natural resources, and this would help promote the necessary environment for peaceful coexistence among its littorals. The Barcelona Convention and its Protocols to abate and combat pollution in the Mediterranean Sea are a significant step forward in that direction.

[47] *Supra*, Ch. 9 s.3(*b*)(i).

[48] Ibid.

[49] *Supra*, Ch. 9 s.2(*f*).

[50] As noted earlier, this argument by Tunisia, however, would be very difficult to substantiate in the light of the Court's finding in the 1982 Judgment. See *supra*, Ch. 9 s.3(*b*)(i).

[51] Ibid.

[52] See e.g. *supra*, Ch. 7, text at nn. 217–20.

Select Bibliography

1. Official Documents and Publications

Al Atlas Al Watani (the National Atlas) (Secretariat of Planning, Tripoli, 1978).

Annuaire statistique de la France (1987) (Institut Nationale de la Statistique et des Études Économiques, Paris, 1987).

Annual Abstract of Statistics (Malta) (Valletta, 1989).

Annuario statistico italiano (Istituto Centrale di Statistica, Rome, 1988).

Anuario estadístico de España (Instituto Nacional de Estadística, Madrid, 1988).

BABINSKI, C., 'Report on the Export Refinery and Ethylene Projects in Libya' (Secretariat of Planning, Tripoli, Oct. 1977).

BASSETT, P., *Australia's Maritime Boundary* (Australian Foreign Affairs Record, Department of Foreign Affairs, 1984).

'Compendium of Negotiations between Italy and France on the Delimitation of the Continental Shelf', meetings of 18–19 Sept. 1973; 25–6 Feb. 1974; note of the French delegation of 6 Sept. 1974 (unpublished documents, but now published in U. Leanza (ed.), *Mediterranean Continental Shelf: Delimitations and Regimes*, 4 vols. (Oceana Publications, New York, 1988), 1613–35).

Hydrographer of the Navy, *Admiralty Tide Tables: European Waters, Including Mediterranean Sea*, i (London, 1989).

—— *Admiralty Tide Tables: Pacific Ocean and Adjacent Seas, Including Tidal Stream Tables*, iii (London, 1989).

—— *Aegean Sea Continental Shelf* case (*Greece* v. *Turkey*), *Pleadings, Oral Arguments, Documents* (1980).

—— *Case Concerning the Continental Shelf* (*Libyan Arab Jamahiriya* v. *Malta*), Application by Italy for Permission to Intervene, *Comptes rendus*, 84/1–84/7.

—— *Case Concerning the Continental Shelf* (*Tunisia* v. *Libyan Arab Jamahiriya*), Request by Malta for Permission to Intervene, *Comptes rendus*, 81/1–81/6.

—— Continental Shelf (*Libyan Arab Jamahiriya* v. *Malta*).

Memorial submitted by the Socialist People's Libyan Arab Jamahiriya, 3 vols., 26 Apr. 1983.

Memorial submitted by the Republic of Malta, 2 vols., 26 Apr. 1983.

Counter-Memorial submitted by the Socialist People's Libyan Arab Jamahiriya, 3 vols., 26 Oct. 1983.

Counter-Memorial submitted by the Republic of Malta, 2 vols., 26 Oct. 1983.

Observations on the Application filed by Italy for permission to intervene, submitted by the Socialist People's Libyan Arab Jamahiriya, 5 Dec. 1983.

Reply to Judge Oda by the Agent of Italy, 30 Jan. 1984.

Reply submitted by the Socialist People's Libyan Arab Jamahiriya, 12, July, 1984.

Reply (with Documentary Annexes) submitted by the Republic of Malta, 12 July 1984.

Observations on the Application filed by Italy for permission to intervene, submitted by the Republic of Malta, 5 Dec. 1983.

—— Continental Shelf (*Tunisia* v. *Libyan Arab Jamahiriya*).

Memorial submitted by the Government of the Republic of Tunisia, 3 vols., 30 May 1980.

Memorial submitted by the Socialist People's Libyan Arab Jamahiriya (with 2 annexes), 30 May 1980.

Counter-Memorial submitted by the Government of the Republic of Tunisia (with 2 annexes), 1 Dec. 1980.

Observations on Malta's 30 Jan. 1981 Application for permission to intervene, submitted by the Socialist People's Libyan Arab Jamahiriya.

Counter-Memorial submitted by the Socialist People's Libyan Arab Jamahiriya, 4 vols., 2 Feb. 1981.

Reply submitted by the Government of the Republic of Tunisia, 3 vols., 15 July 1981.

Reply submitted by the Socialist People's Libyan Arab Jamahiriya, 2 vols., 15 July 1981.

ICJ, *North Sea Continental Shelf* cases (*Federal Republic of Germany* v. *Denmark*; *Federal Republic of Germany* v. *The Netherlands*), *Pleadings, Oral Arguments, Documents*, 2 vols. (1968).

Instrupa Report (Final Report on Results of Test Fishing Program, Gulf of Sirte) (Tennenwalallec, Federal Republic of Germany, July 1977).

International Civil Aviation Organization, *Regional Manual: European-Mediterranean*, Doc. 4600 (Canada, 1951–2).

Revindicación espanola de Gibraltar (Ministerio de Asuntos Exteriores, Oficina de Información Diplomática (Spain), Nov. 1987).

Sogreah (Consulting Engineers), 'Study for a General Masterplan for the Development of Fishing Ports in the Libyan Arab Republic: Final Report' (Grenoble, 1975).

Speerplan/Finnmap (Consulting Engineers), 'El Khalij Region: Master Plan (Sirt)' (Secretariat of Utilities, Tripoli, May 1981).

—— 'El Khalij Region: Regional and Urban Planning: Summary Report' (Secretariat of Utilities, Tripoli, 1982).

Statistical Abstract 1985 and 1986 (Republic of Cyprus, Department of Statistics and Research, Ministry of Finance, 1987).

Statistical Yearbook of Greece (1987) (National Statistical Service of Greece, Athens, 1989).

United Nations, 'Historic Bays: Memorandum by the Secretariat of the United Nations', Doc. A/CONF. 13/1 (preparatory document No. 1) (1957).

—— 'Juridical Regime of Historic Waters, Including Historic Bays: Study Prepared by the Secretariat', Doc. A/CN. 4/143 (1962).

—— FAO (Fisheries Department), *Atlas of Living Resources of the Seas* (Rome, 1981).

—— FAO, *Fish Resources in the Mediterranean and Black Seas (Consultation on the Protection of Living Resources and Fisheries from Pollution in the Mediterranean)* (Rome, 1974).

—— FAO, *Coastal State Requirements for Foreign Fishing*, Legislative Study No.

21 Rev. 2 (by G. Moore) (Rome, 1985).
—— FAO, GFCM, *The State of Marine Pollution in the Mediterranean and Legislative Controls*, No. 51 (Rome, 1972).
—— FAO, GFCM, *Report of the Twelfth Session, Prospectives for Fisheries, Development to 1985* (Rome, 1974).
—— FAO, GFCM, *Report of the Twelfth Session: Plans and Politics for Fisheries Development* (Rome, 1974).
—— FAO, GFCM, *Report of the Thirteenth Session: The Organization of Fisheries Research* (Rome, 1976).
—— FAO, GFCM, *Rapport de la consultation technique pour l'évaluation des stocks dans les divisions statistiques Baléares et Golfe du Lion* (Rome, 1980).
—— FAO, GFCM, *Statistical Bulletin No. 5: Nominal Catches 1972–1982* (Rome, 1984).
—— 'Baselines: An Analysis of the Provisions of the United Nations Convention on the Law of the Sea Dealing with Baselines', a study to be published by Office for Ocean Affairs (second draft dated 1 Sept. 1987).
—— 'Multilateral Treaties in Respect of which the Secretary-General Performs Depositary Functions' (ST/LEG/SER. D/7).
—— Office of the Special Representative of the Secretary-General for the Law of the Sea, *Law of the Sea Bulletin*, 1–2 (1983), 3 (1984), 4–6 (1985), 7–8 (1986).
—— Environment Programme, *Convention for the Protection of the Mediterranean Sea against Pollution and Its Related Protocols* (United Nations, New York, 1982).
—— Conference on the Law of the Sea, 'Scientific Considerations Relating to the Continental Shelf' (Preparatory Document No. 2 UN Doc. A/CONF. 13/2 and Add 1, 20 Sept. 1957).
—— 'Report on Uses of the Sea' (Secretary-General; UN Doc. E/5120, 28 Apr. 1972).
—— A/CONF. 13/37, Preparatory Documents, vol. i.
—— A/CONF. 13/38, Plenary Meetings, vol. ii.
—— A/CONF. 13/39, First Committee (Territorial Sea and Contiguous Zone), vol. iii.
—— A/CONF. 13/41, Third Committee (High Seas: Fishing: Conservation of Living Resources), vol. v.
—— A/CONF. 13/42, Fourth Committee (Continental Shelf), vol. vi.
—— US Department of State, *International Boundary Study*, Series A (*Limits in the Sea*), 1–108 (1970–88).
—— USA, S. Exec. Rep. No. 96, 49, 96th Cong., 2nd Sess. (1980).
—— Presidential Proclamation No. 5,030, The White House, Office of the Press Secretary, 10 Mar. 1983.
Yearbook of the ILC (1951), i, UN Doc. A/CN. 4/SER. A/1951.
Yearbook of the ILC (1951), ii, UN Doc. A/CN. 4/SER. A/1951/Add. 1.
Yearbook of the ILC (1952), i, UN Doc. A/CN. 4/SER. A/1952.
Yearbook of the ILC (1952), ii, UN Doc. A/CN. 4/SER. A/1952/Add. 1.
Yearbook of the ILC (1953), i, UN Doc. A/CN. 4/SER. A/1953.
Yearbook of the ILC (1953), ii, UN Doc. A/CN. 4/SER. A/1953/Add. 1.
Yearbook of the ILC (1954), i, UN Doc. A/CN. 4/SER. A/1954.

Yearbook of the ILC (1954), ii, UN Doc. A/CN. 4/SER. A/1954/Add. 1.
Yearbook of the ILC (1955), i, UN Doc. A/CN. 4/SER. A/1955.
Yearbook of the ILC (1955), ii, UN Doc. A/CN. 4/SER. A/1955/Add. 1.
Yearbook of the ILC (1956), i, UN Doc. A/CN. 4/SER. A/1956.
Yearbook of the ILC (1956), ii, UN Doc. A/CN. 4/SER. A/1956/Add. 1.

2. Books

ABDALLAH, LE PRINCE MOULAY, *Les Nouvelles Règles du droit international de la mer et leur application au Maroc* (Bibliothèque de Droit International, Paris, 1981).

ADAMI, V., *National Frontiers in Relation to International Law*, trans. T. Behrens (Oxford University Press, London, 1927).

ALLAN, J., *Libya: The Experience of Oil* (Croom Helm, London, 1981).

—— (ed.), *Libya since Independence* (Croom Helm, London, 1982).

AMATO, L., DE CESARI, P., GASPARRO, G., and TREVES, T., *La giurisprudenza italiana sul diritto del mare* (Giuffrè Editore, Milan, 1979).

—— FARANDA, T., SCOVAZZI, T., and TREVES, T., *La legislazione italiana sul diritto del mare* (Giuffrè Editore, Milan, 1981).

ATTARD, D., *The Exclusive Economic Zone in International Law* (Clarendon Press, Oxford, 1987).

BARSTON, R., and BIRNIE, P. (eds.), *The Maritime Dimension* (George Allen & Unwin, London, 1980).

BEAZLEY, M., *Atlas of the Oceans: Mediterranean: Living and Mineral Resources* (M. Beazley Publisher, London, 1977).

BLAKE, G. (ed.), *Maritime Boundaries and Ocean Resources* (Croom Helm, London, 1987).

—— and DRYSDALE, A., *The Middle East and North Africa: A Political Geography* (Oxford University Press, New York, 1985).

BLUM, Y., *Historic Titles in International Law* (Martinus Nijhoff, The Hague, 1965).

BOGGS, W., *International Boundaries* (Columbia University Press, New York, 1940).

BOOTH, K., *Law, Force and Diplomacy at Sea* (George Allen & Unwin, London, 1985).

BOUCHEZ, L., *The Regime of Bays in International Law* (A. W. Sythoff, Leyden, 1964).

BOWETT, D., *The Law of the Sea* (Manchester University Press, Manchester, 1967).

—— *The Legal Regime of Islands in International Law* (Oceana Publications, New York, 1979).

BROWN, E., *The Legal Status of Hydrospace* (Stevens & Sons, London, 1971).

BROWNLIE, I., *Principles of Public International Law* (Clarendon Press, Oxford, 1979).

CALAMIA, A., MENGOZZI, P., and RONZITTI, N., *I rapporti di vicinato tra Italia e Jugoslavia* (Giuffrè Editore, Milan, 1984).

CHENG, B., *General Principles of International Law as Applied by International*

Courts and Tribunals (Grotius Publications, Cambridge, 1987).

CHURCHILL, R., and LOWE, A., *The Law of the Sea* (Manchester University Press, Manchester, 1988).

CLINGAN, T. (ed.), *Law of the Sea: State Practice in Zones of Special Jurisdiction (Proceedings of the 13th Annual Conference of the Law of the Sea Institute, Mexico City, 1979)* (Law of the Sea Institute, Honolulu, 1982).

COLOMBOS, C., *The International Law of the Sea* (Longman, London, 1967).

COTTREILI, J., and THEBERGE, J. (eds.), *The Western Mediterranean: Its Political, Economic and Strategic Importance* (Praeger, New York, 1974).

CRONAN, D., *Underwater Minerals* (Academic Press, London, 1980).

DIPLA, H., *Le Régime juridique des îles dans le droit international de la mer* (Presses Universitaires de la France, Paris, 1984).

DUPUY, P., *The Law of the Sea* (Oceana Publications, Leiden, 1974).

DURANTE, F., and RODINO, W., *Western Europe and the Development of the Law of the Sea* (4 Binders) (Oceana Publications, New York, 1983).

EL-HAKIM, A., *The Middle Eastern States and the Law of the Sea* (Manchester University Press, Manchester, 1979).

ELIAS, T., *The International Court of Justice and Some Contemporary Problems* (Martinus Nijhoff, The Hague, 1983).

Essays in Honour of Robert Ago: International Law at the Time of its Codification, 4 vols. (Giuffrè Editore, Milan, 1987).

EXTAVOUR, W., *The Exclusive Economic Zone* (Institut Universitaire de Hautes Études Internationales, Geneva, 1979).

FIELD, D., *Outlines of an International Code* (Baker, Voorhis & Co., New York, 1872).

FRANCALANCI, D., ROMANO, D., and SCOVAZZI, T. (eds.), *Atlas of Baselines* (Giuffrè Editore, Milan, 1986).

GIDEL, G., *Le Droit international public de la mer*, 3 vols. (Sirey, Paris, 1932–4).

GOODCHILD, R., *Libyan Studies*, ed. J. Reynolds (Paul Elek, London, 1976).

GREIG, D., *International Law* (Butterworths, London, 1976).

GROSS, L. (ed.), *The Future of the International Court of Justice*, 2 vols. (Oceana Publications, New York, 1976).

HOLT, S. (ed.), *The Mediterranean Marine Environment and the Development of the Region* (The Royal University of Malta Press, Valletta, 1974).

HYDE, C., *International Law, Chiefly as Interpreted and Applied by the United States*, vol. i, 2nd edn. (Little, Brown & Co., Boston, 1945).

Hydrographer of the Navy, *Mediterranean Pilot* (Taunton), vols. 1, 2, 4 (1978), vol. 3 (1970), vol. 5 (1976).

JAGOTA, S., *Maritime Boundary* (Martinus Nijhoff, Dordrecht, 1985).

JESSUP, P., *The Law of Territorial Waters and Maritime Jurisdiction* (G. A. Jennings, New York, 1927).

JOHNSTON, D., and SAUNDERS, P. (eds.), *Ocean Boundary Making: Regional Issues and Developments* (Croom Helm, Kent, 1988).

KOBAYASHI, T., *The Anglo-Norwegian Fisheries Case of 1951 and the Changing Law of the Territorial Sea* (University of Florida Press, Gainesville, Fla., 1965).

KOH, K., *Straits in International Navigation: Contemporary Issues* (Oceana Publications, London, 1982).

KUWABARA, S., *The Legal Regime of the Protection of the Mediterranean against Pollution from Land-Based Sources* (Tycoolly Int. Publishing, Dublin, 1984).

LAHLOU, A., *Le Maroc et le droit des pêches maritimes* (Librairie Générale de Droit et de Jurisprudence, Paris, 1984).

LAUTERPACHT, H., *The Function of Law in the International Community* (Clarendon Press, Oxford, 1933).

LAZRAK, R., *Les Contentieux territoriaux entre le Maroc et l'Espagne* (Dar El Kitab, Casablanca, 1974).

LEANZA, U. (ed.), *The International Legal Regime of the Mediterranean Sea* (Giuffrè Editore, Milan, 1987).

—— *Nuovi saggi di diritto del mare* (Giappichelli Editore, Turin, 1988).

LEVIE, H., *The Status of Gibraltar* (Bowker Publishing Co., Epping, 1983).

LUCIANI, G. (ed.), *The Mediterranean Region* (Croom Helm, London, 1984).

MA, YING-JEOU, *Legal Problems of Seabed Boundary Delimitation in the East China Sea* (University of Maryland, School of Law, Baltimore, 1984).

MCDOUGAL, M., and BURKE, W., *The Public Order of the Oceans: A Contemporary International Law of the Sea* (Yale University Press, New Haven, Conn., 1962).

MCNAIR, A. (Selector and annolator), *International Law Opinions*, 3 vols. (Cambridge University Press, Cambridge, 1956).

MARSTON, G., *The Marginal Seabed: United Kingdom Legal Practice* (Clarendon Press, Oxford, 1981).

MILLER, H. (ed.), *Treaties and Other International Acts of the United States of America*, ii (Government Printing Office, Washington, DC, 1931).

MOORE, J., *History and Digest of International Arbitrations to which the United States has been a Party*, 5 vols. (Government Printing Office, Washington, DC, 1898).

—— *A Digest of International Law*, 8 vols. (Government Printing Office, Washington, DC, 1909).

MOUSSA, F., *La Tunisie et le droit de la mer* (Imprimerie Officielle de la République Tunisienne, Tunis, 1981).

MOUTON, M., *The Continental Shelf* (Martinus Nijhoff, The Hague, 1952).

NEWMAN, R. (ed.), *Equity in the World's Legal System* (Bruylant, Brussels, 1973).

NORDQUIST, M., and SIMMONDS, K. (eds.), *New Directions in the Law of the Sea*, 11 vols. (Oceana Publications, New York, 1973–81).

O'CONNELL, D., *State Succession*, 2 vols. (Cambridge University Press, Cambridge, 1967).

—— *The International Law of the Sea*, 2 vols. (Clarendon Press, Oxford, 1982, 1984).

—— *International Law*, 2 vols. (Stevens & Sons, London, 1970).

OPPENHEIM, L., and LAUTERPACHT, H., *International Law*, i (Longman, London, 1954).

PRESCOTT, J., *The Maritime Political Boundaries of the World* (Methuen, London, 1985).

—— *Australia's Maritime Boundaries* (Dept. of International Relations,

Canberra, 1985).

RAO, P., *The New Law of Maritime Zones with Special Reference to India's Maritime Zones* (Millind Publications, New Delhi, 1983).

REYNAUD, A., *Le Plateau continental de la France* (Librairie Générale de Droit et de Jurisprudence, Paris, 1984).

ROSSI, E., *Storia di Tripoli e della Tripolitania della Conquista Arabe al 1911*, ed. M. Nallino (Arabic trans.) (Dar Al-Thakhaffa, Beirut, 1974).

ROZAKIS, C., and STEPHANOU, C. (eds.), *The New Law of the Sea: Selected and Edited Papers of the Athens Colloquium on the Law of the Sea* (Elsevier Science, Amsterdam, 1983).

SCOVAZZI, T. (ed.), *Atlas of the Straight Baselines* (Giuffrè Editore, Milan, 1989).

SEBEK, V., *The Eastern European States and the Development of the Law of the Sea* (2 Binders) (Oceana Publications, New York, 1979).

SMITH, G., *Restricting the Concept of Free Seas: Modern Maritime Law Re-evaluated* (Robert Krieger Publishing Co., New York, 1980).

SMITH, H., *Great Britain and the Law of Nations*, 2 vols. (King & Son, London, 1923, 1935).

SMITH, R., *Exclusive Economic Zone: An Analysis and Primary Documents* (Martinus Nijhoff, Dordrecht, 1986).

STANLEY, D. (ed.), *The Mediterranean Sea: A Natural Laboratory* (Hutchinson & Ross, Stroudsburg, Pa., 1971).

STROHL, M., *The International Law of Bays* (Martinus Nijhoff, The Hague, 1963).

SWARZTRAUBER, S., *The Three-Mile Limit of Territorial Seas* (Naval Institute Press, Maryland, 1972).

SYMMONS, C., *The Maritime Zones of Islands in International Law* (Martinus Nijhoff, The Hague, 1979).

TRUVER, S., *The Strait of Gibraltar and the Mediterranean* (Sijhoff & Noordhoff, The Hague, 1980).

TWISS, T., *The Law of Nations*, 2 vols. (Clarendon Press, Oxford, 1861).

VICUNA, F. (ed.), *The Exclusive Economic Zone: A Latin American Perspective* (Westview Press, Boulder, Colo., 1984).

VISSCHER, C. DE, *Theory and Reality in Public International Law*, trans. P. Corbett (Princeton University Press, Princeton, NJ, 1957).

VUKAS, B. (ed.), *Essays on the Law of the Sea* (Zagreb, 1985).

—— (ed.), *The Legal Regime of Enclosed or Semi-enclosed Seas: The Particular Case of the Mediterranean* (Zagreb, 1988).

WESTERMAN, G., *The Juridical Bay: Its Designation and Delimitation in International Law* (Oxford University Press, New York, 1987).

WHITEMAN, M., *Digest of International Law*, 14 vols. (Department of State Publication, Washington, DC, 1963–70).

WICK, L., and SCHMITT, W. (eds.), *Harvesting Ocean Energy* (The UNESCO Press, Paris, 1981).

WRIGHT, L., and MACLEOD, J., *The First Americans in North Africa* (Arabic trans.) (Al Ferjani Bookshop, Tripoli, 1974).

3. Theses

JAYEWARDENE, H., 'The Regime of Islands in International Law', unpublished Ph.D. thesis (Cambridge, 1981).

ORIHUELA, E., 'La delimitación de espacios maritimos', doctoral thesis, ii (University of Murcia, 1986).

4. Articles

ABBADI, A., 'Security and the Co-operation in the Mediterranean Basin', 14 *ODIL* (1984), 55.

ADAM, R., 'Un nuovo provvedimento in materia di linee di base del mare territoriale italiano', 61 *RDI* (1978), 469.

ADEDE, A., 'Toward the Formulation of the Rule of Delimitation of Sea Boundaries between States with Adjacent or Opposite Coasts', 19 *Virginia Journal of International Law* (1978–9), 207.

AKEHURST, M., 'Equity and General Principles of Law', 25 *ICLQ* (1976), 801.

ALEXANDER, L., 'Regionalism and the Law of the Sea: The Case of Semi-enclosed Seas', 2 *ODIL* (1974), 151.

—— and HODGSON, D., *Toward an Objective Analysis of Special Circumstances: Bays, Rivers, Coastal and Oceanic Archipelagos and Atolls*, Occasional Paper No. 13 (Law of the Sea Institute, University of Rhode Island, 1972).

ALLOTT, P., 'Power Sharing in the Law of the Sea', 77 *AJIL* (1983), 1.

AMERASINGHE, C., 'The Problem of Archipelagos in the International Law of the Sea', 23 *ICLQ* (1974), 539.

ANAND, R., 'Mid-ocean Archipelagos in International Law', 19 *IJIL* (1979), 228.

AURRECOCHEA, I., 'Gibraltar', 2 *IJECL* (1987), 36.

AZCÁRRAGA, J. DE, 'Espana subscribe, con Francia e Italia, dos convenios sobre delimitación de sus plataformas submarinas comunes', 28 *REs.DI* (1975), 131.

BASTIANELLI, F., 'Border Delimitation in the Mediterranean Sea', 17 *SI* (1982), 319.

—— 'Boundary Delimitation in the Mediterranean Sea', 5 *MPR* (1983), 1.

BEAZLEY, P., *Maritime Limits and Baselines: A Guide to Their Delineation*, The Hydrographic Society, Special Paper Publication No. 2 (rev. Aug. 1978).

—— 'Half-Effect Applied to Equidistance Lines', 56 *IHR* (1979), 153.

—— 'Maritime Boundaries', 59 *IHR* (1982), 149.

BEELEY, B., 'The Greek–Turkish Boundary: Conflicts at the Interface', 3 *IBGT* (1978), 351.

BENNOUNA, M., 'Le Maroc et le droit de la mer: Évolution des positions marocaines', *Revue Maghreb-Machrék* (1979), 51.

BLAKE, G., 'The 1982 United Nations Convention on the Law of the Sea: Some Implications for Libya', 15 *Society for Libyan Studies* (1984), 129.

BLECHER, M., 'Equitable Delimitation of Continental Shelf', 73 *AJIL* (1979), 60.

BLECHMAN, B., and KAPLAN, S., 'The Political Use of Military Power in the Mediterranean by the United States and the Soviet Union', 13 *SPSI* (1978), 29.

Boggs, W., 'Delimitation of Seaward Areas under National Jurisdiction', 45 *AJIL* (1951), 240.

—— 'Problems of Water-Boundary Definition: Median Line and International Boundaries through Territorial Waters', 27 *Geographical Review* (1937), 445.

Bouchez, L., 'The Fixing of Boundaries in International Boundary Rivers', 12 *ICLQ* (1963), 85.

Bowett, D., 'The Arbitration between the United Kingdom and France Concerning the Continental Shelf Boundary in the English Channel and South-Western Approaches', 49 *BYIL* (1978), 1.

Boyle, A., 'The Law of Treaties and the Anglo-French Continental Shelf Arbitration', 29 *ICLQ* (1980), 498.

Brierly, J., 'Vital Interests and the Law', 21 *BYIL* (1944), 51.

Brown, E., 'The Anglo-French Continental Shelf Case', 33 *Yearbook of World Affairs* (1979), 304.

—— 'The Continental Shelf and the EEZ: The Problem of Delimitation at UNCLOS III', 4 *MPM* (1977), 307.

—— 'The EEZ: Criteria and Machinery for Resolution of International Conflicts between Users of the EEZ', 4 *MPM* (1977), 325.

—— 'The Tunisia–Libya Continental Shelf Case: A Missed Opportunity', 7 *MP* (1983), 142.

Burmester, H., 'The Torres Strait Treaty: Ocean Boundary Delimitation by Agreement', 76 *AJIL* (1982), 321.

Burnett, W., 'Mediterranean Mare Clausum in the Year 2000? An International Analysis of Peacetime Military Navigation in the Mediterranean', 34 *Naval Law Review* (1985), 75.

Charney, J., 'Ocean Boundaries between Nations: A Theory for Progress', 78 *AJIL* (1984), 582.

Cheng, B., 'Justice and Equity in International Law', 8 *CLP* (1955), 185.

Christie, D., 'From the Shoals of Ras Kaboudia to the Shores of Tripoli: The Tunisia–Libya Continental Shelf Boundary Delimitation', 13 *GJICL* (1983), 1.

Churchill, R., 'Maritime Delimitation in the Jan Mayen Area', 9 *MP* (1985), 16.

Collignon, Y., 'Les Côtes et le plateau continental marocain', 13 *Bulletin de L'Institut de Pêches Maritimes du Maroc* (1965), 21.

Collins, E., and Rogoff, M., 'The International Law of Maritime Boundary Delimitation', 34 *MLR* (1982), 1.

Colson, D., 'The United Kingdom–France Continental Shelf Arbitration', 72 *AJIL* (1978), 95.

Conforti, B., 'L'Arrêt de la Cour Internationale de Justice dans l'affaire de la délimitation du plateau continental entre la Libye et Malte', 90 *RGDIP* (1986), 313.

Deloukas, N., 'The Controversy between Greece and Turkey in the Aegean Sea', *Issues and Studies* (Jan. 1980), 70.

Evans, A., 'Aegean Sea Continental Shelf Case (Greece vs. Turkey) (Jurisdiction), 1978', 73 *AJIL* (1979), 493.

Feldman, M., 'The Tunisia/Libya Continental Shelf Case: Geographic Justice or Judicial Compromise?', 77 *AJIL* (1983), 219.

Feldman, M., and Colson, D., 'The Maritime Boundaries of the United States', 75 *AJIL* (1981), 729.

Fitzmaurice, G., 'The Law and Procedure of the International Court of Justice, 1951–54: General Principles and Sources of Law', 30 *BYIL* (1953), 1.

—— 'The Law and Procedure of the International Court of Justice, 1951–54: Points of Substantive Law', 31 *BYIL* (1954), 371.

—— 'The Law and Procedure of the International Court of Justice, 1951–54: Points of Substantive Law', 32 *BYIL* (1955–6), 20.

Florio, F., 'Problemi della frontiera marittima nel Golfo di Trieste', 60, *RDI* (1977), 467.

Fontana, A., 'Le "zone economiche esclusive" nel Mediterraneo', *Rivista marittima* (1977), 21.

Francioni, F., 'The Status of the Gulf of Sirte in International Law', 11 *SJILC* (1984), 311.

Friedmann, W., 'The North Sea Continental Shelf Cases: A Critique', 64 *AJIL* (1970), 229.

Fusillo, M., 'Baselines for Delimiting the Territorial Sea', 3 *IYIL* (1977), 570.

Gaggiano, G., 'Some Reflections on the Treaty of Osimo between Italy and Yugoslavia', 2 *IYIL* (1976), 248.

Gardiner, P., 'Reasons and Methods for Fixing the Outer Limit of the Legal Continental Shelf beyond 200 Nautical Miles', 11/12 *RIRI* (1978), 145.

Georgacopoulos, J., 'The Aegean Sea Continental Shelf Problem: Presentation of the Greek Case', 6 *IBL* (1978), 479.

Gioia, A., 'Tunisia's Claims over Adjacent Seas and the Doctrine of "Historic Rights"', 11 *SJILC* (1984), 327.

Goldie, L., 'Historic Bays in International Law: An Impressionistic Overview', 11 *SJILC* (1984), 211.

—— 'The International Court of Justice's "Natural Prolongation" and the Continental Shelf Problem of Islands', 4 *Netherlands Yearbook of International Law* (1973), 237.

Greig, D., 'The Beagle Channel Arbitration', 7 *AYBIL* (1976–7), 332.

Grisel, E., 'The Lateral Boundaries of the Continental Shelf and the Judgment of the International Court of Justice in the North Sea Continental Shelf Cases', 64 *AJIL* (1970), 562.

Gross, L., 'The Dispute between Greece and Turkey Concerning the Continental Shelf in the Aegean', 71 *AJIL* (1977), 31.

—— 'Under-utilization of the International Court of Justice', 27 *HILJ* (1986), 571.

Guttry, G., 'The Delimitation of Territorial Waters in the Mediterranean Sea', 11 *SJILC* (1984), 377.

Herman, L., 'The Court Giveth and the Court Taketh Away: An Analysis of the Tunisiap/Libya Continental Shelf Case', 33 *ICLQ* (1984), 825.

Hodgson, R., and Cooper, E., 'The Technical Delimitation of a Modern Equidistant Boundary', 3 *ODIL* (1976), 361.

—— 'The Tuniso/Libyan Continental Shelf Case', 16 *Case Western Reserve Journal of International Law* (1984), 1.

Horigan, J., 'Utilization of Petroleum Reservoirs Extending across Sub-sea

Boundary Lines of Bordering States in the North Sea', 7 *NRL* (1974), 67.
HUTCHINSON, D., 'The Concept of Natural Prolongation in the Jurisprudence Concerning Delimitation of Continental Shelf Areas', 55 *BYIL* (1984), 133.
—— 'The Seaward Limit to Continental Shelf Jurisdiction in Customary Law', 56 *BYIL* (1985), 111.
ILUYOMADE, B., 'The Scope and Content of a Complaint of Abuse of Rights in International Law', 16 *HILJ* (1975), 47.
JESSUP, P., 'The Pacific Coast Fisheries', 33 *AJIL* (1939), 129.
JOHNSON, D., 'Acquisitive Prescription in International Law', 27 *BYIL* (1950), 332.
—— 'The Anglo-Norwegian Fisheries Case', 1 *ICLQ* (1952), 145.
KARL, D., 'Islands and the Delimitation of the Continental Shelf: A Framework for Analysis', 71 *AJIL* (1977), 642.
KOYMEN, A., 'The Aegean Continental Shelf Problem: Presentation of the Turkish Case', 6 *IBL* (1978), 459.
LAGONI, R., 'Oil and Gas Deposits across National Frontiers', 73 *AJIL* (1979), 215.
—— 'Interim Measures Pending Maritime Delimitation Agreements', 78 *AJIL* (1984), 345.
LAUTERPACHT, E., 'Equity, Evasion, Equivocation and Evolution in International Law', American Branch of International Law Association, *Proceedings and Committee Reports* (1977–8), 1.
LAUTERPACHT, H., 'Sovereignty over Submarine Areas', 27 *BYIL* (1950), 376.
LEGAULT, L., and HANKEY, B., 'From Sea to Seabed: The Single Maritime Boundary in the Gulf of Maine Case', 79 *AJIL* (1985), 961.
—— and MCRAE, D., 'The Gulf of Maine Case', 22 *CYIL* (1984), 267.
LOWE, A., 'Some Legal Problems Arising from the Use of the Seas for Military Purposes', 10 *MP* (1986), 171.
MACGIBBON, I., 'Customary International Law and Acquiescence', 33 *BYIL* (1957), 115.
—— 'The Scope of Acquiescence in International Law', 31 *BYIL* (1954), 134.
MCGINLEY, G., 'Intervention in the International Court: The Libya/Malta Shelf Case', 34 *ICLQ* (1985), 671.
MCRAE, D., 'Adjudication of Maritime Boundary in the Gulf of Maine', 15 *CYIL* (1979), 292.
—— 'Delimitation of the Continental Shelf between the United Kingdom and France: The Channel Arbitration', 15 *CYIL* (1977), 173.
MARSTON, G., 'Delimitation of Maritime Zones: Recent Commonwealth Decisions', 51 *BYIL* (1980), 263.
—— 'Low-Tide Elevations and Straight Baselines', 46 *BYIL* (1972–3), 405.
NWEIHED, K., 'EZ (Uneasy) Delimitation in the Semi-enclosed Caribbean Sea: Recent Agreements between Venezuela and Her Neighbours', 8 *ODIL* (1980), 1.
ONORATO, W., 'Apportionment of an International Common Petroleum Deposit', 26 *ICLQ* (1977), 324.
OSTRENG, W., 'Delimitation Arrangements in Arctic Seas', 1 *MP* (1986), 132.
OXMAN, B., 'The Third United Nations Conference on the Law of the Sea: The

1976 New York Session', 71 *AJIL* (1977), 247.

Oxman, B., 'The Third United Nations Conference on the Law of the Sea: The Tenth Session (1981)', 76 *AJIL* (1982), 1.

Padwa, D., 'Submarine Boundaries', 9 *ICLQ* (1960), 628.

Peters, P., and Tanja, G., 'Lateral Delimitation of Continental Shelf and Exclusive Economic Zone', 86 *Diritto marittimo* (1984), 463.

Phylactopoulos, A., 'Mediterranean Discord: Conflicting Greek–Turkish Claims on the Aegean Seabed', 8 *IL* (1974), 431.

Quéneudec, J., 'La Loi Française du 30 décembre 1968 sur le plateau continental', 15 *AFDI* (1969), 737.

Ratner, S., 'The Gulf of Sidra Incident of 1981: A Study of the Lawfulness of Peacetime Aerial Engagements', 10 *YJIL* (1984), 59.

Rhee, Sang-Myon, 'Equitable Solutions to the Maritime Boundary Dispute between the United States and Canada in the Gulf of Maine', 75 *AJIL* (1981), 590.

—— 'Sea Boundary Delimitation between States before World War II', 76 *AJIL* (1982), 555.

Ronzitti, N., 'Is the Gulf of Taranto an Historic Bay?', 11 *SJILC* (1984), 275.

Rousseau, C., 'Évolution du différend relatif à la délimitation du plateau continental dans la Mer Égée', 80 *RGDIP* (1976), 1252.

—— 'Grèce et Turquie: Différend entre les deux États au sujet de la délimitation du plateau continental dans la Mer Égée', 79 *RGDIP* (1975), 508.

—— 'Signature de l'accord du 24 mai 1977 concernant la délimitation plateau continental dans la Mer Ionienne', 82 *RGDIP* (1978), 293.

Rozakis, C., *The Greek–Turkish Dispute over the Aegean Continental Shelf*, Occasional Paper No. 27 (Law of the Sea Institute, University of Rhode Island, Kingston, Dec. 1975).

Ryan, K., and White, M., 'The Torres Strait Treaty', 7 *AYBIL* (1976–7), 87.

Schmitt, K., 'The Problem of Maritime Boundaries in U.S.–Mexican Relations', 22 *NRJ* (1982), 139.

Schneider, J., 'The Gulf of Maine Case: The Nature of an Equitable Result', 79 *AJIL* (1985), 539.

Scovazzi, T., 'Implications of the New Law of the Sea for the Mediterranean', 5 *MP* (1981), 302.

Sharma, S., 'The Single Maritime Boundary Regime and the Relationship between the Continental Shelf and the Exclusive Economic Zone', 2 *IJECL* (1987), 203.

Spinnato, J., 'Historic and Vital Bays: An Analysis of Libya's Claim to the Gulf of Sidra', 13 *ODIL* (1983), 65.

Stevenson, J., and Oxman, B., 'The Third United Nations Conference on the Law of the Sea: The 1974 Caracas Session; the Geneva Session', 69 *AJIL* (1977), 1 and 763, respectively.

Symonides, J., 'Delimitation of Maritime Areas between States with Opposite or Adjacent Coasts', 13 *PYIL* (1984), 19.

Tanja, G., 'A New Treaty Regime for the Ems–Dollard Region', 2 *IJECL* (1987), 123.

Visscher, C. de, 'Reflections on Present Prospects of International Adjudication', 50 *AJIL* (1956), 467.

VOELCKEL, M., 'Les Lignes de base dans la Convention de Genève sur la mer territoriale', 19 *AFDI* (1973), 820.

VUKAS, B., 'Enclosed and Semi-enclosed Seas', 11/12 *RIRI* (1978), 171.

—— 'The Mediterranean and the New International Law of the Sea', *Jugaslovenska Revija za Medjunarodno Pravo* (1977), 115.

WALDOCK, C., 'The Legal Basis of Claims to the Continental Shelf', 36 *Grotius Trans.* (1961), 115.

WEISSBERG, G., 'Maps as Evidence in International Boundary Disputes: A Reappraisal', 57 *AJIL* (1963), 781.

WILSON, A., *The Aegean Sea Dispute*, Adelphi Papers No. 155 (Institute for Strategic Studies, London, 1979–80).

YOUNG, R., 'Equitable Solutions for Offshore Boundaries: The 1968 Saudi Arabia–Iran Agreement', 64 *AJIL* (1970), 152.

ZARTMAN, I., 'The Mediterranean: Bridge or Barrier?', 93 *US Naval Institute Proceedings* (1967), 63.

ZIMMERMAN, P., 'The Doctrine of Historic Bays: Applying an Anachronism in the Alabama and Mississippi Boundary Case', 23 *SLR* (1986), 763.

ZOPPO, C., *Naval Arms Control in the Mediterranean*, California Seminar on Arms Control and Foreign Policy, Research Paper No. 57 (Santa Monica, Calif., 1975).

5. Conference Papers

BOWETT, D., 'Exploitation of Mineral Resources and the Continental Shelf', paper given at Conference on the International Juridical Regime of the Mediterranean Sea (Rome, 18–19 Oct. 1985) (photocopy of a typescript).

CHARNEY, J., 'Ocean Boundaries in the Aegean Sea', paper given at the International Symposium on Aegean Issues (Cesme, İzmir, 15–17 Oct. 1987) (photocopy of a typescript).

COLSON, D., 'Environmental Factors: Are They Relevant to Delimitation?', paper given at the 19th Annual Conference of the Law of the Sea Institute (Cardiff, 1985) (photocopy of a typescript).

DYKE, J., 'The Role of Islands in Delimiting Maritime Zones: The Boundary between Turkey and Greece', paper given at the International Symposium on Aegean Issues (Cesme, İzmir, 15–17 Oct. 1987) (photocopy of a typescript).

KARL, D., 'The Delimitation of the Aegean Continental Shelf: Equitable Principles and the Problem of Islands', paper given at the International Symposium on Aegean Issues (Cesme, İzmir, 15–17 Oct. 1987) (photocopy of a typescript).

KWIATKOWSKA, B., 'Maritime Boundary Delimitation between Opposite and Adjacent States in the New Law of the Sea: Some Implications for the Aegean', paper given at the International Symposium on Aegean Issues (Cesme, İzmir, 15–17 Oct. 1987) (photocopy of a typescript).

MCRAE, D., 'The Single Maritime Boundary: Problems in Theory and Practice', paper given at the 19th Annual Conference of the Law of the Sea Institute (Cardiff, 1985) (photocopy of a typescript).

PRESCOTT, J., 'Straight Baselines: Theory and Practice', paper given at the 19th

Annual Conference of the Law of the Sea Institute (Cardiff, 1985) (photocopy of a typescript).

RONZITTI, N., 'The Aegean Demilitarizations, Greek–Turkish Relations and Mediterranean Security', paper given at the International Symposium on Aegean Issues (Cesme, İzmir, 15–17 Oct. 1987) (photocopy of a typescript).

SCOTTO, R., 'Status of Petroleum Exploration in Malta', paper given at International Conference on 'The Mediterranean in the New Law of the Sea' (Malta, Feb. 1987) (photocopy of a typescript).

SHARMA, S., 'Relevance of Economic Factors to the Law of Maritime Delimitation between Neighbouring States', paper given at the 19th Annual Conference of the Law of the Sea Institute (Cardiff, 1985) (photocopy of a typescript) (the Cardiff Conference papers are now published in E. D. Brown and R. R. Churchill (eds.), *The UN Convention on the Law of the Sea: Impact and Implementations* (Law of the Sea Institute, University of Hawaii, Honolulu, 1987), pts. III and IV).

VUKAS, B., 'The Protection of the Mediterranean Sea against Pollution', paper given at Conference on the International Juridical Regime of the Mediterranean Sea (Rome, 18–19 Oct. 1985) (photocopy of a typescript).

Index

UNIV DUNELM

DURHAM UNIVERSITY LIBRARY
3 0104 00665190 9

1999
1999
2000
2001